T0202981

Marco Corazza

Claudio Pizzi

Mathematical and Statistical Methods for Actuarial Sciences and Finance

Marco Corazza (Editor)
Claudio Pizzi (Editor)

Mathematical and Statistical Methods for Actuarial Sciences and Finance

 Springer

Marco Corazza
Department of Applied Mathematics
University Ca' Foscari Venice
Venice, Italy

Claudio Pizzi
Department of Statistics
University Ca' Foscari Venice
Venice, Italy

ISBN 978-88-470-3906-3

ISBN 978-88-470-1481-7 (eBook)

DOI 10.1007/978-88-470-1481-7

Springer Dordrecht Heidelberg London Milan New York

9 8 7 6 5 4 3 2 1

Cover-Design: Simona Colombo, Milan
Typesetting with LaTeX: PTP-Berlin, Protago TeX-Production GmbH, Germany (www.ptp-berlin.eu)

Springer-Verlag Italia srl – Via Decembrio 28 – 20137 Milano
Springer is a part of Springer Science+Business Media (www.springer.com)

Preface

This volume collects a selection of refereed papers of the more than one hundred presented at the *International Conference MAF 2008 – Mathematical and Statistical Methods for Actuarial Sciences and Finance*.

The conference was organised by the Department of Applied Mathematics and the Department of Statistics of the University Ca' Foscari Venice (Italy), with the collaboration of the Department of Economics and Statistical Sciences of the University of Salerno (Italy). It was held in Venice, from March 26 to 28, 2008, at the prestigious Cavalli Franchetti palace, along Grand Canal, of the *Istituto Veneto di Scienze, Lettere ed Arti*.

This conference was the first international edition of a biennial national series begun in 2004, which was born of the brilliant belief of the colleagues – and friends – of the Department of Economics and Statistical Sciences of the University of Salerno: the idea following which the cooperation between mathematicians and statisticians in working in actuarial sciences, in insurance and in finance can improve research on these topics. The proof of this consists in the wide participation in these events. In particular, with reference to the 2008 international edition:

– More than 150 attendants, both academicians and practitioners;
– More than 100 accepted communications, organised in 26 parallel sessions, from authors coming from about twenty countries (namely: Canada, Colombia, Czech Republic, France, Germany, Great Britain, Greece, Hungary, Ireland, Israel, Italy, Japan, Poland, Spain, Sweden, Switzerland, Taiwan, USA);
– two plenary guest-organised sessions; and
– a prestigious keynote lecture delivered by Professor Wolfgang Härdle of the Humboldt University of Berlin (Germany).

The papers published in this volume cover a wide variety of subjects: actuarial models; ARCH and GARCH modelling; artificial neural networks in finance; copulæ; corporate finance; demographic risk; energy markets; insurance and reinsurance; interest rate risk; longevity risk; Monte Carlo approaches; mutual fund analysis; non-parametric testing; option pricing models; ordinal models; probability distributions and stochastic processes in finance; risk measures; robust estimation in finance;

solvency analysis; static and dynamic portfolio management; time series analysis; volatility term structure; and trading systems.

Of course, the favourable outcome of this conference would not have been possible without the precious help of our sponsors (in alphabetical order): *Banca d'Italia*; *Casinò Municipale di Venezia*; *Cassa di Risparmio di Venezia*; *Istituto Veneto di Scienze, Lettere ed Arti*; *Provincia di Venezia*; and *VENIS – Venezia Informatica e Sistemi*. We truly thank them all.

Moreover, we also express our gratitude to the members of the Scientific and the Organising Committees, and to all the people whose collaboration contributed to the success of the MAF 2008 conference.

Finally, we would like to report that the organization of the next conference has already begun: the MAF 2010 conference will be held in Ravello (Italy), on the Amalfitan Coast, from April 7 to 9, 2010 (for more details visit the website http://maf2010.unisa.it/). We anticipate your attendance.

Venezia, August 2009   *Marco Corazza and Claudio Pizzi*

Contents

List of Contributors

Laura Ballester
Department of Economic Analysis
and Finance,
University of Castilla – La Mancha,
Cuenca, Spain

Diana Barro
Department of Applied Mathematics,
University Ca' Foscari Venice,
Venice, Italy

Cristina Bencivenga
Department of Economic Theory
and Quantitative Methods for
Political Choices,
"La Sapienza" University of Rome,
Rome, Italy

Michele Leonardo Bianchi
Specialized Intermediaries
Supervision Department,
Bank of Italy,
Rome, Italy

Catalina Bolancé
Department Econometrics,
RFA-IREA,
University of Barcelona,
Barcelona, Spain

Antonella Campana
Department SEGeS,
University of Molise,
Campobasso, Italy

Elio Canestrelli
Department of Applied Mathematics,
University Ca' Foscari Venice,
Venice, Italy

Marta Cardin
Department of Applied Mathematics,
University Ca' Foscari Venice,
Venice, Italy

Paola Cerchiello
Department of Statistics and Applied
Economics "L.Lenti",
University of Pavia,
Pavia, Italy

Rosa Cocozza
Department of Business Administration,
University of Naples Federico II,
Naples, Italy

Marco Corazza
Department of Applied Mathematics,
University Ca' Foscari Venice,
Venice, Italy

Pietro Coretto
Department of Economics and Statistics,
University of Salerno,
Fisciano, Italy

Valeria D'Amato
Department of Economics and Statistics,
University of Salerno,
Fisciano, Italy

Rita L. D'Ecclesia
Department of Economic Theory
and Quantitative Methods for
Political Choices,
"La Sapienza" University of Rome,
Rome, Italy

Fernanda D'Ippoliti
Dipartimento di Scienze,
University of Chieti-Pescara,
Pescara, Italy

Antonio Díaz
Department of Economic Analysis
and Finance,
University of Castilla – La Mancha,
Albacete, Spain

Emilia Di Lorenzo
Department of Business Administration,
University of Naples Federico II,
Naples, Italy

Andrea Ellero
Department of Applied Mathematics,
University Ca' Foscari Venice,
Venice, Italy

Frank J. Fabozzi
Yale School of Management,
New Haven, CT USA

Román Ferrer
Department of Economic Analysis
and Finance,
University of Castilla – La Mancha,
Cuenca, Spain

Paola Ferretti
Department of Applied Mathematics,
University Ca' Foscari Venice,
Venice, Italy

Cinzia Franceschini
Facoltà di Economia,
University of Urbino
"Carlo Bo",
Urbino, Italy

Cristóbal González
Department of Financial Economics,
University of Valencia,
Valencia, Spain

Luca Grilli
Dipartimento di Scienze Economiche,
Matematiche e Statistiche,
University of Foggia,
Foggia, Italy

Montserrat Guillén
Department Econometrics,
RFA-IREA,
University of Barcelona,
Barcelona, Spain

Maria Iannario
Department of Statistical Sciences,
University of Naples Federico II,
Naples, Italy

Francisco Jareño
Department of Economic Analysis
and Finance,
University of Castilla – La Mancha,
Albacete, Spain

Young Shin (Aaron) Kim
School of Economics and Business
Engineering,
University of Karlsruhe and KIT,
Karlsruhe, Germany

Michele La Rocca
Department of Economics and Statistics,
University of Salerno,
Fisciano, Italy

Susanna Levantesi
"La Sapienza" University of Rome,
Rome, Italy

Francesco Lisi
Department of Statistical Sciences,
University of Padova,
Padova, Italy

Nicola Maria Rinaldo Loperfido
Facoltà di Economia,
University of Urbino
"Carlo Bo",
Urbino, Italy

Elisa Luciano
University of Torino,
ICER & Collegio Carlo Alberto,
Torino, Italy

Marco Marozzi
Dipartimento di Economia e Statistica,
University of Calabria,
Rende (CS), Italy

Massimiliano Menzietti
University of Calabria,
Rende (CS), Italy

Enrico Moretto
Facoltà di Economia,
University of Insubria,
Varese, Italy

Silvia Muzzioli
Department of Economics and CEFIN,
University of Modena and
Reggio Emilia,
Modena, Italy

Martina Nardon
Department of Applied Mathematics,
University Ca' Foscari Venice,
Venice, Italy

Eliseo Navarro
Department of Economic Analysis
and Finance,
University of Castilla – La Mancha,
Albacete, Spain

Jens Perch Nielsen
Cass Business School,
City University London,
London, UK

Marcella Niglio
Department of Economics and Statistics,
University of Salerno,
Fisciano, Italy

Włodzimierz Ogryczak
Warsaw University of Technology,
Institute of Control & Computation
Engineering,
Warsaw, Poland

Albina Orlando
National Research Council,
Istituto per le Applicazioni del Calcolo
"Mauro Picone",
Naples, Italy

Edoardo Otranto
Dipartimento di Economia, Impresa
e Regolamentazione,
University of Sassari,
Sassari, Italy

Elisa Pagani
Department of Economics,
University of Verona,
Verona, Italy

Maria Lucia Parrella
Department of Economics and Statistics,
University of Salerno,
Fisciano, Italy

Sara Pasquali
CNR-IMATI,
Milan, Italy

Danilo Pelusi
Dipartimento di Scienze della
Comunicazione,
University of Teramo,
Teramo, Italy

Cira Perna
Department of Economics and Statistics,
University of Salerno,
Fisciano, Italy

Paolo Pianca
Department of Applied Mathematics,
University Ca' Foscari Venice,
Venice, Italy

Domenico Piccolo
Department of Statistical Sciences,
University of Naples Federico II,
Naples, Italy

Claudio Pizzi
Department of Statistics,
University Ca' Foscari Venice,
Venice, Italy

Gianni Pola
Crédit Agricole Asset Management
SGR,
Quantitative Research,
Milan, Italy

François Quittard-Pinon
Institut de Science Financière
et d'Assurances,
Lyon, France
and
Université de Lyon
Lyon, France
and
EMLyon Business School,
Lyon, France

Svetlozar (Zari) T. Rachev
School of Economics and Business
Engineering,
University of Karlsruhe and KIT,
Karlsruhe, Germany
and
Department of Statistics and
Applied Probability,
University of California, Santa Barbara
and
Chief-Scientist, FinAnalytica, INC

Rivo Randrianarivony
Institut de Science Financière
et d'Assurances,
Lyon, France
and
Université de Lyon,
Lyon, France
and
EMLyon Business School,
Lyon, France

Jean Roy
HEC-Montreal
Canada

Massimo Alfonso Russo
Dipartimento di Scienze Economiche,
Matematiche e Statistiche,
University of Foggia,
Foggia, Italy

Maria Russolillo
Department of Economics and Statistics,
University of Salerno,
Fisciano, Italy

Luigi Santamaria
Dipartimento di Scienze Statistiche,
Cattolica University,
Milan, Italy

Giulia Sargenti
Department of Economic Theory
and Quantitative Methods for
Political Choices,
"La Sapienza" University of Rome,
Rome, Italy

Patrizia Semeraro
Department of Applied Mathematics
D. De Castro,
University of Turin,
Turin, Italy

Angelo Sfrecola
Dipartimento di Scienze Economiche,
Matematiche e Statistiche,
University of Foggia,
Foggia, Italy

Marilena Sibillo
Department of Economics and Statistics,
University of Salerno,
Fisciano (Salerno), Italy

Tomasz Śliwiński
Warsaw University of Technology,
Institute of Control & Computation
Engineering,
Warsaw, Poland

Barbara Trivellato
CNR-IMATI,
Milan, Italy
and
Polytechnic University of Turin,
Turin, Italy

Giovanni Villani
Dipartimento di Scienze Economiche,
Matematiche e Statistiche,
University of Foggia
Foggia, Italy

Domenico Vistocco
Dipartimento di Scienze Economiche,
University of Cassino,
Cassino, Italy

Alberto Zorzi
Department of Applied Mathematics,
University Ca' Foscari Venice,
Venice, Italy

Impact of interest rate risk on the Spanish banking sector

Laura Ballester, Román Ferrer, and Cristóbal González

Abstract. This paper examines the exposure of the Spanish banking sector to interest rate risk. With that aim, a univariate GARCH-M model, which takes into account not only the impact of interest rate changes but also the effect of their volatility on the distribution of bank stock returns, is used. The results show that both changes and volatility of interest rates have a negative and significant impact on the stock returns of the Spanish banking industry. Moreover, there seems to be a direct relationship between the size of banking firms and their degree of interest rate sensitivity.

Key words: interest rate risk, banking firms, stocks, volatility

1 Introduction

Interest rate risk (IRR) is one of the key forms of financial risk faced by banks. This risk stems from their role as financial intermediaries and it has been attributed to two major reasons. First, in their balance sheets, banks primarily hold financial assets and liabilities contracted in nominal terms. Second, banks traditionally perform a maturity transformation function using short-term deposits to finance long-term loans. The resulting maturity mismatch introduces volatility into banks' income and net worth as interest rates change, and this is often seen as the main source of bank IRR. In recent years, IRR management has gained prominence in the banking sector due to the fact that interest rates have become substantially more volatile and the increasing concern about this topic under the new Basel Capital Accord (Basel II). The most common approach to measuring bank interest rate exposure has consisted of estimating the sensitivity of bank stock returns to interest rate fluctuations. The knowledge of the effect of interest rate variability on bank stocks is important for bank managers to adequately manage IRR, investors for hedging and asset allocation purposes, and banking supervisors to guarantee the stability of the banking system. The main objective of this paper is to investigate the interest rate exposure of the Spanish banking industry at a portfolio level by using the GARCH (generalised autoregressive conditional heteroskedasticity) methodology. Its major contribution is to examine for the first time in the Spanish case the joint impact of interest rate changes and interest rate volatility on the distribution of bank stock returns. The rest of the paper is organised as follows. Section 2 contains a review of the relevant literature. The methodology employed and data used are described in Sections 3 and 4, respectively. Section 5 reports the empirical results. Finally, Section 6 concludes.

2 Literature review

The influence of IRR on bank stocks is an issue addressed by a considerable amount of literature. The bulk of the research has focused on the two-index model postulated by [18] and several general findings can be stated. First, most of the papers document a significant negative relationship between interest rate movements and bank stock returns. This result has been mainly attributed to the typical maturity mismatch between banks' assets and liabilities. Banks are normally exposed to a positive duration gap because the average duration of their assets exceeds the average duration of their liabilities. Thus, the net interest income and the bank value are negatively affected by rising interest rates and vice versa. Second, bank stocks tend to be more sensitive to changes in long-term interest rates than to short-term rates. Third, interest rate exposure of banks has declined over recent years, probably due to the development of better systems for managing IRR.

Early studies on bank IRR were based on standard regression techniques under the restrictive assumptions of linearity, independence and constant conditional variance of stock returns (see, e.g., [1, 10]). Later on, several studies (see, e.g., [14, 15]) provided evidence against constant conditional variance. A more recent strand of literature attempts to capture the time-varying nature of the interest rate sensitivity of bank stock returns by using GARCH-type methodology. Specifically, [17] led the way in the application of ARCH methodology in banking, showing its suitability for bank stock analysis. Subsequent studies have used different types of GARCH processes to examine interest rate exposure of banks. For example, [5] and [16] have employed univariate GARCH-M (GARCH in mean) models to examine both the effect of changes in interest rates and their volatility on bank stock returns, whereas [6] and [9] have used multivariate GARCH-M models.

3 Methodology

The model proposed can be viewed as an extended version of a univariate GARCH(1,1)-M model similar to the formulations by [5] and [16]. It is as follows:

$$R_{it} = \omega_i + \lambda_i R_{mt} + \theta_i \Delta I_t + \gamma_i \log h_{it} + \epsilon_{it} \qquad (1)$$

$$h_{it} = \alpha_0 + \alpha_1 \epsilon_{it-1}^2 + \beta h_{it-1} + \delta_i V C I_{t-1} \qquad (2)$$

$$\epsilon_{it} | \Omega_{t-1} \sim N(0, h_{it}) \qquad (3)$$

where R_{it} denotes the return on bank i's stock in period t, R_{mt} the return on the market portfolio in period t, ΔI_t the change in the interest rate in period t, ϵ_{it} an error term with zero mean and conditional variance h_{it}, which is dependent on the information set Ω_{t-1}, and VCI_{t-1} the interest rate volatility in period $t-1$. Moreover, ω_i, λ_i, θ_i, γ_i, α_0, α_1, β and δ_i are the parameters to be estimated. In particular, λ_i describes the sensitivity of the return on ith bank stock to general market fluctuations and it can be seen as a measure of market risk. In turn, θ_i reflects the sensitivity of the return on ith bank stock to movements in interest rates controlling for changes in the market return. Hence, it is a measure of ith bank IRR. As usual, to preserve the non-negativity requirement for the conditional variance $\alpha_0, \alpha_1, \beta \geq 0$, whereas $\alpha_1 + \beta < 1$ for stability to hold.

The GARCH-M approach is consistent with the patterns of leptokurtosis and volatility clustering frequently observed in stock markets and allows for the consideration of time-varying risk premia and an empirical assessment of the relationship between risk and return. Some features of the model should be highlighted. First, it incorporates the conditional variance h_{it} as an additional explanatory variable in (1). The specification of volatility in logarithmic form is based on [7]. Second, the typical structure of GARCH processes has been extended in (2) by modelling the conditional variance as a function of the conditional interest rate volatility lagged in one period. In this respect, even though the effect of interest rate volatility on stock returns has been considered in the literature to a lesser extent than the impact of interest rate changes, the interest rate volatility is important enough to be taken into account. As [5] points out, this variable conveys critical information about the overall volatility of the financial markets and it influences the volatility of bank stock returns also at the micro level.

There are also several critical aspects regarding the model estimation. The first issue has to do with the possible multicolinearity between the series of market portfolio return and interest rate changes, which could generate serious estimation problems. Due to the significant negative correlation typically observed in the Spanish case between these two variables, an orthogonalisation procedure has been used. Since the central aim of this study is to analyse the banks' IRR, the market portfolio return has been orthogonalised as in [10] or [11]. Thus, the residuals from an auxiliary regression of the market return series on a constant and the interest rate fluctuations series, by construction uncorrelated with the interest rate changes, have replaced the original market portfolio returns in (1).

A second issue concerns the choice of the interest rate proxy to be used. In this sense, long-term interest rates are the proxy most employed in the literature, since they are assumed to exert great influence on corporate decisions and overall economic activity. Nevertheless, in order to enhance the robustness of the results, short-term interest rates and the spread between long- and short-term rates have been used as well. With regard to the short-term rates, an interbank rate has been chosen since the money market has become a key reference for Spanish banks during recent years. In turn, the interest rate spread is considered a good proxy for the slope of the yield curve.

4 Data

The sample consists of all commercial banks listed on the Spanish Stock Exchange for at least one year over the period January 1993–December 2005 (23 banks in total). Monthly stock returns have been obtained from the *Bolsa de Madrid* database. The market portfolio used is a modified version of the *Indice General de la Bolsa de Madrid (IGBM)*, the widest Spanish stock market index. Due to the major relevance of bank stocks in the IGBM, an alternative index where banks are excluded has been constructed in order to obtain a series of market returns as exogenous as possible. Market interest rates have been proxied by the monthly average yield on 10-year Spanish government bonds and the 3-month average rate of the Spanish interbank market, whereas the interest rate spread has been computed as the difference between them. Following [5] and [16], interest rate volatility has been measured by the conditional variance of interest rates, which is generated using a GARCH process.

To check whether there is a relationship between bank size and IRR, bank stocks have been sorted by size into three portfolios – large (L), medium (M) and small (S) banks. This classification (see Table 1) is based on the three categories of commercial banks typically distinguished in the Spanish banking industry. Thus, the L portfolio includes the banks that have given rise to the two currently multinational Spanish banking conglomerates (B. Santander and BBVA). The M portfolio is made up of a group of medium-sized Spanish banks that operate in national territory. Finally, the S portfolio is constituted by a broad set of small banks that operate mostly at a regional

Table 1. List of banks and composition of bank portfolios

Portfolios	Asset Volume (€ × 10³)	Obs.	Portfolios	Asset Volume (€ × 10³)	Obs.
Portfolio L					
BSCH	396,124,995	81	B. Bilbao Vizcaya	100,026,979	85
BBVA	297,433,664	71	Argentaria	69,998,972	80
B. Santander	113,404,303	75	B. Central Hispano	68,793,146	75
Portfolio M					
Banesto	42,332,585	156	Bankinter	15,656,910	156
B. Exterior	32,130,967	51	B. Pastor	8,789,945	156
B. Popular	29,548,620	156	B. Atlántico	7,591,378	138
B. Sabadell	26,686,670	56			
Portfolio S					
B. Zaragozano	4,597,099	130	B. Galicia	1,726,563	156
B. Valencia	4,213,420	156	B. de Vasconia	1,330,458	156
B. Guipuzcoano	4,082,463	156	B. de Vitoria	875,974	62
B. Andalucía	3,521,838	156	B. Crédito Balear	854,972	156
B. Herrero	2,624,824	95	B. Alicante	835,576	64
B. de Castilla	2,151,742	156	B. Simeón	686,451	67

This table displays the list of Spanish commercial banks considered and their distribution in portfolios according to size criteria (portfolios L, M and S).

level.[1] The formation of portfolios has a twofold advantage. First, it is an efficient way of condensing substantial amounts of information. Second, it helps to smooth out the noise in the data due to shocks to individual stocks. On the contrary, portfolios can mask the dissimilarities among banks within each portfolio. In this case, the mentioned advantages seem to outweigh this inconvenience, according to the number of papers based on bank stock portfolios (see, e.g., [5,6,17]). Monthly value-weighted portfolio returns have been obtained using year-end market capitalisation as the weight factor for each individual bank stock.

5 Empirical results

Table 2 contains the descriptive statistics of bank stock portfolio returns. They suggest that the data series are skewed and leptokurtic relative to the normal distribution. In addition, there is evidence of nonlinear dependence, possibly due to autoregressive heteroskedasticity. Overall, these diagnostics indicate that a GARCH-type process is appropriate for studying the IRR of bank stocks. Table 3 reports the parameters of the GARCH models estimated using the three alternative interest rate proxies.[2] The coefficient on the market return, λ_i, is highly significant, positive and less than unity in all cases. Further, its absolute value increases as the portfolio size increases, indicating that market risk is directly related to bank size. This is a relevant and unambiguous result, because it is not affected by the weight of banks in the market index since they have been explicitly excluded from the market portfolio. The fact that $\lambda_i < 1$ suggests that bank stock portfolios have less market risk than the overall stock market.

Table 2. Descriptive statistics of bank portfolio stock returns

	Mean	Variance	Skewness	Kurtosis	JB	Q(12)	Q(24)	$Q^2(12)$	$Q^2(24)$
L	0.016	0.006	−0.44***	5.15***	35.41***	9.63	12.55	49.59***	61.6***
M	0.011	0.002	−0.002	5.34***	35.82***	9.89	19.51	95.92***	109.5***
S	0.013	0.001	2.20***	13.42***	833.6***	29.28***	35.63*	25.93**	28.35

JB is the Jarque-Bera test statistic which tests the null hypothesis of normality of returns. Q(n) is the Ljung-Box statistic at a lag of n, which tests the presence of serial correlation. As usual ***, ** and * denote significance at the 1%, 5% and 10% levels, respectively.

---

[1] The composition of bank portfolios is fixed for the whole sample period. Alternatively, we have also considered an annual restructuring of the portfolios according to their volume of total assets, and the results obtained in that case were very similar to those reported in this paper.

[2] The final model to be estimated for portfolio S does not include the conditional variance of interest rates since its inclusion would generate serious problems in the estimation of the model due to the small variability of the returns on that portfolio.

Table 3. Maximum likelihood estimates of the GARCH-M extended model

ω_i	λ_i	θ_i	γ_i	α_0	α_1	β	δ_i
			3-month interest rate changes				
Portfolio L							
−0.01***	0.96***	−1.12	0.004***	0.0003***	0.09***	0.82***	−15.04***
(4.99)	(17.60)	(−0.09)	(9.47)	(10.22)	(5.67)	(54.94)	(−8.56)
Portfolio M							
0.02***	0.50***	−1.31	0.002***	0.0004***	0.15***	0.66***	−13.88***
(11.04)	(10.16)	(−1.17)	(8.87)	(18.63)	(4.74)	(27.83)	(−12.98)
Portfolio S							
−0.15***	0.27***	−1.31	−0.02***	0.00009***	0.03***	0.89***	—
(−53.73)	(5.12)	(−1.17)	(−58.56)	(12.89)	(6.33)	(148.20)	—
			10-year interest rate changes				
ω_i	λ_i	θ_i	γ_i	α_0	α_1	β	δ_i
Portfolio L							
0.03***	0.89***	−6.80***	0.003***	0.0004***	0.14***	0.79***	−45.34***
(9.08)	(15.38)	(−7.25)	(6.41)	(11.02)	(6.65)	(43.33)	(−8.97)
Portfolio M							
0.04***	0.48***	−3.19***	0.005***	0.0003***	0.11***	0.78***	−30.49***
(14.29)	(8.82)	(−3.04)	(12.18)	(19.48)	(5.19)	(48.76)	(−10.36)
Portfolio S							
−0.11***	0.25***	−3.28***	−0.01***	0.00009***	0.04***	0.87***	—
(−42.01)	(4.26)	(−3.37)	(−46.50)	(13.17)	(6.80)	(138.17)	—
			Interest rate spread				
ω_i	λ_i	θ_i	γ_i	α_0	α_1	β	δ_i
Portfolio L							
0.13***	0.95***	−0.32	0.018***	0.0003***	0.10**	0.80***	−10.00***
(3.24)	(15.82)	(−0.90)	(3.10)	(2.17)	(2.08)	(11.16)	(−3.79)
Portfolio M							
0.05***	0.51***	0.03	0.007***	0.0001***	0.08***	0.83***	−9.10***
(18.84)	(9.71)	(0.18)	(16.78)	(12.45)	(5.39)	(66.60)	(−5.40)
Portfolio S							
−0.18***	0.26***	−0.61***	−0.03***	0.00008***	0.02***	0.89***	—
(−66.93)	(5.20)	(−3.23)	(−75.23)	(12.67)	(5.90)	(155.75)	—

This table shows the maximum likelihood estimates of the GARCH(1,1)-M extended model for the different interest rate proxies based on equations (1)–(3). Values of t-statistics are in parentheses and ***, ** and * denote statistical significance at the 1%, 5% and 10% levels, respectively.

Concerning the impact of interest rate changes, θ_i is always negative and statistically significant when long-term rates are used. Long-term rates exert the strongest influence on bank stock portfolio returns, consistent with previous research (see, e.g., [3,5,6]).

The IRR of the Spanish banking industry also seems to be directly related to bank size. This finding may be attributed to three factors. First, the aggressive pricing policies – especially on the asset side – introduced by larger banks over recent years aimed to increase their market share in an environment of a sharp downward trend of interest rates and intense competition have led to an extraordinary increase of adjustable-rate products tied to interbank market rates. Second, the more extensive engagement of large banks in derivative positions. Third, large banks may have an incentive to assume higher risks induced by a moral hazard problem associated to their *too big to fail* status. As a result, the revenues and stock performance of bigger banks are now much more affected by market conditions. In contrast, more conservative pricing strategies of small banks, together with a minor use of derivatives and a heavier weight of idiosyncratic factors (e.g., rumours of mergers and acquisitions), can justify their lower exposure to IRR.

To provide greater insight into the relative importance of both market risk and IRR for explaining the variability of bank portfolio returns, a complementary analysis has been performed. A two-factor model as in [18] is the starting point:

$$R_{it} = \omega_i + \lambda_i R_{mt} + \theta_i \Delta I_t + \epsilon_{it} \qquad (4)$$

Since both explanatory variables are linearly independent by construction, the variance of the return of each bank stock portfolio, $Var(R_{it})$, can be written as:

$$Var(R_{it}) = \hat{\lambda}_i^2 Var(R_{mt}) + \hat{\theta}_i^2 Var(\Delta I_t) + Var(\epsilon_{it}) \qquad (5)$$

To compare both risk factors, equation (5) has been divided by $Var(R_{it})$. Thus, the contribution of each individual factor can be computed as the ratio of its variance over the total variance of the bank portfolio return. As shown in Table 4, the market risk is indisputably the most important determinant of bank returns. IRR is comparatively less relevant, long-term rates being the ones which show greater incidence.

Table 4. Relative importance of risk factors

		Interest rate changes								
		3 months			10 years			Spread		
		ΔI_t	R_{mt}	Total	ΔI_t	R_{mt}	Total	ΔI_t	R_{mt}	Total
Portfolio L	$R^2(\%)$	0.85	53.84	54.69	2.81	51.77	54.58	1.22	53.47	54.69
Portfolio M	$R^2(\%)$	1.30	34.21	35.52	2.74	32.78	35.52	1.19	34.83	36.02
Portfolio S	$R^2(\%)$	1.24	15.19	16.42	5.59	12.40	17.99	1.08	15.35	16.43

This table shows the contribution of interest rate and market risks, measured through the factor R^2 obtained from equation (5) in explaining the total variance of bank portfolio returns.

Turning to the mean equation of the GARCH-M model, the parameter γ_i has usually been interpreted as the compensation required to invest in risky assets by risk-averse investors. Since volatility as measured in GARCH models is not a measure of systematic risk, but total risk, γ_i does not necessarily have to be positive because increases of total risk do not always imply higher returns.[3] For our case, the estimated values for γ_i differ in sign across bank portfolios (positive for portfolios L and M and negative for portfolio S). This heterogeneity among banks may be basically derived from differences in product and client specialisation, interest rate hedging strategies, etc. The absence of a conclusive result concerning this parameter is in line with the lack of consensus found in prior research. In this sense, whereas [12] and [4] detected a positive relationship between risk and return ($\gamma_i > 0$), [5,9,13] suggested a negative relationship ($\gamma_i < 0$). In turn, [2] and [16] found an insignificant γ_i.

With regard to the conditional variance equation, α_1 and β are positive and significant in the majority of cases. In addition, the volatility persistence ($\alpha_1 + \beta$) is always less than unity, consistent with the stationarity conditions of the model. This implies that the traditional constant-variance capital asset pricing models are inappropriate for describing the distribution of bank stock returns in the Spanish case.

The parameter δ_i, which measures the effect of interest rate volatility on bank portfolio return volatility, is negative and significant for portfolios L and M.[4] A possible explanation suggested by [5] is that, in response to an increase in interest rate volatility, L and M banks seek shelter from IRR and are able to reduce their exposure within one month, e.g., by holding derivatives and/or reducing the duration gap of their assets and liabilities. Hence, this generates a lower bank stock volatility in the following period. Moreover, a direct relationship seems to exist between the absolute value of δ_i, the bank size and the term to maturity of interest rates. Thus, analogously to the previous evidence with interest rate changes, interest rate volatility has a greater negative effect on bank return volatility as the bank size increases. Further, interest rate volatility has a larger impact when long-term rates are considered. In sum, it can be concluded that the Spanish bank industry does show a significant interest rate exposure, especially with respect to long-term interest rates.

In addition, the proposed GARCH model has been augmented with the purpose of checking whether the introduction of the euro as the single currency within the Monetary European Union from January 1, 1999 has significantly altered the degree of IRR of Spanish banks.[5] Thus, the following extended model has been esti-

---

[3] [13] indicates several reasons for the relationship between risk and return being negative. In the framework of the financial sector, [5] also suggests an explanation to get a negative trade-off coefficient between risk and return.

[4] Recall that this parameter does not appear in the model for portfolio S.

[5] Since the GARCH model estimation requires a considerable number of observations, a dummy variable procedure has been employed instead of estimating the model for each subperiod.

mated:

$$R_{it} = \omega_i + \lambda_i R_{mt} + \theta_i \Delta I_t + \eta_i D_t \Delta I_t + \gamma_i \log h_{it} + \epsilon_{it} \tag{6}$$
$$h_{it} = \alpha_0 + \alpha_1 \epsilon_{it-1}^2 + \beta h_{it-1} + \delta_i V C I_{t-1} \tag{7}$$
$$\epsilon_{it} | \Omega_{t-1} \sim N(0, h_{it}) \tag{8}$$

where $D_t = 1$ if $t \leq$ January 1999 and $D_t = 0$ if $t >$ January 1999. Its associated coefficient, η_i, reflects the differential impact in terms of exposure to IRR during the pre-euro period. The results are reported in Table 5.

Table 5. Maximum likelihood estimates of the GARCH-M extended model with dummy variable

	Portfolio L			Portfolio M			Portfolio S		
	3 month	10 years	Spread	3 months	10 years	Spread	3 month	10 years	Spread
θ_i	1.94	−3.44	−1.88***	2.42	2.59	±0.73***	−1.69	−0.67	−0.17
η_i	−6.52**	−4.69**	1.52**	−4.57***	−6.53***	0.95***	0.13	−3.43***	−0.61**

This table shows the IRR estimated parameters in the GARCH-M model following (6)–(8). *** ,** and * denote statistical significance at the 1%, 5% and 10% levels, respectively.

The coefficient η_i is negative and significant at the usual levels in most cases with the long- and short-term interest rate changes, whereas the results are not totally conclusive with the spread series. This finding shows that the IRR is substantially higher during the pre-euro period, in line with prior evidence (see [15]) and indicating that interest rate sensitivity of bank stock returns has decreased considerably since the introduction of the euro. The declining bank IRR during the last decade can be basically attributed to the adoption of a more active role in asset-liability management by the banks in response to the increase of volatility in financial markets, which has led to more effective IRR management.

6 Conclusions

This paper examines the interest rate exposure of the Spanish banking sector within the framework of the GARCH-M. In particular, the analysis has been carried out on bank stock portfolios constructed according to size criteria. Following the most recent strand of research, this study investigates the impact of both interest rate changes and interest rate volatility on the distribution of bank stock returns.

The results confirm the common perception that interest rate risk is a significant factor to explain the variability in bank stock returns but, as expected, it plays a secondary role in comparison with market risk. Consistent with previous work, bank stock portfolio returns are negatively correlated with changes in interest rates, the

long-term rates being the ones which exert greater influence. This negative relationship has been mostly attributed to the typical maturity mismatch between banks'
assets and liabilities. Another explanation is closely linked to the expansion phase
of the Spanish economy since the mid-1990s. Specifically, bank profits did increase
dramatically, reaching their greatest figures ever, with the subsequent positive effect
on stock prices, in a context of historically low interest rates within the framework
of the Spanish housing boom. Further, interest rate volatility is also found to be a
significant determinant of bank portfolio return volatility, with a negative effect.

Another major result refers to the direct relationship found between bank size and
interest rate sensitivity. This size-based divergence could be the result of differences
between large and small banks in terms of bank pricing policy, extent of use of
derivative instruments or product and client specialisation. Thus, larger banks have
a stock performance basically driven by market conditions, whereas smaller banks
are influenced more heavily by idiosyncratic risk factors. Finally, a decline of bank
interest rate sensitivity during recent years has been documented, which may be linked
to the greater availability of systems and instruments to manage and hedge interest
rate risk.

Acknowledgement. The authors are grateful for the financial support from the Spanish Ministry
of Education and Science and FEDER Funds, grant number SEJ2005-08931-C02-02/ECON.

References

1. Bae, S.C.: Interest rate changes and common stock returns of financial institutions: revisited. J. Finan. Res. 13, 71–79 (1990)
2. Baillie, R.T., DeGennaro, R.P.: Stock returns and volatility. J. Finan. Quant. Anal. 25,
   203–214 (1990)
3. Bartram, S.M.: The interest rate exposure of nonfinancial corporations. Eur. Finan. Rev.
   6, 101–125 (2002)
4. Campbell, J.Y., Hentschel, L.: No news is good news: An asymmetric model of changing
   volatility in stocks returns. J. Finan. Econ. 31, 281–318 (1992)
5. Elyasiani, E., Mansur, I.: Sensitivity of the bank stock returns distribution to changes
   in the level and volatility of interest rate: A GARCH-M model. J. Banking Finan. 22,
   535–563 (1998)
6. Elyasiani, E., Mansur, I.: Bank stock return sensitivities to the long-term and short-term
   interest Rate: A multivariate GARCH approach. Managerial Finan. 30, 32–45 (2004)
7. Engle, R.F., Lilien, D.M., Robins, R.P.: Estimating time varying risk premia in the term
   structure: The ARCH-M model. Econometrica 55, 391–407 (1987)
8. Engle, R.F., Ng, V.K.: Measuring and testing the impact of news on volatility. J. Finan.
   48, 1749–1778 (1993)
9. Faff, R.W., Hodgson, A., Kremmer, M.L.: An investigation of the impact of interest rates
   and interest rate volatility on Australian financial sector stock return distributions. J. Bus.
   Finan. Acc. 32, 1001–1032 (2005)
10. Flannery, M.J., James, C.M.: The effect of interest rate changes on the common stock
    returns of financial institutions. J. Finan. 39, 1141–1153 (1984)

11. Fraser, D.R., Madura, J., Weigand, R.A.: Sources of bank interest rate risk. Finan. Rev. 37, 351–368 (2002)
12. French, K.R., Schwert, G.W., Stambaugh, R.F.: Expected stock returns and volatility. J. Finan. Econ. 19, 3–29 (1987)
13. Glosten, L.R., Jagannathan, R., Runkle, D.: On the relationship between the expected value and the volatility on the nominal excess returns on stocks. J. Finan. 48, 1779–1801 (1993)
14. Kane, E.J., Unal, H.: Change in market assessment of deposit institution riskiness. J. Finan. Serv. Res. 2, 201–229 (1988)
15. Kwan, S.H.: Reexamination of interest rate sensitivity of commercial bank stock returns using a random coefficient model. J. Finan. Serv. Res. 5, 61–76 (1991)
16. Ryan, S.K., Worthington, A.C.: Market, interest rate and foreign exchange rate risk in Australian banking: A GARCH-M approach. Int. J. Appl. Bus. Econ. Res. 2, 81–103 (2004)
17. Song, F.: A two factor ARCH model for deposit-institution stock returns. J. Money Credit Banking 26, 323–340 (1994)
18. Stone, B.K.: Systematic interest rate risk in a two-index model of returns. J. Finan. Quant. Anal. 9, 709–721 (1974)

Tracking error with minimum guarantee constraints

Diana Barro and Elio Canestrelli

Abstract. In recent years the popularity of indexing has greatly increased in financial markets and many different families of products have been introduced. Often these products also have a minimum guarantee in the form of a minimum rate of return at specified dates or a minimum level of wealth at the end of the horizon. Periods of declining stock market returns together with low interest rate levels on Treasury bonds make it more difficult to meet these liabilities. We formulate a dynamic asset allocation problem which takes into account the conflicting objectives of a minimum guaranteed return and of an upside capture of the risky asset returns. To combine these goals we formulate a double tracking error problem using asymmetric tracking error measures in the multistage stochastic programming framework.

Key words: minimum guarantee, benchmark, tracking error, dynamic asset allocation, scenario

1 Introduction

The simultaneous presence of a benchmark and a minimum guaranteed return characterises many structured financial products. The objective is to attract potential investors who express an interest in high stock market returns but also are not risk-seeking enough to fully accept the volatility of this investment and require a cushion. This problem is of interest also for the asset allocation choices for pension funds both in the case of defined benefits (which can be linked to the return of the funds) and defined contribution schemes in order to be able to attract members to the fund. Moreover, many life insurance products include an option on a minimum guaranteed return and a minimum amount can be guaranteed by a fund manager for credibility reasons. Thus the proper choice of an asset allocation model is of interest not only for investment funds or insurance companies that offer products with investment components, but also for pension fund industry.

In the literature there are contributions which discuss the two components separately, and there are contributions which discuss the tracking error problem when a Value at Risk (VaR), Conditional Value at Risk (CVaR) or Maximum Drawdown (MD) constraint is introduced mainly in a static framework, but very few contributions address the dynamic portfolio management problem when both a minimum guarantee and a tracking error objective are present; see for example [14]. To jointly model these goals we work in the stochastic programming framework as it has proved to be flexible enough to deal with many different issues which arise in the formulation and solution of these problems. We do not consider the point of view of an investor who wants to maximise the expected utility of his wealth along the planning horizon or at the end of the investment period. Instead we consider the point of view of a manager of a fund, thus representing a collection of investors, who is responsible for the management of a portfolio connected with financial products which offer not only a minimum guaranteed return but also an upside capture of the risky portfolio returns. His goals are thus conflicting since in order to maximise the upside capture he has to increase the total riskiness of the portfolio and this can result in a violation of the minimum return guarantee if the stock market experiences periods of declining returns or if the investment policy is not optimal. On the other hand a low risk profile on the investment choices can assure the achievement of the minimum return guarantee, if properly designed, but leaves no opportunity for upside capture.

2 Minimum guaranteed return and constraints on the level of wealth

The relevance of the introduction of minimum guaranteed return products has grown in recent years due to financial market instability and to the low level of interest rates on government (sovereign) and other bonds. This makes it more difficult to fix the level of the guarantee in order to attract potential investors. Moreover, this may create potential financial instability and defaults due to the high levels of guarantees fixed in the past for contracts with long maturities, as the life insurance or pension fund contracts. See, for example, [8, 20, 31].

A range of guarantee features can be devised such as rate-of-return guarantee, including the principal guarantee, i.e., with a zero rate of return, minimum benefit guarantee and real principal guarantee. Some of them are more interesting for participants in pension funds while others are more relevant for life insurance products or mutual funds. In the case of minimum return guarantee, we ensure a deterministic positive rate of return (given the admissibility constraints for the attainable rate of returns); in the minimum benefit a minimum level of payments are guaranteed, at retirement date, for example. In the presence of nominal guarantee, a fixed percentage of the initial wealth is usually guaranteed for a specified date while real or flexible guarantees are usually connected to an inflation index or a capital market index.

The guarantee constraints can be chosen with respect to the value of terminal wealth or as a sequence of (possibly increasing) guaranteed returns. This choice may be led by the conditions of the financial products linked to the fund. The design of the guarantee is a crucial issue and has a consistent impact on the choice of management strategies.

Not every value of minimum guarantee is reachable; no arbitrage arguments can be applied. The optimal design of a minimum guarantee has been considered and discussed in the context of pension fund management in [14]. Muermann et al. [26] analyses the willingness of participants in a defined contribution pension fund to pay for a guarantee from the point of view of regret analysis.

Another issue which has to be tackled in the formulation is the fact that policies which give a minimum guaranteed return usually provide to policyholders also a certain amount of the return of the risky part of the portfolio invested in the equity market. This reduces the possibility of implementing a portfolio allocation based on Treasury bonds since no upside potential would be captured. The main objective is thus a proper combination of two conflicting goals, namely a guaranteed return, i.e., a low profile of risk, and at least part of the higher returns which could be granted by the equity market at the cost of a high exposure to the risk of not meeting the minimum return requirement.

The first possibility is to divide the investment decision into two steps. In the first the investor chooses the allocation strategy without taking care of the guarantee, while in the second step he applies a dynamic insurance strategy (see for example [15]).

Consiglio et al. [9] discuss a problem of asset and liability management for UK insurance products with guarantees. These products offer the owners both a minimum guaranteed rate of return and the possibility to participate in the returns of the risky part of the portfolio invested in the equity market. The minimum guarantee is treated as a constraint and the fund manager maximises the certainty equivalent excess return on equity (CEexROE). This approach is flexible and allows one to deal also with the presence of bonuses and/or target terminal wealth.

Different contributions in the literature have tackled the problem of optimal portfolio choices with the presence of a minimum guarantee both in continuous and discrete time also from the point of view of portfolio insurance strategies both for a European type guarantee and for an American type guarantee, see for example [10, 11].

We consider the problem of formulating and solving an optimal allocation problem including minimum guarantee requirements and participation in the returns generated from the risky portfolio. These goals can be achieved both considering them as constraints or including them in the objective function. In the following we will analyse in more detail the second case in the context of dynamic tracking error problems, which in our opinion provide the more flexible framework.

3 Benchmark and tracking error issues

The introduction of benchmarks and of indexed products has greatly increased since the Capital Asset Pricing Model (see [23,25,28]) promoted a theoretical basis for index funds. The declaration of a benchmark is particularly relevant in the definition of the risk profile of the fund and in the evaluation of the performance of funds' managers. The analysis of the success in replicating a benchmark is conducted through tracking error measures.

Considering a given benchmark, different sources of tracking error can be analysed and discussed, see, for example [19]. The introduction of a liquidity component in the management of the portfolio, the choice of a partial replication strategy, and management expenses, among others, can lead to tracking errors in the replication of the behaviour of the index designed as the benchmark. This issue is particularly relevant in a pure passive strategy where the goal of the fund manager is to perfectly mime the result of the benchmark, while it is less crucial if we consider active asset allocation strategies in which the objective is to create overperformance with respect to the benchmark. For instance, the choice of asymmetric tracking error measures allows us to optimise the portfolio composition in order to try to maximise the positive deviations from the benchmark. For the use of asymmetric tracking error measures in a static framework see, for example, [16,22,24,27].

For a discussion on risk management in the presence of benchmarking, see Basak et al. [4]. Alexander and Baptista [1] analyse the effect of a drawdown constraint, introduced to control the shortfall with respect to a benchmark, on the optimality of the portfolios in a static framework.

We are interested in considering dynamic tracking error problems with a stochastic benchmark. For a discussion on dynamic tracking error problems we refer to [2,5,7, 13,17].

4 Formulation of the problem

We consider the asset allocation problem for a fund manager who aims at maximising the return on a risky portfolio while preserving a minimum guaranteed return. Maximising the upside capture increases the total risk of the portfolio. This can be balanced by the introduction of a second goal, i.e., the minimisation of the shortfall with respect to the minimum guarantee level.

We model the first part of the objective function as the maximisation of the overperformance with respect to a given stochastic benchmark. The minimum guarantee itself can be modelled as a, possibly dynamic, benchmark. Thus the problem can be formalized as a double tracking error problem where we are interested in maximising the positive deviations from the risky benchmark while minimising the downside distance from the minimum guarantee. The choice of asymmetric tracking error measures allows us to properly combine the two goals.

To describe the uncertainty, in the context of a multiperiod stochastic programming problem, we use a scenario tree. A set of scenarios is a collection of paths from $t = 0$ to T, with probabilities π_{k_t} associated to each node k_t in the path: according to the information structure assumed, this collection can be represented as a scenario tree where the current state corresponds to the root of the tree and each scenario is represented as a path from the origin to a leaf of the tree.

If we fix it as a minimal guaranteed return, without any requirement on the upside capture we obtain a problem which fits the portfolio insurance framework, see, for example, [3,6,18,21,29]. For portfolio insurance strategies there are strict restrictions on the choice of the benchmark, which cannot exceed the return on the risk-free security for no arbitrage conditions.

Let x_{k_t} be the value of the risky benchmark at time t in node k_t; z_t is the value of the lower benchmark, the minimum guarantee, which can be assumed to be constant or have deterministic dynamics, thus it does not depend on the node k_t. We denote with y_{k_t} the value of the managed portfolio at time t in node k_t. Moreover let $\phi_{k_t}(y_{k_t}, x_{k_t})$ be a proper tracking error measure which accounts for the distance between the managed portfolio and the risky benchmark, and $\psi_{k_t}(y_{k_t}, z_t)$ a distance measure between the risky portfolio and the minimum guarantee benchmark. The objective function can be written as

$$\max_{y_{k_t}} \sum_{t=0}^{T} \left[\alpha_t \sum_{k_t=K_{t-1}+1}^{K_t} \phi_{k_t}(y_{k_t}, x_{k_t}) - \beta_t \sum_{k_t=K_{t-1}+1}^{K_t} \psi_{k_t}(y_{k_t}, z_t) \right] \quad (1)$$

where α_t and β_t represent sequences of positive weights which can account both for the relative importance of the two goals in the objective function and for a time preference of the manager. For example, if we consider a pension fund portfolio management problem we can assume that the upside capture goal is preferable at the early stage of the investment horizon while a more conservative strategy can be adopted at the end of the investment period. A proper choice of ϕ_t and ψ_t allows us to define different tracking error problems.

The tracking error measures are indexed along the planning horizon in such a way that we can monitor the behaviour of the portfolio at each trading date t. Other formulations are possible. For example, we can assume that the objective of a minimum guarantee is relevant only at the terminal stage where we require a minimum level of wealth z_T

$$\max_{y_{k_t}} \sum_{t=0}^{T} \left[\alpha_t \sum_{k_t=K_{t-1}+1}^{K_t} \phi_{k_t}(y_{k_t}, x_{k_t}) \right] - \beta_T \sum_{k_T=K_{T-1}+1}^{K_T} \psi_{k_T}(y_{k_T}, z_T). \quad (2)$$

The proposed model can be considered a generalisation of the tracking error model of Dembo and Rosen [12], who consider as an objective function a weighted average of positive and negative deviations from a benchmark. In our model we consider two different benchmarks and a dynamic tracking problem.

The model can be generalised in order to take into account a monitoring of the shortfall more frequent than the trading dates, see Dempster et al. [14].

We consider a decision maker who has to compose and manage his portfolio using $n = n_1 + n_2$ risky assets and a liquidity component. In the following $q_{i\,k_t}$, $i = 1, \ldots, n_1$, denotes the position in the ith stock and $b_{j\,k_t}, j = 1, \ldots, n_2$ denotes the position in the jth bond while c_{k_t} denotes the amount of cash.

We denote with $r_{k_t} = (r_{1\,k_t}, \ldots, r_{n\,k_t})$ the vector of returns of the risky assets for the period $[t-1, t]$ in node k_t and with $r_{c\,k_t}$ the return on the liquidity component in node k_t. In order to account for transaction costs and liquidity component in the portfolio we introduce two vector of variables $a_{k_t} = (a_{1\,k_t}, \ldots, a_{n\,k_t})$ and $v_{k_t} =$

$(v_{1k_t}, \ldots, v_{nk_t})$ denoting the value of each asset purchased and sold at time t in node k_t, while we denote with κ^+ and κ^- the proportional transaction costs for purchases and sales.

Different choices of tracking error measures are possible and different trade-offs between the goals on the minimum guarantee side and on the enhanced tracking error side, for the risky benchmark, are possible, too. In this contribution we do not tackle the issue of comparing different choices for tracking error measures and trade-offs in the goals with respect to the risk attitude of the investor. Among different possible models, we propose the absolute downside deviation as a measure of tracking error between the managed portfolio and the minimum guarantee benchmark, while we consider only the upside deviations between the portfolio and the risky benchmark

$$\phi_{k_t}(y_{k_t}, x_{k_t}) = [y_{k_t} - x_{k_t}]^+ = \theta_{k_t}^+; \tag{3}$$

$$\psi_{k_t}(y_{k_t}, z_t) = [y_{k_t} - z_t]^- = \gamma_{k_t}^-, \tag{4}$$

where $[y_{k_t} - x_{k_t}]^+ = \max[y_{k_t} - x_{k_t}, 0]$ and $[y_{k_t} - z_t]^- = -\min[y_{k_t} - z_t, 0]$. The minimum guarantee can be assumed constant over the entire planning horizon or it can follow a deterministic dynamics, i.e, it is not scenario dependent. Following [14] we assume that there is an annual guaranteed rate of return denoted with ρ. If the initial wealth is $W_0 = \sum_{i=1}^{n+1} x_{i0}$, then the value of the guarantee at the end of the planning horizon is $W_T = W_0(1 + \rho)^T$. At each intermediate date the value of the guarantee is given by $z_t = e^{\delta(t,T-t)(T-t)} W_0(1 + \rho)^T$, where $e^{\delta(t,T-t)(T-t)}$ is a discounting factor, i.e., the price at time t of a zcb which pays 1 at terminal time T.

The objective function becomes a weighted trade-off between negative deviations from the minimum guarantee and positive deviations from the risky benchmark. Given the choice for the minimum guarantee, the objective function penalises the negative deviations from the risky benchmark only when these deviations are such that the portfolio values are below the minimum guarantee and penalises them for the amounts which are below the minimum guarantee. Thus, the choice of the relative weights for the two goals is crucial in the determination of the level of risk of the portfolio strategy.

The obtained dynamic tracking error problem in its arborescent form is

$$\max_{q_{k_t}, b_{k_t}, c_{k_t}} \sum_{t=1}^{T} \left[\alpha_t \sum_{k_t=K_{t-1}+1}^{K_t} \theta_{k_t}^+ - \beta_t \sum_{k_t=K_{t-1}+1}^{K_t} \gamma_{k_t}^- \right] \tag{5}$$

$$\theta_{k_t}^+ - \theta_{k_t}^- = y_{k_t} - x_{k_t} \tag{6}$$

$$-\gamma_{k_t}^- \leq y_{k_t} - z_t \tag{7}$$

$$y_{k_t} = c_{k_t} + \sum_{i=1}^{n_1} q_{i\,k_t} + \sum_{j=1}^{n_2} b_{j\,k_t} \tag{8}$$

$$q_{i\,k_t} = (1 + r_{i\,k_t})\left[q_{i\,f(k_t)} + a_{i\,f(k_t)} - v_{i\,f(k_t)}\right] \quad i = 1, \ldots, n_1 \qquad (9)$$

$$b_{j\,k_t} = (1 + r_{j\,k_t})\left[b_{j\,f(k_t)} + a_{j\,f(k_t)} - v_{j\,f(k_t)}\right] \quad j = 1, \ldots, n_2 \quad (10)$$

$$c_{k_t} = (1 + r_{c\,k_t})\left[c_{f(k_t)} - \sum_{i=1}^{n_1}(\kappa^+)a_{i f(k_t)} + \sum_{i=1}^{n_1}(\kappa^-)v_{i f(k_t)}\right.$$

$$\left. + \sum_{j=1}^{n_2}(\kappa^+)a_{j f(k_t)} + \sum_{j=1}^{n_2}(\kappa^-)v_{j f(k_t)} + \sum_{j=1}^{n_2}g_{k_t}b_{j f(k_t)}\right] \qquad (11)$$

$$a_{i\,k_t} \geq 0 \ v_{i\,k_t} \geq 0 \ i = 1, \ldots, n_1 \qquad (12)$$

$$a_{j\,k_t} \geq 0 \ v_{j\,k_t} \geq 0 \ j = 1, \ldots, n_2 \qquad (13)$$

$$q_{i\,k_t} \geq 0 \ i = 1, \ldots, n_1 \qquad (14)$$

$$b_{j\,k_t} \geq 0 \ j = 1, \ldots, n_2 \qquad (15)$$

$$\theta_{k_t}^+ \theta_{k_t}^- = 0 \qquad (16)$$

$$\theta_{k_t}^+ \geq 0 \ \theta_{k_t}^- \geq 0 \qquad (17)$$

$$\gamma_{k_t}^- \geq 0 \qquad (18)$$

$$c_{k_t} \geq 0 \qquad (19)$$

$$q_{i0} = \bar{q}_i \ i = 1, \ldots, n_1 \qquad (20)$$

$$b_{j0} = \bar{b}_j \ j = 1, \ldots, n_2 \qquad (21)$$

$$c_0 = \bar{c} \qquad (22)$$

$$k_t = K_{t-1} + 1, \ldots, K_t$$

$$t = 1, \ldots, T$$

where equation (8) represents the portfolio composition in node k_t; equations (9)–(11) describe the dynamics of the amounts of stocks, bonds and cash in the portfolio moving from the ancestor node $f(k_t)$, at time $t-1$, to the descendent nodes k_t, at time t, with $K_0 = 0$. In equation (11), with g_{k_t} we denote the inflows from the bonds in the portfolio. Equation (16) represents the complementarity conditions which prevent positive and negative deviations from being different from zero at the same time. Equations (20)–(22) give the initial endowments for stocks, bonds and cash.

We need to specify the value of the benchmark and the value of the minimum guarantee at each time and for each node. The stochastic benchmark y_{k_t} and the prices of the risky assets in the portfolio must be simulated according to given stochastic processes in order to build the corresponding scenario trees. Other dynamics for the minimum guaranteed level of wealth can be designed. In particular, we can discuss a time-varying rate or return ρ_t along the planning horizon, or we can include the accrued bonuses as in [8].

A second approach to tackle the problem of the minimum return guarantee is to introduce probabilistic constraints in the dynamic optimisation problem. Denoting with θ the desired confidence level, we can formulate the shortfall constraints both on the level of wealth at an intermediate time t and on the terminal wealth as

follows
$$Pr(W_t \leq z_t) \leq 1 - \theta \quad Pr(W_T \leq z_T) \leq 1 - \theta$$

where W_t is the random variable representing the level of wealth. Under the assumption of a discrete and finite number of realisations we can compute the shortfall probability using the values of the wealth in each node $W_{k_t} = \sum_{i=1}^{n+1} x_{i\,k_t}$. This gives rise to a chance constrained stochastic optimisation problem which can be extremely difficult to solve due to non-convexities which may arise, see [14].

5 Conclusions

We discuss the issue of including in the formulation of a dynamic portfolio optimisation problem both a minimum return guarantee and the maximisation of the potential returns from a risky portfolio. To combine these two conflicting goals we formulate them in the framework of a double dynamic tracking error problem using asymmetric tracking measures.

References

1. Alexander, G.J., Baptista, A.M.: Portfolio selection with a drawdown constraint. J. Banking Finan. 30, 3171–3189 (2006)
2. Barro, D., Canestrelli, E.: Tracking error: a multistage portfolio model. Ann. Oper. Res. 165(1), 44–66 (2009)
3. Basak, S.: A general equilibrium of portfolio insurance. Rev. Finan. Stud. 8, 1059–1090 (1995)
4. Basak, S., Shapiro, A., Tepla, L.: Risk management with benchmarking. Working paper series asset management SC-AM-03-16, Salomon Center for the Study of Financial Institutions, NYU (2003)
5. Boyle, P., Tian, W.: Optimal portfolio with constraints. Math. Finan. 17, 319–343 (2007)
6. Brennan, M., Solanki, R.: Optimal portfolio insurance. J. Finan. Quant. Anal. 16, 279–300 (1981)
7. Browne, S.: Beating a moving target: optimal portfolio strategies for outperforming a stochastic benchmark. Finan. Stochastics 3, 255–271 (1999)
8. Consiglio, A., Cocco, F., Zenios, S.A.: The Prometeia model for managing insurance policies with guarantees. In: Zenios, S.A. and Ziemba, W.T. (Eds.) Handbook of Asset and Liability Management, vol. 2, pp. 663–705. North-Holland (2007)
9. Consiglio, A., Saunders, D., Zenios, S.A.: Asset and liability management for insurance products with minimum guarantees: The UK case. J. Banking Finan. 30, 645–667 (2006)
10. Deelstra, G., Grasselli, M., Koehl, P.-F.: Optimal investment strategies in the presence of a minimum guarantee. Ins. Math. Econ. 33, 189–207 (2003)
11. Deelstra, G., Grasselli, M., Koehl, P.-F.: Optimal design of the guarantee for defined contribution funds. J. Econ. Dyn. Control 28, 2239–2260 (2004)
12. Dembo, R., Rosen, D.: The practice of portfolio replication, a practical overview of forward and inverse probems. Ann. Oper. Res. 85, 267–284 (1999)

13. Dempster, M.H.A., Thompson, G.W.P.: Dynamic portfolio replication usign stochastic programming. In: Dempster, M.H.A. (ed.) Risk Management: Value at Risk and Beyond, pp. 100–128. Cambridge Univeristy Press (2001)

14. Dempster, M.A.H., Germano, M., Medova, E.A., Rietbergen, M.I., Sandrini, F., and Scrowston, M.: Designing minimum guaranteed return funds. Research Papers in Management Studies, University of Cambridge, WP 17/2004 (2004)

15. El Karoui, N., Jeanblanc, M., Lacoste, V.: Optimal portfolio management with American capital guarantee. J. Econ. Dyn. Control 29, 449–468 (2005)

16. Franks, E.C.: Targeting excess-of-benchmark returns. J. Portfolio Man. 18, 6–12 (1992)

17. Gaivoronski, A., Krylov, S., van der Vijst, N.: Optimal portfolio selection and dynamic benchmark tracking. Eur. J. Oper. Res. 163, 115–131 (2005)

18. Grossman, S.J., Zhou, Z.: Equilibrium analysis of portfolio insurance. J. Finan. 51, 1379–1403 (1996)

19. Halpern, P., Kirzner, E.: Quality standards for index products. Working paper. Rotman School of Management, University of Toronto (2001)

20. Hochreiter, R., Pflug, G., Paulsen, V.: Design and management of Unit-linked Life insurance Contracts with guarantees. In: Zenios, S.A. and Ziemba W.T. (eds.) Handbook of asset and liability management vol. 2, pp. 627–662. North-Holland (2007)

21. Jensen, B.A., Sorensen, C.: Paying for minimal interest rate guarantees: Who should compensate whom? Eur. Finan. Man. 7, 183–211 (2001)

22. Konno, H., Yamazaki, H.: Mean absolute deviation portfolio optimization model and its applications to Tokyo stock market. Man. Sci. 37, 519–531 (1991)

23. Lintner, J.: The valuation of risky assets and the selection of risky investments in stock portfolios and capital budget. Rev. Econ. Stat. 47, 13–37 (1965)

24. Michalowski, W., Ogryczak, W.: Extending the MAD portfolio optimization model to incorporate downside risk aversion. Naval Res. Logis. 48, 185–200 (2001)

25. Mossin, J.: Equilibrium in a Capital Asset Market. Econ. 34, 768–783 (1966)

26. Muermann, A., Mitchell, O.S., Volkman, J.M.: Regret, portfolio choice, and guarantees in defined contribution schemes. Insurance: Math. Econ. 39, 219–229 (2006)

27. Rudolf, M., Wolter, H.-J., Zimmermann, H.: A linear model for tracking error minimization. J. Banking Finan. 23, 85–103 (1999)

28. Sharpe, W.F.: Capital asset prices: a theory of market equilibrium under conditions of risk. J. Finan. 19(3), 425–442 (1964)

29. Sorensen, C.: Dynamic asset allocation and fixed income management. J. Finan. Quant. Anal. 34, 513–531 (1999)

30. Zenios, S.A., Ziemba, W.T. (eds.): Handbook of Asset and Liability Management, vol. 2. North-Holland (2007)

31. Ziemba, W.T.: The Russel-Yasuda Kasai, InnoALM and related models for pensions, Insurance companies and high net worth individuals. In: Zenios, S.A., Ziemba, W.T. (eds.) Handbook of Asset and Liability Management, vol. 2, pp. 861–962. North-Holland (2007)

Energy markets: crucial relationship between prices

Cristina Bencivenga, Giulia Sargenti, and Rita L. D'Ecclesia

Abstract. This study investigates the relationship between crude oil, natural gas and electricity prices. A possible integration may exist and it can be measured using a cointegration approach. The relationship between energy commodities may have several implications for the pricing of derivative products and for risk management purposes. Using daily price data for Brent crude oil, NBP UK natural gas and EEX electricity we analyse the short- and long-run relationship between these markets. An unconditional correlation analysis is performed to study the short-term relationship, which appears to be very unstable and dominated by noise. A long-run relationship is analysed using the Engle-Granger cointegration framework. Our results indicate that gas, oil and electricity markets are integrated. The framework used allows us to identify a short-run relationship.

Key words: energy commodities, correlation, cointegation, market integration

1 Introduction

Energy commodities have been a leading actor in the economic and financial scene in the last decade. The deregulation of electricity and gas markets in western countries caused a serious change in the dynamic of electricity and gas prices and necessitated the adoption of adequate risk management strategies. The crude oil market has also been also experiencing serious changes over the last decade caused by economic and political factors. The deregulation of gas and electricity markets should cause, among other things, more efficient price formation of these commodities. However their dependence on oil prices is still crucial. An analysis of how these commodities are related to each other represents a milestone in the definition of risk measurement and management tools.

For years natural gas and refined petroleum products have been used as close substitutes in power generation and industry. As a consequence, movements of natural gas prices have generally tracked those of crude oil. This brought academics and practitioners to use a simple rule of thumb to relate natural gas prices to crude oil prices according to which a simple deterministic function may be able to explain the relationships between them (see, e.g., [7]). Recently the number of facilities able to switch between natural gas and residual fuel oil has declined, so gas prices seem to move more independently from oil prices. However, to a certain extent, oil prices are expected to remain the main drivers of energy prices through inter-fuel competition and price indexation clauses in some long-term gas contracts.

Finally, the high price volatility in the energy commodity markets boosted the development of energy derivative instruments largely used for risk management. In particular, spread options have been largely used, given that the most useful and important structure in the world of energy is represented by the spread.[1] The joint behaviour of commodity prices as well as gas, oil and electricity, is crucial for a proper valuation of spread contracts. This requires a real understanding of the nature of volatility and correlation in energy markets.

The aim of this paper is twofold. First, to investigate the short-run relationship between oil, natural gas and electricity in the European energy markets. Second, to identify possible long-run equilibrium relationships between these commodities. In particular we test for shared price trends, or common trends, in order to detect if natural gas and electricity are driven by a unique source of randomness, crude oil. In a financial context the existence of cointegrating relationships implies no arbitrage opportunity between these markets as well as no leading market in the price discovery process. This is going to be a key feature for the definition of hedging strategies also for energy markets, given the recent deregulation process of the gas and the electricity market in Europe.

The paper is organised as follows. Section 2 provides an overview of the relevant literature on this topic. Section 3 describes the data set given by daily prices of electricity, oil and natural gas for the European market over the period 2001–2007 and examines the annualised quarterly volatilities of each time series. In Sections 4 and 5 current state of the art methodologies are used to analyse the short- and long-run relationships, as well as a rolling correlation and cointegration approach. Section 6 draws some preliminary conclusions.

2 Relevant literature

Economic theory suggests the existence of a relationship between natural gas and oil prices. Oil and natural gas are competitive substitutes and complements in the electricity generation and in industrial production. Due to the asymmetric relationship in the relative size of each market, past changes in the price of oil caused changes in the natural gas market, but the converse did not hold [17].

The relationship between natural gas and crude oil has been largely investigated. In [14] UK gas and Brent oil prices over the period 1996–2003 have been analysed. In [3] the degree of market integration both among and between the crude oil, coal, and natural gas markets in the US has been investigated. A longer time period 1989–2005, is used in [17] where a cointegration relationship between oil and natural gas

---

[1] Spreads are price differentials between two commodities and are largely used to describe power plant refineries, storage facilities and transmission lines. For an extensive description of energy spread options, see [6].

prices has been found despite periods where they may have appeared to decouple. A cointegration relationship between the prices of West Texas Intermediate (WTI) crude oil and Henry Hub (HH) natural gas has been examined in [4] and [9].

Analysis of the relationship between electricity and fossil fuel prices has only been performed at regional levels and on linked sets of data given the recent introduction of spot electricity markets. Serletis and Herbert [15] used the dynamics of North America natural gas, fuel oil and power prices from 1996 to 1997 to find that the HH and Transco Zone 6 natural gas prices and the fuel oil price are cointegrated, whereas power prices series appears to be stationary. In [8] the existence of a medium- and long-term correlation between electricity and fuel oil in Europe is analysed. [2] investigates the dynamic of gas, oil and electricity during an interim period 1995–1998: deregulation of the UK gas market (1995) and the opening up of the Interconnector (1998). Cointegration between natural gas, crude oil and electricity prices is found and a leading role of crude oil is also identified. More recently, using a multivariate time series framework, [13] interrelationships among electricity prices from two diverse markets, Pennsylvania, New Jersey, Maryland Interconnection (PJM) and Mid-Columbia (Mid-C), and four major fuel source prices, natural gas, crude oil, coal and uranium, in the period 2001–2008, are examined.

To the best of our knowledge the level of integration between the gas, oil and electricity markets in the European market has not been investigated. The purpose of this study is mainly to perform such analysis in order to verify if an integrated energy market can be detected.

3 The data set

Time series for the daily prices of ICE Brent crude oil,[2] natural gas at the National Balancing Point (NBP) UK[3] and European Energy Exchange (EEX) electricity[4] are used for the period September 2001 – December 2007.

Oil prices are expressed in US$/barrel per day (bd), gas in UK p/therm and electricity prices in €/Megawatt hour (MWh). We convert all prices into €/MWh using the conversion factors for energy content provided by the Energy Information Administration (EIA).[5] The dynamics of the energy prices are represented into Figure 1.

Following the standard literature we perform a finer analysis of the volatility of each price series by estimating the annualised quarterly volatilities $\sigma_i = \sigma_{i,N}\sqrt{250}$,

---

[2] Brent blend is the reference crude oil for the North Sea and is one of the three major benchmarks in the international oil market [7].

[3] The NBP is the most liquid gas trading point in Europe. The NBP price is the reference for many forward transactions and for the International Petroleum Exchange (IPE) Future contracts [7].

[4] EEX is one of the leading energy exchanges in central Europe [7]. For the purpose of our analysis peak load prices have been used.

[5] According to EIA conversion factors, 1 barrel of crude oil is equal to 1.58 MWh.

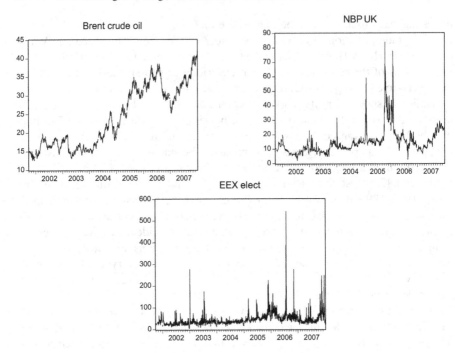

Fig. 1. Crude oil, natural gas and electricity prices, 2001–2007

$i = 1, \ldots, 25$, $N = 60$, where

$$\sigma_{i,N} = \sqrt{\frac{1}{N-1} \sum_{\tau=(i-1)N+1}^{iN} \left(\ln \frac{S_\tau}{S_{\tau-1}} - E\left(\ln \frac{S_\tau}{S_{\tau-1}} \right) \right)^2}.$$

The oil price volatility swings between 21% and 53%, confirming the non-stationarity of the data. The same non-stationarity characterises the data of natural gas, fluctuating between 65% and 330%. Electricity prices, as expected, were far more volatile than oil and gas prices,[6] with a range of quarterly volatility which swings between around 277% and 868%.

A preliminary analysis is going to be performed on the stationarity of the time series. In line with most of the recent literature we transform the original series in logs. First we test the order of integration of a time series using the Augmented Dickey-Fuller (ADF) type regression:

$$\Delta y_t = \alpha_0 + \alpha_1 t + \gamma \, y_{t-1} + \sum_{j=1}^{k} \beta_j \, \Delta y_{t-j} + \epsilon_t \tag{1}$$

---

[6] Seasonality and mean reversion are common features in commodity price dynamics; in addition a jump component has to be included when describing electricity prices.

where $\Delta y_t = y_t - y_{t-1}$ and the lag length k is automatic based on Scharwz information criterion (SIC). The results of the unit root test are reported in Table 1.

Table 1. Unit root test results for the logged price series

Series	t_y	τ_0	τ_1	τ_d	Decision
Oil	1.17 (1)	−0.70 (1)	−2.73 (1)	−43.2 (1)	$I(1)$
Gas	−0.30 (6)	−3.37* (6)	−5.58** (2)	−21.1 (1)	$I(1)$
Elect	−0.06 (14)	−3.41 (14)	−4.36** (14)	−18.4 (1)	$I(1)$

The 5% significance levels are −1.94 for ADF without exogenous variables, −2.86 for ADF with a constant and −3.41 for ADF with a constant and trend. (∗) denotes acceptance of the null at 1%, (∗∗) denotes rejection of the null at the conventional test sizes. The SIC-based optimum lag lengths are in parentheses.

We run the test without any exogenous variable, with a constant and a constant plus a linear time trend as exogenous variables in equation (1). The reported t-statistics are t_y, τ_0 and τ_1, respectively. τ_d is the t-statistic for the ADF tests in first-differenced data. t_y is greater than the critical values but we reject the hypothesis in first-difference, hence we conclude that the variables are first-difference stationary (i.e., all the series are $I(1)$).

4 The short-run relationship

Alexander [1] presents the applications of correlation analysis to the crude oil and natural gas markets. Correlation measures co-movements of prices or returns and can be considered a short-term measure. It is essentially a static measure, so it cannot reveal any dynamic causal relationship. In addition, estimated correlations can be significantly biased or nonsense if the underlying variables are polynomials of time or when the two variables are non-stationary [18].

To analyse a possible short-run relationship among the variables, we estimate a rolling correlation over $\tau_j = 100$ days[7] according to:

$$\rho_s[x, y] = \frac{\frac{1}{\tau_j - 1} \sum_{i=s}^{s+\tau_j} (x_i - \widehat{x})(y_i - \widehat{y})}{\widehat{\sigma}_x \widehat{\sigma}_y}, \quad s = 1, \ldots, T - \tau_j, \quad (2)$$

where $T = 1580$ (the entire period 2001–2007), and $\widehat{\sigma}_x$ and $\widehat{\sigma}_y$ are the standard deviations of x and y, estimated on the corresponding time window.

Correlation changes over time, as expected, given the non-stationarity of the underlying processes. Volatilities of commodity prices are time dependent and so are the covariance and the unconditional correlation. This means that we can only attempt

---

[7] This window period is suggested in [6]. We also perform the analysis with larger windows ($\tau_j = 100, 150, 200$ days), getting similar results.

to catch seasonal changes in correlations when interpreting the rolling correlation coefficient. The unconditional correlation coefficients,[8] ρ_T, together with the main statistical features of the rolling correlations, ρ_s, $s = 1, \ldots, T$, between the energy price series are reported in Table 2. It is interesting to notice that the rolling correlations between gas and oil show some counterintuitive behaviour.

Table 2. Unconditional correlation and rolling correlations between log prices

Matrices	ρ_T	$E(\rho_s)$	$\sigma(\rho_s)$	Max(ρ_s)	Min(ρ_s)
Oil/Elect	0.537	0.0744	0.260	0.696	−0.567
Gas/Elect	0.515	0.119	0.227	0.657	−0.280
Oil/Gas	0.590	-0.027	0.426	0.825	−0.827

These results do not provide useful insights into the real nature of the relationship between the main commodities of the energy markets.

5 The long-run relationship

Table 1 confirms a stochastic trend for all the price series; a possible cointegration relationship among the energy commodity prices may therefore be captured (i.e., the presence of a shared stochastic trend or common trend). Two non-stationary series are cointegrated if a linear combination of them is stationary. The vector which realises such a linear combination is called the cointegrating vector.

We examine the number of cointegrating vectors by using the Johansen method (see [10] and [11]). For this purpose we estimate a vector error correction model (VECM) based on the so-called reduced rank regression method (see [12]). Assume that the n-vector of non-stationary $I(1)$ variables Y_t follows a vector autoregressive (VAR) process of order p,

$$Y_t = A_1 Y_{t-1} + A_2 Y_{t-2} + \ldots + A_p Y_{t-p} + \epsilon_t \qquad (3)$$

with ϵ_t as the corresponding n-dimensional white noise, and $n \times n$ A_i, $i = 1, \ldots, p$ matrices of coefficients.[9] Equation (3) is equivalently written in a VECM framework,

$$\Delta Y_t = D_1 \Delta Y_{t-1} + D_2 \Delta Y_{t-2} + \cdots + D_p \Delta Y_{t-p+1} + D Y_{t-1} + \epsilon_t \qquad (4)$$

where $D_i = -(A_{i+1} + \cdots + A_p)$, $i = 1, 2, \ldots, p-1$ and $D = (A_1 + \cdots + A_p - I_n)$. The Granger's representation theorem [5] asserts that if D has reduced rank $r \in (0, n)$, then $n \times r$ matrices Γ and B exist, each with rank r, such that $D = -\Gamma B'$ and $B' Y_t$ is $I(0)$. r is the number of cointegrating relations and the coefficients of the cointegrating vectors are reported in the columns of B.

The cointegration results for the log prices are shown in Table 3.

---

[8] The unconditional correlation for the entire period is given by $\rho_T = \frac{cov(x,y)}{\sigma_x \sigma_y}$.

[9] In the following, for the VAR(p) model we exclude the presence of exogenous variables.

Table 3. Cointegration rank test (trace and maximum eigenvalue)

Nr. of coint. vec.	Eigenvalue	λ_{trace}	$\lambda_{trace}^{0.05}$	λ_{max}	$\lambda_{max}^{0.05}$
$r = 0$	0.121	237.2	29.79	240.3	21.13
$r \leq 1$	0.020	32.91	15.49	32.24	14.26
$r \leq 2$	0.000	0.672	3.841	0.672	3.841

A rejection of the null 'no cointegrated' relationship and 'r at most 1' in favour of 'r at most 2' at the 5% significance level is provided. This provides evidence of the existence of two cointegrating relationships among the three commodity price series. In a VECM framework, the presence of two cointegrating vectors, $r = 2$, on a set of $n = 3$ variables allows the estimation of a $n - r = 1$ common (stochastic) trend [16]. The common trend may be interpreted as a source of randomness which affects the dynamics of the commodity prices. In this case we may assume oil prices represent the leading risk factor in the energy market as a whole.

To better analyse the dynamics of the markets we use the Engle-Granger [5] two-step methodology. This method consists in estimating each cointegrating relationship individually using ordinary least squares (OLS) and then including the errors from those cointegrating equations in short-run dynamic adjustment equations which allow the explanation of adjustment to the long-run equilibrium. The first step is to estimate the so-called cointegrating regression

$$y_{1,t} = \alpha + \beta y_{2,t} + z_t \tag{5}$$

where $y_{1,t}$ and $y_{2,t}$ are two price series, both integrated of order one, and z_t denotes the OLS regression residuals. We perform the test twice for each couple of time series using as dependent variable both of the series. For each couple of time series, using both of the series as dependent variables. The results are reported in Table 4. The null hypothesis of no cointegration is rejected at the 8% significance level for the regression oil vs electricity, at the 1 % level in all the other cases. The coefficients β in equation (5), which represent the factors of proportionality for the common trend, are estimated by OLS.

According to the Granger representation theorem, if two series cointegrate, the short-run dynamics can be described by the ECM. The basic ECM proposed in [5] can be written as follows:

$$\Delta y_{1,t} = \phi \Delta y_{2,t} + \theta(y_{1,t-1} - \alpha - \beta y_{2,t-1}) + \epsilon_t \tag{6}$$

where $(y_{1,t-1} - \alpha - \beta y_{2,t-1})$ represents the error correction term z_{t-1} of equation (5), ϕ measures the contemporaneous price response,[10] θ represents the speed of the adjustment towards the long-term cointegrating relationship, and $\epsilon_t \sim i.i.d.(0, \Sigma)$.

---

[10] The parameter ϕ approximates the correlation coefficient between first differences in prices ($\Delta y_{i,t}$ and $\Delta y_{j,t}$) and it will be close to 1 when the two commodities are in the same market. Therefore, a higher value of ϕ is a sign of a stronger integration of the market [3].

Table 4. Engle and Granger cointegration test

Dep. variable	Indep. variable	β	t_β	p-value
Elect	Gas	0.514	23.90	0.00
Gas	Elect	0.516		0.00
Elect	Oil	0.732	25.33	0.00
Oil	Elect	0.394		0.08
Oil	Gas	0.432	29.03	0.00
Gas	Oil	0.805		0.00

t_β are the t-statistics for the coefficients β in equation (5). The last column reports the p-values for the unit root tests on the regression residuals.

Cointegration tests per se do not focus on the economically interesting parameters α, β, ϕ and θ [3]. The ECM highlights that the deviations from the long-run cointegrating relationship are corrected gradually through a series of partial short-run adjustments. In the long run equilibrium the error correction term will be equal to zero. However, if the variables $y_{1,t}$ and $y_{2,t}$ deviate from the long-run equilibrium, the error correction term will be different from zero and each variable adjusts to restore the equilibrium relation whose speed of adjustment is represented by θ.

The results reported in Table 5 highlight no significant value for coefficient ϕ in any cases. Therefore we apply an ECM using a different lag for the independent variable.

Table 5. Estimated speed of adjustment parameters for the ECM

Dep. variable	Indep. variable	ϕ	t_ϕ	p-value	θ	t_θ	p-value
Δ Elect	Δ Gas	0.010	0.150	0.880	−0.452	−21.54	0.00
Δ Elect	Δ Oil	−0.427	−1.059	0.289	−0.461	−21.71	0.00
Δ Gas	Δ Oil	0.028	0.189	0.849	−0.053	−6.553	0.00

For electricity and gas, significant coefficients ϕ and θ are found ($\phi = 0.25$, $\theta = 0.46$) with a lag of two days, indicating that in addition to a long-run relationship a short-run influence exists among the two series. For the pair electricity/oil, considering the independent variable with a five-day lag, a significant coefficient, $\phi = 0.68$ (9% level), is found whereas $\theta = 0.452$; also in this case, the price adjustment in the short run is detected with a lag of five days. For the pair gas/oil a significant coefficient ϕ is found ($\phi = 0.29$) at the 5% level with a lag of six days. θ is equal to 0.05, showing that the speed adjustment to the long-run equilibrium is particularly low. The presence of a short-run relationship among the various commodities may also be explained by the fact that the analysis refers to European markets where deregulation has not been completely performed yet. Some of the markets still experience market power and in this context the oil prices may still represent the leading actor for the electricity and gas price formation. The misalignment between oil and gas in the short run may depend on different forces (i.e., the supply of gas from Algeria or Russia) that may provide some independent source of randomness for natural gas prices. This may explain why, especially in turbulent periods, gas and oil tend to have different dynamics, while natural gas prices follow crude oil in the long run.

6 Conclusions

This paper analyses the dynamics of the prices of oil, electricity and natural gas in the European markets in order to estimate the nature of the existing relationship among them. The simple correlation analysis among the various time series is non-effective given the non-stationarity of the data. A cointegration approach is chosen to measure a possible integration among the markets.

A cointegration relationship among each pair of commodities is found using the Engle-Granger approach. The Johansen cointegration test reports that oil, gas and electricity prices are all cointegrated. Two further integrating equations are found, implying that one common trend is present in the energy market. From an economic point of view this can be interpreted as a simple source of risk (the oil market), which affects the dynamics of the two other commodities (electricity and gas).

References

1. Alexander, C.: Correlation and cointegration in energy markets. In: Kaminski, V. (ed.) Managing Energy Price Risk, pp. 291–304. RISK Publications (1999)
2. Asche, F., Osmundsen, P., Sandsmark, M.: The UK market for natural gas, oil and electricty: are the prices decoupled? Energ. J. 27, 27–40 (2006)
3. Bachmeier, L.J., Griffin, J.M.: Testing for market integration: crude oil, coal, and natural gas. The Energy Journal 27, 55–72 (2006)
4. Brown, S.P.A., Yücel, M.K.: What drives natural gas prices?. Research Department Working Paper 0703, Federal Reserve Bank of Dallas (2007)
5. Engle, R.F., Granger, C.W.J.: Co-integration and error correction: representation, estimation, and testing. Econometrica 55, 251–276 (1987)
6. Eydeland, A., Wolyniec, K.: Energy and Power Risk Management. New Development in Modeling, Pricing, and Hedging. Wiley (2003)
7. Geman, H.: Commodity and Commodity Derivatives. Modeling and Pricing for Agriculturals, Metals and Energy. Wiley (2005)
8. Gjølberg, O.: When (and how) will the markets for oil and electricity become integrated? Econometric evidence and trends 1993–99. Discussion paper D:20, (2001)
9. Hartley, P.R., Medlock, III K.B.:, Rosthal J.: The relationship between crude oil and natural gas prices. The James A. Baker III Institute for Public Policy, Rice University (2007)
10. Johansen, S.: Statistical analysis of cointegration vectors. J. Econ. Dyn. Control 12, 231–254 (1988)
11. Johansen, S.: The role of constant and linear terms in cointegration of nonstationary variables. Econometric Rev. 13, 205–209 (1991)

12. Johansen, S.: Likelihood-based Inference in Cointegrated Vector Autoregressive Models. Oxford University Press, Oxford (1995)
13. Mjelde, J.W., Bessler, D.A.: Market integration among electricity markets and their major fuel source markets. Energy Econ. 31, 482–491 (2009)
14. Panagiotidis, T., Rutledge, E.: Oil and gas markets in the UK: evidence from a cointegrating approach. Energy Econ. 29, 329–347 (2006)
15. Serletis, A., Herbert, J.: The message in North America energy prices. Energy Econ. 21, 471–483 (1999)
16. Stock, J.H., Watson, M.W.: Testing for common trends. J. Am. Stat. Assoc. 83, 1097–1107 (1988)
17. Villar, J.A., Joutz, F.L.: The relationship between crude oil and natural gas prices. Energy Information Administration, Office of Oil and Natural Gas. Washington, D.C. (2006)
18. Yule, U.: Why do we sometimes get nonsense-correlations between time series? A study in sampling and the nature of time series. J. R. Stat. Soc. 89, 1–63 (1926)

Tempered stable distributions and processes in finance: numerical analysis

Michele Leonardo Bianchi*, Svetlozar T. Rachev, Young Shin Kim, and Frank J. Fabozzi

Abstract. Most of the important models in finance rest on the assumption that randomness is explained through a normal random variable. However there is ample empirical evidence against the normality assumption, since stock returns are heavy-tailed, leptokurtic and skewed. Partly in response to those empirical inconsistencies relative to the properties of the normal distribution, a suitable alternative distribution is the family of tempered stable distributions. In general, the use of infinitely divisible distributions is obstructed the difficulty of calibrating and simulating them. In this paper, we address some numerical issues resulting from tempered stable modelling, with a view toward the density approximation and simulation.

Key words: stable distribution, tempered stable distributions, Monte Carlo

1 Introduction

Since Mandelbrot introduced the α-stable distribution in modelling financial asset returns, numerous empirical studies have been done in both natural and economic sciences. The works of Rachev and Mittnik [19] and Rachev et al. [18] (see also references therein), have focused attention on a general framework for market and credit risk management, option pricing, and portfolio selection based on the α-stable distribution. While the empirical evidence does not support the normal distribution, it is also not always consistent with the α-stable distributional hypothesis. Asset returns time series present heavier tails relative to the normal distribution and thinner tails than the α-stable distribution. Moreover, the stable scaling properties may cause problems in calibrating the model to real data. Anyway, there is a wide consensus to assume the presence of a leptokurtic and skewed pattern in stock returns, as showed by the α-stable modelling. Partly in response to the above empirical inconsistencies, and to maintain suitable properties of the stable model, a proper alternative to the α-stable distribution is the family of tempered stable distributions.

Tempered stable distributions may have all moments finite and exponential moments of some order. The latter property is essential in the construction of tempered

---
* The views expressed in this paper are those of the author and should not be attributed to the institution to which he belongs.

stable option pricing models. The formal definition of tempered stable processes has been proposed in the seminal work of Rosiński [21]. The KoBol (Koponen, Boyarchenko, Levendorskiĭ) [4], CGMY (Carr, Geman, Madan, Yor) [5], Inverse Gaussian (IG) and the tempered stable of Tweedie [22] are only some of the parametric examples in this class that have an infinite-dimensional parametrisation by a family of measures [24]. Further extensions or limiting cases are also given by the fractional tempered stable framework [10], the bilateral gamma [15] and the generalised tempered stable distribution [7] and [16]. The general formulation is difficult to use in practical applications, but it allows one to prove some interesting results regarding the calculus of the characteristic function and the random number generation. The infinite divisibility of this distribution allows one to construct the corresponding Lévy process and to analyse the change of measure problem and the process behaviour as well.

The purpose of this paper is to show some numerical issues arising from the use of this class in applications to finance with a look at the density approximation and random number generation for some specific cases, such as the CGMY and the Kim-Rachev (KR) case. The paper is related to some previous works of the authors [13, 14] where the exponential Lévy and the tempered stable GARCH models have been studied. The remainder of this paper is organised as follows. In Section 2 we review the definition of tempered stable distributions and focus our attention on the CGMY and KR distributions. An algorithm for the evaluation of the density function for the KR distribution is presented in Section 3. Finally, Section 4 presents a general random number generation method and an option pricing analysis via Monte Carlo simulation.

2 Basic definitions

The class of infinitely divisible distribution has a large spectrum of applications and in recent years, particularly in mathematical finance and econometrics, non-normal infinitely divisible distributions have been widely studied. In the following, we will refer to the Lévy-Khinchin representation with Lévy triplet (a_h, σ, ν) as in [16]. Let us now define the Lévy measure of a TS_α distribution.

Definition 1 *A real valued random variable X is TS_α if is infinitely divisible without a Gaussian part and has Lévy measure ν that can be written in polar coordinated*

$$\nu(dr, dw) = r^{-\alpha-1} q(r, w) dr \, \sigma(dw), \tag{1}$$

where $\alpha \in (0, 2)$ and σ is a finite measure on S^{d-1} and

$$q : (0, \infty) \times S^{d-1} \mapsto (0, \infty)$$

is a Borel function such that $q(\cdot, w)$ is completely monotone with $q(\infty, w) = 0$ for each $w \in S^{d-1}$. A TS_α distribution is called a proper TS_α distribution if

$$\lim_{r \to 0^+} q(r, w) = 1$$

for each $w \in S^{d-1}$.

Furthermore, by theorem 2.3 in [21], the Lévy measure ν can be also rewritten in the form

$$\nu(A) = \int_{\mathbb{R}_0^d} \int_0^\infty I_A(tx) \alpha t^{-\alpha-1} e^{-t} \, dt \, R(dx), \qquad A \in \mathcal{B}(\mathbb{R}^d), \qquad (2)$$

where R is a unique measure on $\mathbb{R}^d$ such that $R(\{0\}) = 0$

$$\int_{\mathbb{R}^d} (\|x\|^2 \wedge \|x\|^\alpha) R(dx) < \infty, \qquad \alpha \in (0, 2). \qquad (3)$$

Sometimes the only knowledge of the Lévy measure cannot be enough to obtain analytical properties of tempered stable distributions. Therefore, the definition of Rosiński measure R allows one to overcome this problem and to obtain explicit analytic formulas and more explicit calculations. For instance, the characteristic function can be rewritten by directly using the measure R instead of ν (see theorem 2.9 in [21]). Of course, given a measure R it is always possible to find the corresponding tempering function q; the converse is true as well. As a consequence of this, the specification of a measure R satisfying conditions (3), or the specification of a completely monotone function q, uniquely defines a TS_α distribution.

Now, let us define two parametric examples. In the first example the measure R is the sum of two Dirac measures multiplied for opportune constants, while the spectral measure R of the second example has a nontrivial bounded support. If we set

$$q(r, \pm 1) = e^{-\lambda_{\pm} r}, \qquad \lambda > 0, \qquad (4)$$

and the measure

$$\sigma(\{-1\}) = c_- \quad \text{and} \quad \sigma(\{1\}) = c_+, \qquad (5)$$

we get

$$\nu(dr) = \frac{c_-}{|r|^{1+\alpha_-}} e^{-\lambda_- r} I_{\{r<0\}} + \frac{c_+}{|r|^{1+\alpha_+}} e^{-\lambda_+ r} I_{\{r>0\}}. \qquad (6)$$

The measures Q and R are given by

$$Q = c_- \delta_{-\lambda_-} + c_+ \delta_{\lambda_+} \qquad (7)$$

and

$$R = c_- \lambda_-^\alpha \delta_{-\frac{1}{\lambda_-}} + c_+ \lambda_+^\alpha \delta_{\frac{1}{\lambda_+}}, \qquad (8)$$

where δ_λ is the Dirac measure at λ (see [21] for the definition of the measure Q).

Then the characteristic exponent has the form

$$\psi(u) = iub + \Gamma(-\alpha)c_+((\lambda_+ - iu)^\alpha - \lambda_+^\alpha + i\alpha\lambda_+^{\alpha-1}u)$$
$$+ \Gamma(-\alpha)c_-((\lambda_- + iu)^\alpha - \lambda_-^\alpha - i\alpha\lambda_-^{\alpha-1}u), \qquad (9)$$

where we are considering the Lévy-Khinchin formula with truncation function $h(x) = x$. This distribution is usually referred to as the KoBoL or generalised tempered stable (GTS) distribution. If we take $\lambda_+ = M$, $\lambda_- = G$, $c_+ = c_- = C$, $\alpha = Y$ and $m = b$, we obtain that X is CGMY distributed with expected value m. The definition of the corresponding Lévy process follows.

Definition 2 *Let X_t be the process such that $X_0 = 0$ and $E[e^{iuX_t}] = e^{t\psi(u)}$ where*

$$\psi(u) = ium + \Gamma(-Y)C((M - iu)^Y - M^Y + iYM^{Y-1}u)$$
$$+ \Gamma(-Y)C((G + iu)^Y - G^Y - iYG^{Y-1}u).$$

We call this process the CGMY process with parameter (C, G, M, Y, m) where $m = E[X_1]$.

A further example is given by the KR distribution [14], with a Rosiński measure of the following form

$$R(dx) = (k_+ r_+^{-p_+} I_{(0,r_+)}(x)|x|^{p_+ - 1} + k_- r_-^{-p_-} I_{(-r_-,0)}(x)|x|^{p_- - 1})\, dx, \qquad (10)$$

where $\alpha \in (0, 2)$, $k_+, k_-, r_+, r_- > 0$, $p_+, p_- \in (-\alpha, \infty) \setminus \{-1, 0\}$, and $m \in \mathbb{R}$. The characteristic function can be calculated by theorem 2.9 in [21] and is given in the following result [14].

Definition 3 *Let X_t be a process with $X_0 = 0$ and corresponding to the spectral measure R defined in (10) with conditions $p \neq 0$, $p \neq -1$ and $\alpha \neq 1$, and let $m = E[X_1]$. By considering the Lévy-Khinchin formula with truncation function $h(x) = x$, we have $E[e^{iuX_t}] = e^{t\psi(u)}$ with*

$$\psi(u) = \frac{k_+ \Gamma(-\alpha)}{p_+} \left({}_2F_1(p_+, -\alpha; 1 + p_+; ir_+ u) - 1 + \frac{i\alpha p_+ r_+ u}{p_+ + 1} \right)$$
$$\frac{k_- \Gamma(-\alpha)}{p_-} \left({}_2F_1(p_-, -\alpha; 1 + p_-; -ir_- u) - 1 - \frac{i\alpha p_- r_- u}{p_- + 1} \right) + ium, \qquad (11)$$

where ${}_2F_1(a, b; c; x)$ is the hypergeometric function [1]. We call this process the KR process with parameter $(k_+, k_-, r_+, r_-, p_+, p_-, \alpha, m)$.

3 Evaluating the density function

In order to calibrate asset returns models through an exponential Lévy process or tempered stable GARCH model [13, 14], one needs a correct evaluation of both the pdf and cdf functions. With the pdf function it is possible to construct a maximum likelihood estimator (MLE), while the cdf function allows one to assess the goodness of fit. Even if the MLE method may lead to a local maximum rather than to a global one due to the multidimensionality of the optimisation problem, the results obtained seem to be satisfactory from the point of view of goodness-of-fit tests. Actually, an analysis of estimation methods for this kind of distribution would be interesting, but it is far from the purpose of this work.

Numerical methods are needed to evaluate the pdf function. By the definition of the characteristic function as the Fourier transform of the density function [8], we consider the inverse Fourier transform that is

$$f(x) = \frac{1}{2\pi} \int_{\mathbb{R}} e^{-iux} E[e^{iuX}] du \qquad (12)$$

where $f(x)$ is the density function. If the density function has to be calculated for a large number of x values, the fast Fourier Transform (FFT) algorithm can be employed as described in [23]. The use of the FFT algorithm largely improves the speed of the numerical integration above and the function f is evaluated on a discrete and finite grid; consequently a numerical interpolation is necessary for x values out of the grid. Since a personal computer cannot deal with infinite numbers, the integral bounds $(-\infty, \infty)$ in equation (12) are replaced with $[-M, M]$, where M is a large value. We take $M \sim 2^{16}$ or 2^{15} in our study and we have also noted that smaller values of M generate large errors in the density evaluation given by a wave effect in both density tails. We have to point out that the numerical integration as well as the interpolation may cause some numerical errors. The method above is a general method that can be used if the density function is not known in closed form.

While the calculus of the characteristic function in the CGMY case involves only elementary functions, more interesting is the evaluation of the characteristic function in the KR case that is connected with the Gaussian hypergeometric function. Equation (11) implies the evaluation of the hypergeometric $_2F_1(a, b; c; z)$ function only on the straight line represented by the subset $I = \{iy \mid y \in \mathbb{R}\}$ of the complex plane $\mathbb{C}$. We do not need a general algorithm to evaluate the function on the entire complex plane $\mathbb{C}$, but just on a subset of it. This can be done by means of the analytic continuation, without having recourse either to numerical integration or to numerical solution of a differential equation [17] (for a complete table of the analytic continuation formulas for arbitrary values of $z \in \mathbb{C}$ and of the parameters a, b, c, see [3] or [9]). The hypergeometric function belongs to the special function class and often occurs in many practical computational problems. It is defined by the power series

$$_2F_1(a, b, c; z) = \sum_{n=0}^{\infty} \frac{(a)_n (b)_n}{(c)_n} \frac{z^n}{n!}, \quad |z| < 1, \tag{13}$$

where $(a)_n := \Gamma(a + n)/\Gamma(n)$ is the Ponchhammer symbol (see [1]). By [1] the following relations are fulfilled

$$_2F_1(a, b, c; z) = (1 - z)^{-b}{_2F_1}\left(b, c - a, c, \frac{z}{z - 1}\right) \quad \text{if} \quad \left|\frac{z}{z - 1}\right| < 1$$

$$_2F_1(a, b, c; z) = (-z)^{-a} \frac{\Gamma(c)\Gamma(b - a)}{\Gamma(c - a)\Gamma(b)} {_2F_1}\left(a, a - c + 1, a - b + 1, \frac{1}{z}\right)$$

$$+ (-z)^{-b} \frac{\Gamma(c)\Gamma(a - b)}{\Gamma(c - b)\Gamma(a)} {_2F_1}\left(b, b - c + 1, b - a + 1, \frac{1}{z}\right)$$

$$\text{if} \quad \left|\frac{1}{z}\right| < 1$$

$$_2F_1(a, b, c; -iy) = \overline{_2F_1(a, b, c; iy)} \quad \text{if} \quad y \in \mathbb{R}. \tag{14}$$

First by the last equality of (14), one can determine the values of $_2F_1(a, b, c; z)$ only for the subset $I_+ = \{iy \mid y \in \mathbb{R}_+\}$ and then simply consider the conjugate for the set $I_- = \{iy \mid y \in \mathbb{R}_-\}$, remembering that $_2F_1(a, b, c; 0) = 1$. Second, in order to obtain a fast convergence of the series (13), we split the positive imaginary line into three subsets without intersection,

$$I_+^1 = \{iy \mid 0 < y \leq 0.5\}$$
$$I_+^2 = \{iy \mid 0.5 < y \leq 1.5\}$$
$$I_+^3 = \{iy \mid y > 1.5\},$$

then we use (13) to evaluate $_2F_1(a, b, c; z)$ in I_+^1. Then, the first and the second equalities of (14) together with (13) are enough to evaluate $_2F_1(a, b, c; z)$ in I_+^2 and I_+^3 respectively. This subdivision allows one to truncate the series (13) to the integer $N = 500$ and obtain the same results as Mathematica. We point out that the value of y ranges in the interval $[-M, M]$. This method, together with the MATLAB vector calculus, considerably increases the speed with respect to algorithms based on the numerical solution of the differential equation [17]. Our method is grounded only on basic summations and multiplication. As a result the computational effort in the KR density evaluation is comparable to that of the CGMY one. The KR characteristic function is necessary also to price options, not only for MLE estimation. Indeed, by using the approach of Carr and Madan [6] and the same analytic continuation as above, risk-neutral parameters may be directly estimated from option prices, without calibrating the underlying market model.

4 Simulation of TS_α processes

In order to generate random variate from TS_α processes, we will consider the general shot noise representation of proper TS_α laws given in [21]. There are different methods to simulate Lévy processes, but most of these methods are not suitable for the simulation of tempered stable processes due to the complicated structure of their Lévy measure. As emphasised in [21], the usual method of the inverse of the Lévy measure [20] is difficult to implement, even if the spectral measure R has a simple form. We will apply theorem 5.1 from [21] to the previously considered parametric examples.

Proposition 1 *Let $\{U_j\}$ and $\{T_j\}$ be i.i.d. sequences of uniform random variables in $(0, 1)$ and $(0, T)$ respectively, $\{E_j\}$ and $\{E_j'\}$ i.i.d. sequences of exponential variables of parameter 1 and $\{\Gamma_j\} = E_1' + \ldots + E_j'$, $\{V_j\}$ an i.i.d. sequence of discrete random variables with distribution*

$$P(V_j = -G) = P(V_j = M) = \frac{1}{2},$$

a positive constant $0 < Y < 2$ (with $Y \neq 1$), and $\|\sigma\| = \sigma(S^{d-1}) = 2C$. Furthermore, $\{U_j\}$, $\{E_j\}$, $\{E_j'\}$ and $\{V_j\}$ are mutually independent. Then

$$X_t \overset{d}{=} \sum_{j=1}^{\infty} \left[\left(\frac{Y\Gamma_j}{2C} \right)^{-1/Y} \wedge E_j U_j^{1/Y} |V_j|^{-1} \right] \frac{V_j}{|V_j|} I_{\{T_j \leq t\}} + tb_T \quad t \in [0, T], \quad (15)$$

where

$$b_T = -\Gamma(1 - Y)C(M^{Y-1} - G^{Y-1}) \qquad (16)$$

and γ is the Euler constant [1, 6.1.3], converges a.s. and uniformly in $t \in [0, T]$ to a CGMY process with parameters $(C, G, M, Y, 0)$.

This series representation is not new in the literature, see [2] and [12]. It is a slight modification of the series representation of the stable distribution [11], but here big jumps are removed. The shot noise representation for the KR distribution follows.

Proposition 2 *Let $\{U_j\}$ and $\{T_j\}$ be i.i.d. sequences of uniform random variables in $(0, 1)$ and $(0, T)$ respectively, $\{E_j\}$ and $\{E_j'\}$ i.i.d. sequences of exponential variables of parameter 1 and $\{\Gamma_j\} = E_1' + \ldots + E_j'$, and constants $\alpha \in (0, 2)$ (with $\alpha \neq 1$), $k_+, k_-, r_+, r_- > 0$ and, $p_+, p_- \in (-\alpha, \infty) \setminus \{-1, 0\}$. Let $\{V_j\}$ be an i.i.d. sequence of random variables with density*

$$f_V(r) = \frac{1}{\|\sigma\|} \left(k_+ r_+^{-p_+} I_{\{r > \frac{1}{r_+}\}} r^{-\alpha - p_+ - 1} + k_- r_-^{-p_+} I_{\{r < -\frac{1}{r_-}\}} |r|^{-\alpha - p_- - 1} \right)$$

where

$$\|\sigma\| = \frac{k_+ r_+^\alpha}{\alpha + p_+} + \frac{k_- r_-^\alpha}{\alpha + p_-}.$$

Furthermore, $\{U_j\}$, $\{E_j\}$, $\{E_j'\}$ and $\{V_j\}$ are mutually independent. If $\alpha \in (0, 1)$, or if $\alpha \in (1, 2)$ with $k_+ = k_-$, $r_+ = r_-$ and $p_+ = p_-$, then the series

$$X_t = \sum_{j=1}^{\infty} I_{\{T_j \leq t\}} \left(\left(\frac{\alpha \Gamma_j}{T \|\sigma\|} \right)^{-1/\alpha} \wedge E_j U_j^{1/\alpha} |V_j|^{-1} \right) \frac{V_j}{|V_j|} + t b_T \qquad (17)$$

converges a.s. and uniformly in $t \in [0, T]$ to a KR tempered stable process with parameters $(k_+, k_+, r_+, r_+, p_+, p_+, \alpha, 0)$ with

$$b_T = -\Gamma(1 - \alpha) \left(\frac{k_+ r_+}{p_+ + 1} - \frac{k_- r_-}{p_- + 1} \right).$$

If $\alpha \in (1, 2)$ and $k_+ \neq k_-$ (or $r_+ \neq r_-$ or alternatively $p_+ \neq p_-$), then

$$X_t = \sum_{j=1}^{\infty} \left[I_{\{T_j \leq t\}} \left(\left(\frac{\alpha \Gamma_j}{T \|\sigma\|} \right)^{-1/\alpha} \wedge E_j U_j^{1/\alpha} |V_j|^{-1} \right) \frac{V_j}{|V_j|} \right.$$
$$\left. - \frac{t}{T} \left(\frac{\alpha j}{T \|\sigma\|} \right)^{-1/\alpha} x_0 \right] + t b_T, \qquad (18)$$

converges a.s. and uniformly in $t \in [0, T]$ to a KR tempered stable process with parameters $(k_+, k_-, r_+, r_-, p_+, p_-, \alpha, 0)$, where we set

$$b_T = \alpha^{-1/\alpha} \zeta \left(\frac{1}{\alpha} \right) T^{-1} (T \|\sigma\|)^{1/\alpha} x_0 - \Gamma(1 - \alpha) x_1$$

with

$$x_0 = \|\sigma\|^{-1}\left(\frac{k_+ r_+^\alpha}{\alpha + p_+} - \frac{k_- r_-^\alpha}{\alpha + p_-}\right),$$

$$x_1 = \frac{k_+ r_+}{p_+ + 1} - \frac{k_- r_-}{p_- + 1},$$

where ζ denotes the Riemann zeta function [1, 23.2], γ is the Euler constant [1, 6.1.3].

4.1 A Monte Carlo example

In this section, we assess the goodness of fit of random number generators proposed in the previous section. A brief Monte Carlo study is performed and prices of European put options with different strikes are calculated. We take into consideration a CGMY process with the same artificial parameters as [16], that is, $C = 0.5$, $G = 2$, $M = 3.5$, $Y = 0.5$, interest rate $r = 0.04$, initial stock price $S_0 = 100$ and annualised maturity $T = 0.25$. Furthermore we consider also a GTS process defined by the characteristic exponent (9) and parameters $c_+ = 0.5$, $c_- = 1$, $\lambda_+ = 3.5$, $\lambda_- = 2$ and $\alpha = 0.5$, interest rate r, initial stock price S_0 and maturity T as in the CGMY case.

Monte Carlo prices are obtained through 50,000 simulations. The Esscher transform with $\theta = -1.5$ is considered to reduce the variance [12]. We want to emphasise that the Esscher transform is an exponential tilting [21], thus if applied to a CGMY or a GTS process, it modifies only parameters but not the form of the characteristic function.

In Table 1 simulated prices and prices obtained by using the Fourier transform method [6] are compared. Even if there is a competitive CGMY random number generator, where a time changed Brownian motion is considered [16], we prefer to use an algorithm based on series representation. Contrary to the CGMY case, in

Table 1. European put option prices computed using the Fourier transform method (price) and by Monte Carlo simulation (Monte Carlo)

	CGMY			GTS	
Strike	Price	Monte Carlo	Strike	Price	Monte Carlo
80	1.7444	1.7472	80	3.2170	3.2144
85	2.3926	2.3955	85	4.2132	4.2179
90	3.2835	3.2844	90	5.4653	5.4766
95	4.5366	4.5383	95	7.0318	7.0444
100	6.3711	6.3724	100	8.9827	8.9968
105	9.1430	9.1532	105	11.3984	11.4175
110	12.7632	12.7737	110	14.3580	14.3895
115	16.8430	16.8551	115	17.8952	17.9394
120	21.1856	21.2064	120	21.9109	21.9688

general there is no constructive method to find the subordinator process that changes the time of the Brownian motion; that is we do not know the process T_t such that the TS_α process X_t can be rewritten as $W_{T(t)}$ [7]. The shot noise representation allows one to generate any TS_α process.

5 Conclusions

In this work, we have focused our attention on the practical implementation of numerical methods involving the use of TS_α distributions and processes in the field of finance. Basic definitions are given and a possible algorithm to approximate the density function is proposed. Furthermore, a general Monte Carlo method is developed with a look at option pricing.

References

1. Abramowitz, M., Stegun, I.A.: Handbook of Mathematical Functions, with Formulas, Graphs, and Mathematical Tables. Dover Publications, Mineola, NY (1974)
2. Asmussen, S., Glynn, P.W.: Stochastic Simulation: Algorithms and Analysis. Springer, New York (2007)
3. Becken, W., Schmelcher, P.: The analytic continuation of the Gaussian hypergeometric function 2F1 (a, b; c; z) for arbitrary parameters. J. Comp. Appl. Math. 126(1–2), 449–478 (2000)
4. Boyarchenko, S.I., Levendorskiĭ, S.: Non-Gaussian Merton-Black-Scholes Theory. World Scientific, River Edge, NJ (2002)
5. Carr, P., Geman, H., Madan, D.B., Yor, M.: The Fine Structure of Asset Returns: An Empirical Investigation. J. Bus. 75(2), 305–332 (2002)
6. Carr, P., Madan, D.: Option valuation using the fast Fourier transform. J. Comput. Finan. 2(4), 61–73 (1999)
7. Cont, R., Tankov, P.: Financial Modelling with Jump Processes. Chapman & Hall / CRC Press, London (2004)
8. Feller, W.: An Introduction to Probability Theory and Its Applications. Vol. 2. Wiley, Princeton, 3rd edition (1971)
9. Gil, A., Segura, J., Temme, N.M.: Numerical Methods for Special Functions. Siam, Philadelphia (2007)
10. Houdré, C., Kawai, R.: On fractional tempered stable motion. Stoch. Proc. Appl. 116(8), 1161–1184 (2006)
11. Janicki, A., Weron, A.: Simulation and Chaotic Behavior of α-stable Stochastic Processes. Marcel Dekker, New York (1994)
12. Kawai, R.: An importance sampling method based on the density transformation of Lévy processes. Monte Carlo Meth. Appl. 12(2), 171–186 (2006)
13. Kim, Y.S., Rachev, S.T., Bianchi, M.L., Fabozzi, F.J.: Financial market models with lévy processes and time-varying volatility. J. Banking Finan. 32(7), 1363–1378 (2008)
14. Kim, Y.S., Rachev, S.T., Bianchi, M.L., Fabozzi, F.J.: A new tempered stable distribution and its application to finance. In: Bol, G., Rachev, S.T., Wuerth, R. (eds.) Risk Assessment: Decisions in Banking and Finance, pp. 77–119. Physika Verlag, Springer, Heidelberg (2009)

15. Küchler, U., Tappe, S.: Bilateral gamma distributions and processes in financial mathematics. Stochastic Proc. Appl. 118, 261–283 (2008)
16. Poirot J., Tankov, P.: Monte Carlo Option Pricing for Tempered Stable (CGMY) Processes. Asia Pac. Finan. Markets 13(4), 327–344 (2006)
17. Press, W.H., Teukolsky, S.A., Vetterling, W.T., Flannery, B.P.: Numerical Recipes in C: The Art of Scientific Computing. Cambridge University Press, Cambridge (2002)
18. Rachev, S.T., Menn, C., Fabozzi, F.J.: Fat-tailed and Skewed Asset Return Distributions: Implications for Risk Management, Portfolio Selection, and Option Pricing. Wiley, New York (2005)
19. Rachev, S.T., Mittnik, S.: Stable Paretian Models in Finance. Wiley, New York (2000)
20. Rosiński, J.: Series representations of Lévy processes from the perspective of point processes. In: Barndorff-Nielsen, O.E., Mikosch, T., Resnick, S.I. (eds) Lévy Processes – Theory and Applications, pp. 401–415. Birkhäuser, Boston (2001)
21. Rosiński, J.: Tempering stable processes. Stoch. Proc. Appl. 117, 677–707 (2007)
22. Schoutens, W.: Lévy Processes in Finance: Pricing Financial Derivatives. Wiley, Chichester (2003)
23. Stoyanov, S., Racheva-Iotova, B.: Univariate stable laws in the field of finance–parameter estimation. J. Concrete Appl. Math. 2(4), 24–49 (2004)
24. Terdik, G., Woyczyński, W.A.: Rosiński measures for tempered stable and related Ornstein-Uhlenbeck processes. Prob. Math. Stat. 26(2), 213–243 (2006)

Transformation kernel estimation of insurance claim cost distributions

Catalina Bolancé, Montserrat Guillén, and Jens Perch Nielsen

Abstract. A transformation kernel density estimator that is suitable for heavy-tailed distributions is discussed. Using a truncated beta transformation, the choice of the bandwidth parameter becomes straightforward. An application to insurance data and the calculation of the value-at-risk are presented.

Key words: non-parametric statistics, actuarial loss models, extreme value theory

1 Introduction

The severity of claims is measured in monetary units and is usually referred to as insurance loss or claim cost amount. The probability density function of claim amounts is usually right skewed, showing a big bulk of small claims and some relatively infrequent large claims. For an insurance company, density tails are therefore of special interest due to their economic magnitude and their influence on re-insurance agreements.

It is widely known that large claims are highly unpredictable while they are responsible for financial instability and so, since solvency is a major concern for both insurance managers and insurance regulators, there is a need to estimate the density of claim cost amounts and to include the extremes in all the analyses.

This paper is about estimating the density function nonparametrically when data are heavy-tailed. Other approaches are based on extremes, a subject that has received much attention in the economics literature. Embrechts et al., Coles, and Reiss and Thomas [8, 11, 15] have discussed extreme value theory (EVT) in general. Chavez-Demoulin and Embrechts [6], based on Chavez-Demoulin and Davison [5], have discussed smooth extremal models in insurance. They focused on highlighting nonparametric trends, as a time dependence is present in many catastrophic risk situations (such as storms or natural disasters) and in the financial markets. A recent work by Cooray and Ananda [9] combines the lognormal and the Pareto distribution and derives a distribution which has a suitable shape for small claims and can handle heavy tails. Others have addressed this subject with the g-and-h distribution, like Dutta and Perry [10] for operation risk analysis. The g-and-h distribution [12] can be formed by two nonlinear transformations of the standard normal distribution and has two parameters, skewness and kurtosis.

In previous papers, we have analysed claim amounts in a one-dimensional setting and we have proved that a nonparametric approach that accounts for the asymmetric nature of the density is preferred for insurance loss distributions [2, 4]. Moreover, we have applied the method to a liability data set and compared the nonparametric kernel density estimation procedure to classical methods [4]. Several authors [7] have devoted much interest to transformation kernel density estimation, which was initially proposed by Wand et al. [21] for asymmetrical variables and based on the shifted power transformation family. The original method provides a good approximation for heavy-tailed distributions. The statistical properties of the density estimators are also valid when estimating the cumulative density function (cdf). Transformation kernel estimation turns out to be a suitable approach to estimate quantiles near 1 and therefore it can be used to estimate Value-at-Risk (VaR) in financial and insurance-related applications.

Buch-Larsen et al. [4] proposed an alternative transformation based on a generalisation of the Champernowne distribution; simulation studies have shown that it is preferable to other transformation density estimation approaches for distributions that are Pareto-like in the tail. In the existing contributions, the choice of the bandwidth parameter in transformation kernel density estimation is still a problem. One way of undergoing bandwidth choice is to implement the transformation approach so that it leads to a beta distribution, then use existing theory to optimise bandwidth parameter selection on beta distributed data and backtransform to the original scale. The main drawback is that the beta distribution may be very steep in the domain boundary, which causes numerical instability when the derivative of the inverse distribution function is needed for the backward transformation. In this work we propose to truncate the beta distribution and use the truncated version at transformation kernel density estimation. The results on the optimal choice of the bandwidth for kernel density estimation of beta density are used in the truncated version directly. In the simulation study we see that our approach produces very good results for heavy-tailed data. Our results are particularly relevant for applications in insurance, where the claims amounts are analysed and usually small claims (low cost) coexist with only a few large claims (high cost).

Let $f_\mathbf{x}$ be a density function. Terrell and Scott [19] and Terrell [18] analysed several density families that minimise functionals $\int \left\{ f_\mathbf{x}^{(p)}(x) \right\}^2 dx$, where superscript (p) refers to the pth derivative of the density function. We will use these families in the context of transformed kernel density estimation. The results for those density families are very useful to improve the properties of the transformation kernel density estimator.

Given a sample $X_1, \ldots, X_n$ of independent and identically distributed (iid) observations with density function $f_{\mathbf{x}}$, the classical kernel density estimator is:

$$\hat{f}_{\mathbf{x}}(x) = \frac{1}{n} \sum_{i=1}^{n} K_b (x - X_i), \tag{1}$$

where b is the bandwidth or smoothing parameter and $K_b(t) = K(t/b)/b$ is the kernel. In Silverman [16] or Wand and Jones [20] one can find an extensive revision of classical kernel density estimation.

An error distance between the estimated density $\hat{f}_{\mathbf{x}}$ and the theoretical density $f_{\mathbf{x}}$ that has widely been used in the analysis of the optimal bandwidth b is the mean integrated squared error ($MISE$):

$$E \left\{ \int \left(f_{\mathbf{x}}(x) - \hat{f}_{\mathbf{x}}(x) \right)^2 dx \right\}. \tag{2}$$

It has been shown (see, for example, Silverman [16], chapter 3) that the $MISE$ is asymptotically equivalent to $A - MISE$:

$$\frac{1}{4} b^4 (k_2)^2 \int \left\{ f_{\mathbf{X}}''(x) \right\}^2 dx + \frac{1}{nb} \int K(t)^2 dt, \tag{3}$$

where $k_2 = \int t^2 K(t) dt$. If the second derivative of $f_{\mathbf{x}}$ exists (and we denote it by $f_{\mathbf{x}}''$), then $\int \left\{ f_{\mathbf{x}}''(x) \right\}^2 dx$ is a measure of the degree of smoothness because the smoother the density, the smaller this integral is. From the expression for $A - MISE$ it follows that the smoother $f_{\mathbf{x}}$, the smaller the value of $A - MISE$.

Terrell and Scott (1985, Lemma 1) showed that $Beta(3, 3)$ defined on the domain $(-1/2, 1/2)$ minimises the functional $\int \left\{ f_{\mathbf{x}}''(x) \right\}^2 dx$ within the set of beta densities with the same support. The $Beta(3, 3)$ distribution will be used throughout our work. Its pdf and cdf are:

$$g(x) = \frac{15}{8} \left(1 - 4x^2 \right)^2, \quad -\frac{1}{2} \leq x \leq \frac{1}{2}, \tag{4}$$

$$G(x) = \frac{1}{8} \left(4 - 9x + 6x^2 \right) (1 + 2x)^3. \tag{5}$$

We assume that a transformation exists so that $T(X_i) = Z_i$ $(i = 1, \ldots, n)$ is assumed from a $Uniform(0, 1)$ distribution. We can again transform the data so that $G^{-1}(Z_i) = Y_i$ is a random sample from a random variable $\mathbf{y}$ with a $Beta(3, 3)$ distribution, whose pdf and cdf are defined respectively in (4) and (5).

In this work, we use a parametric transformation $T(\cdot)$, namely the modified Champernowne cdf, as proposed by Buch-Larsen et al. [4].

Let us define the kernel estimator of the density function for the transformed variable:

$$\hat{g}(y) = \frac{1}{n} \sum_{i=1}^{n} K_b (y - Y_i), \tag{6}$$

which should be as close as possible to a $Beta(3, 3)$. We can obtain an exact value for the bandwidth parameter that minimizes $A - MISE$ of $\hat{g}$. If $K(t) = (3/4)\left(1 - t^2\right) 1\left(|t| \leq 1\right)$ is the Epanechnikov kernel, where $1(\cdot)$ equals one when the condition is true and zero otherwise, then we show that the optimal smoothing parameter for $\hat{g}$ if $\mathbf{y}$ follows a $Beta(3, 3)$ is:

$$b = \left(\frac{1}{5}\right)^{-\frac{2}{5}} \left(\frac{3}{5}\right)^{\frac{1}{5}} (720)^{-\frac{1}{5}} n^{-\frac{1}{5}}. \tag{7}$$

Finally, in order to estimate the density function of the original variable, since $y = G^{-1}(z) = G^{-1}\{T(x)\}$, the transformation kernel density estimator is:

$$\hat{f}_{\mathbf{x}}(x) = \hat{g}(y)\left[G^{-1}\{T(x)\}\right]' T'(x) =$$

$$= \frac{1}{n}\sum_{i=1}^{n} K_b\left(G^{-1}\{T(x)\} - G^{-1}\{T(X_i)\}\right)\left[G^{-1}\{T(x)\}\right]' T'(x). \tag{8}$$

The estimator in (8) asymptotically minimises $MISE$ and the properties of the transformation kernel density estimation (8) are studied in Bolancé et al. [3]. Since we want to avoid the difficulties of the estimator defined in (8), we will construct the transformation so as to avoid the extreme values of the beta distribution domain.

2 Estimation procedure

Let $\mathbf{z} = \mathbf{T}(\mathbf{x})$ be a $Uniform(0, 1)$; we define a new random variable in the interval $[1 - l, l]$, where $1/2 < l < 1$. The values for l should be close to 1. The new random variable is $z^* = T^*(x) = (1 - l) + (2l - 1)T(x)$. We will discuss the value of l later.

The pdf of the new variable $y^* = G^{-1}(z^*)$ is proportional to the $Beta(3, 3)$ pdf, but it is in the $[-a, a]$ interval, where $a = G^{-1}(l)$. Finally, our proposed transformation kernel density estimation is:

$$\hat{f}_{\mathbf{x}}(x) = \frac{\hat{g}(y^*)\left[G^{-1}\{T^*(x)\}\right]' T^{*'}(x)}{(2l - 1)} = \hat{g}(y^*)\left[G^{-1}\{T^*(x)\}\right]' T'(x)$$

$$= \frac{1}{n}\sum_{i=1}^{n} K_b\left(G^{-1}\{T^*(x)\} - G^{-1}\{T(X_i)\}\right)\left[G^{-1}\{T^*(x)\}\right]' T'(x). \tag{9}$$

The value of $A - MISE$ associated to the kernel estimation $\hat{g}(y^*)$, where the random variable y^* is defined on an interval that is smaller than $Beta(3, 3)$ domain is:

$$A - MISE_a = \frac{1}{4}b^4(k_2)^2\int_{-a}^{a}\{g''(y)\}^2 dy + \frac{1}{nb}\int_{-a}^{a}g(y)dy\int K(t)^2 dt. \tag{10}$$

And finally, the optimal bandwidth parameter based on the asymptotic mean integrated squared error measure for $\hat{g}\left(y^{*}\right)$ is:

$$
\begin{aligned}
b_g^{opt} &= k_2^{-\frac{2}{5}}\left(\int_{-1}^{1} K\left(t\right)^2 dt \int_{-a}^{a} g\left(y\right) dy\right)^{\frac{1}{5}}\left(\int_{-a}^{a}\left\{g''\left(y\right)\right\}^2 dy\right)^{-\frac{1}{5}} n^{-\frac{1}{5}} \\
&= \left(\frac{1}{5}\right)^{-\frac{2}{5}}\left(\frac{3}{5}\left(\frac{1}{4}a\left(-40a^2+48a^4+15\right)\right)\right)^{\frac{1}{5}} \\
&\quad \times \left(360a\left(-40a^2+144a^4+5\right)\right)^{-\frac{1}{5}} n^{-\frac{1}{5}},
\end{aligned} \tag{11}
$$

The difficulty that arises when implementing the transformation kernel estimation expressed in (9) is the selection of the value of l. This value can be chosen subjectively as discussed in the simulation results by Bolancé et al. [3]. Let $X_i, i = 1, \ldots, n$, be iid observations from a random variable with an unknown density $f_{\mathbf{x}}$. The transformation kernel density estimator of $f_{\mathbf{x}}$ is called KIBMCE (kernel inverse beta modified Champernowne estimator).

3 VaR estimation

In finance and insurance, the VaR represents the magnitude of extreme events and therefore it is used as a risk measure, but VaR is a quantile. Let $\mathbf{x}$ be a loss random variable with distribution function $F_{\mathbf{x}}$; given a probability level p, the VaR of $\mathbf{x}$ is $VaR\left(\mathbf{x}, p\right) = \inf\left\{x, F_{\mathbf{x}}\left(x\right) \geq p\right\}$. Since $F_{\mathbf{x}}$ is a continuous and nondecreasing function, then $VaR\left(\mathbf{x}, p\right) = F_{\mathbf{x}}^{-1}\left(p\right)$, where p is a probability near 1 (0.95, 0.99,...). One way of approximating $VaR\left(\mathbf{x}, p\right)$ is based on the empirical distribution function, but this has often been criticised because the empirical estimation is based only on a limited number of observations, and even np may not be an integer number. As an alternative to the empirical distribution approach, classical kernel estimation of the distribution function can be useful, but this method will be very imprecise for asymmetrical or heavy-tailed variables.

Swanepoel and Van Graan [17] propose to use a nonparametric transformation of the data, which is equal to a classical kernel estimation of the distribution function. We propose to use a parametric transformation based on a distribution function.

Given a transformation function $Tr\left(\mathbf{x}\right)$, it follows that $F_{\mathbf{x}}\left(x\right) = F_{Tr(\mathbf{x})}\left(Tr\left(x\right)\right)$. So, the computation of $VaR\left(\mathbf{x}, p\right)$ is based on the kernel estimation of the distribution function of the transformed variable.

Kernel estimation of the distribution function is [1, 14]:

$$
\hat{F}_{Tr(\mathbf{x})}\left(Tr\left(x\right)\right) = \frac{1}{n}\sum_{i=1}^{n}\int_{-1}^{\frac{Tr(x)-Tr(X_i)}{b}} K\left(t\right) dt. \tag{12}
$$

Therefore, the $VaR\left(\mathbf{x}, p\right)$ can be found as:

$$
VaR\left(\mathbf{x}, p\right) = Tr^{-1}\left[VaR\left(Tr\left(\mathbf{x}\right), p\right)\right] = Tr^{-1}\left[\hat{F}_{Tr(\mathbf{x})}^{-1}\left(p\right)\right]. \tag{13}
$$

4 Simulation study

This section presents a comparison of our inverse beta transformation method with the results presented by Buch-Larsen et al. [4] based only on the modified Champernowne distribution. Our objective is to show that the second transformation, which is based on the inverse of a beta distribution, improves density estimation.

In this work we analyse the same simulated samples as in Buch-Larsen et al. [4], which were drawn from four distributions with different tails and different shapes near 0: *lognormal, lognormal-Pareto, Weibull* and *truncated logistic*. The distributions and the chosen parameters are listed in Table 1.

Table 1. Distributions in simulation study

Distribution	Density	Parameters
Lognormal(μ, σ)	$f(x) = \dfrac{1}{\sqrt{2\pi\sigma^2}x} e^{-\frac{(\log x - \mu)^2}{2\sigma^2}}$	$(\mu, \sigma) = (0, 0.5)$
Weibull(γ)	$f(x) = \gamma x^{(\gamma-1)} e^{-x^\gamma}$	$\gamma = 1.5$
Mixture of pLognormal(μ, σ) and $(1-p)$Pareto(λ, ρ, c)	$f(x) = p \dfrac{1}{\sqrt{2\pi\sigma^2}x} e^{-\frac{(\log x - \mu)^2}{2\sigma^2}}$ $+ (1-p)(x-c)^{-(\rho+1)} \rho \lambda^\rho$	$(p, \mu, \sigma, \lambda, \rho, c)$ $= (0.7, 0, 1, 1, 1, -1)$ $= (0.3, 0, 1, 1, 1, -1)$
Tr. Logistic	$f(x) = \dfrac{2}{s} e^{\frac{x}{s}} \left(1 + e^{\frac{x}{s}}\right)^{-2}$	$s = 1$

Buch-Larsen et al. [4] evaluate the performance of the KMCE estimators compared to the estimator described by Clements et al. [7], the estimator described by Wand et al. [21] and the estimator described by Bolancé et al. [2]. The Champernowne transformation substantially improves the results from previous authors. Here we see that if the second transformation based on the inverse beta transformation improves the results presented in Buch-Larsen et al. [4], this means that the double-transformation method presented here is a substantial gain with respect to existing methods.

We measure the performance of the estimators by the error measures based on L_1 norm, L_2 norm and $WISE$. The last one weighs the distance between the estimated and the true distribution with the squared value of x. This results in an error measure that emphasises the tail of the distribution, which is very relevant in practice when dealing with income or cost data:

$$\left(\int_0^\infty \left(\widehat{f}(x) - f(x)\right)^2 x^2 \, dx \right)^{1/2}. \tag{14}$$

The simulation results can be found in Table 2. For every simulated density and for sample sizes $N = 100$ and $N = 1000$, the results presented here correspond to

Table 2. Estimated error measures (L_1, L_2 and $WISE$) for KMCE and KIBMCE $l = 0.99$ and $l = 0.98$ for sample size 100 and 1000 based on 2000 replications

				Lognormal	Log-Pareto		Weibull	Tr. Logistic
					$p = 0.7$	$p = 0.3$		
N=100	L1	KMCE		0.1363	0.1287	0.1236	0.1393	0.1294
		KIBMCE	$l = 0.99$	0.1335	0.1266	0.1240	0.1374	0.1241
			$l = 0.98$	0.1289	0.1215	0.1191	0.1326	0.1202
	L2	KMCE		0.1047	0.0837	0.0837	0.1084	0.0786
		KIBMCE	$l = 0.99$	0.0981	0.0875	0.0902	0.1085	0.0746
			$l = 0.98$	0.0956	0.0828	0.0844	0.1033	0.0712
	WISE	KMCE		0.1047	0.0859	0.0958	0.0886	0.0977
		KIBMCE	$l = 0.99$	0.0972	0.0843	0.0929	0.0853	0.0955
			$l = 0.98$	0.0948	0.0811	0.0909	0.0832	0.0923
N=1000	L1	KMCE		0.0659	0.0530	0.0507	0.0700	0.0598
		KIBMCE	$l = 0.99$	0.0544	0.0509	0.0491	0.0568	0.0497
			$l = 0.98$	0.0550	0.0509	0.0522	0.0574	0.0524
	L2	KMCE		0.0481	0.0389	0.0393	0.0582	0.0339
		KIBMCE	$l = 0.99$	0.0394	0.0382	0.0393	0.0466	0.0298
			$l = 0.98$	0.0408	0.0385	0.0432	0.0463	0.0335
	WISE	KMCE		0.0481	0.0384	0.0417	0.0450	0.0501
		KIBMCE	$l = 0.99$	0.0393	0.0380	0.0407	0.0358	0.0393
			$l = 0.98$	0.0407	0.0384	0.0459	0.0369	0.0394

the following error measures: L_1, L_2 and $WISE$ for different values of the trimming parameter $l = 0.99, 0.98$. The benchmark results are labelled KMCE and they correspond to those presented in Buch-Larsen et al. [4].

In general, we can conclude that after a second transformation based on the inverse of a modified beta distribution cdf, the error measures diminish with respect to the KMCE method. In some situations the errors diminish quite substantially with respect to the existing approaches.

We can see that the error measure that shows improvements when using the KIBMCE estimator is the $WISE$, which means that this new approach fits the tail of positive distributions better than existing alternatives. The $WISE$ error measure is always smaller for the KIBMCE than for the KMCE, at least for one of the two possible values of l that have been used in this simulation study. This would make the KIBMCE estimator specially suitable for positive heavy-tailed distributions. When looking more closely at the results for the mixture of a lognormal distribution and a Pareto tail, we see that larger values of l are needed to improve the error measures that were encountered with the KMCE method only for $N = 1000$. For $N = 100$, a contrasting conclusion follows.

We can see that for the truncated logistic distribution, the lognormal distribution and the Weibull distribution, the method presented here is clearly better than the existing KMCE. We can see in Table 2 that for $N = 1000$, the KIBMCE $WISE$ is about 20 % lower than the KMCE $WISE$ for these distributions. A similar behaviour is shown by the other error measures, L_1 and L_2, which for $N = 1000$, are about 15 % lower for the KIBMCE.

Note that the KMCE method was studied in [4] and the simulation study showed that it improved on the error measures for the existing methodological approaches [7, 21].

5 Data study

In this section, we apply our estimation method to a data set that contains automobile claim costs from a Spanish insurance company for accidents that occurred in 1997. This data set was analysed in detail by Bolancé et al. [2]. It is a typical insurance claims amount data set, i.e., a large sample that looks heavy-tailed. The data are divided into two age groups: claims from policyholders who are less than 30 years old and claims from policyholders who are 30 years old or older. The first group consists of 1061 observations in the interval [1;126,000] with mean value 402.70. The second group contains 4061 observations in the interval [1;17,000] with mean value 243.09. Estimation of the parameters in the modified Champernowne distribution function for the two samples is, for young drivers $\widehat{\alpha}_1 = 1.116$, $\widehat{M}_1 = 66$, $\widehat{c}_1 = 0.000$ and for older drivers $\widehat{\alpha}_2 = 1.145$, $\widehat{M}_2 = 68$, $\widehat{c}_2 = 0.000$. We notice that $\alpha_1 < \alpha_2$, which indicates that the data set for young drivers has a heavier tail than the data set for older drivers.

For small costs, the KIBMCE density in the density peak is greater than for the KMCE approach proposed by Buch-Larsen et al. [4] both for young and older drivers. For both methods, the tail in the estimated density of young policyholders is heavier than the tail of the estimated density of older policyholders. This can be taken as evidence that young drivers are more likely to claim a large amount than older drivers. The KIBMCE method produces lighter tails than the KMCE methods. Based on the results in the simulation study presented in Bolancé et al. [3], we believe that the KIBMCE method improves the estimation of the density in the extreme claims class.

Table 3. Estimation of VaR at the 95% level, in thousands

	Empirical	KMCE	KIBMCE	
			$l = 0.99$	$l = 0.98$
Young	1104	2912	1601	1716
Older	1000	1827	1119	1146

Table 3 presents the VaR at the 95% level, which is obtained from the empirical distribution estimation and the computations obtained with the KMCE and KIBMCE. We believe that the KIBMCE provides an adequate estimation of the VaR and it seems a recommendable approach to be used in practice.

Acknowledgement. The Spanish Ministry of Education and Science support SEJ2007-63298 is acknowledged. We thank the reviewers for helpful comments.

References

1. Azzalini, A.: A note on the estimation of a distribution function and quantiles by kernel method. Biometrika 68, 326–328 (1981)
2. Bolancé, C., Guillén, M., Nielsen, J.P.: Kernel density estimation of actuarial loss functions. Ins. Math. Econ. 32, 19–36 (2003)
3. Bolancé, C., Guillén, M., Nielsen, J.P.: Inverse Beta transformation in kernel density estimation. Stat. Prob. Lett. 78, 1757–1764 (2008)
4. Buch-Larsen, T., Guillen, M., Nielsen, J.P., Bolancé, C.: Kernel density estimation for heavy-tailed distributions using the Champernowne transformation. Statistics 39, 503–518 (2005)
5. Chavez-Demoulin, V., Davison, P.: Generalized additive modelling of sample extremes. Appl. Stat. 54, 207–222 (2005)
6. Chavez-Demoulin, V., Embrechts, P.: Smooth extremal models in finance and insurance. J. Risk Ins. 71, 183–199 ((2004)
7. Clements, A.E., Hurn, A.S., Lindsay, K.A.: Möbius-like mappings and their use in kernel density estimation. J. Am. Stat. Assoc. 98, 993–1000 (2003)
8. Coles, S.: An Introduction to Statistical Modeling of Extreme Values. Springer, London (2001)
9. Cooray, K., Ananda, M.M.A.: Modeling actuarial data with a composite lognormal-Pareto model. Scand. Act. J. 5, 321–334 (2005)
10. Dutta, K., Perry, J.: A tale of tails: an empirical analysis of loss distribution models for estimating operational risk capital. FRB of Boston Working Paper No. 06-13 (2006) Available at SSRN: http://ssrn.com/abstract=918880
11. Embrechts, P., Klüppelberg, C., Mikosch, T.: Modelling Extremal Events. Springer, London (1999)
12. Hoaglin, D.C.: g-and-h distributions. In: Kotz, S., Johnson, N.L. (eds.) Encyclopedia of Statistical Sciences, pp. 298–301, vol. 3. Wiley, New York (1983)
13. Reiss, R.D.: Nomparametric estimation of smooth distribution functions. Scand. J. Stat. 8, 116–119 (1981)
14. Reiss, R.D., Thomas, M.: Statistical Analysis of Extreme Values: with Applications to Insurance, Finance, Hydrology and other Fields, 2nd Edition. Birkhäuser (2001)
15. Silverman, B.W.: Density Estimation for Statistics and Data Analysis. Chapman & Hall, London (1986)
16. Swanepoel, J.W.H., Van Graan F.C.: A new kernel distribution function estimator based on a non-parametric transformation of the data. Scand. J. Stat. 32, 551–562 (2005)
17. Terrell, G.R.: The maximal smoothing principle in density estimation. J. Am. Stat. Assoc. 85, 270–277 (1990)
18. Terrell, G.R., Scott, D.W.: Oversmoothed nonparametric density estimates. J. Am. Stat. Assoc. 80, 209–214 (1985)
19. Wand, M.P., Jones, M.C.: Kernel Smoothing. Chapman & Hall, London (1995)
20. Wand, P., Marron, J.S., Ruppert, D.: Transformations in density estimation. J. Am. Stat. Assoc. 86, 343–361 (1991)

What do distortion risk measures tell us on excess of loss reinsurance with reinstatements?

Antonella Campana and Paola Ferretti

Abstract. In this paper we focus our attention on the study of an excess of loss reinsurance with reinstatements, a problem previously studied by Sundt and, more recently, by Mata and Hürlimann. It is well known that the evaluation of pure premiums requires knowledge of the claim size distribution of the insurance risk: in order to face this question, different approaches have been followed in the actuarial literature. In a situation of incomplete information in which only some characteristics of the involved elements are known, it appears to be particularly interesting to set this problem in the framework of risk-adjusted premiums. It is shown that if risk-adjusted premiums satisfy a generalised expected value equation, then the initial premium exhibits some regularity properties as a function of the percentages of reinstatement.

Key words: excess of loss reinsurance, reinstatements, distortion risk measures

1 Introduction

In recent years the study of excess of loss reinsurance with reinstatements has become a major topic, in particular with reference to the classical evaluation of pure premiums, which is based on the collective model of risk theory.

The problem, previously studied by Sundt [5] and, more recently, by Mata [4] and Hürlimann [3], requires the evaluation of pure premiums given the knowledge of the claim size distribution of the insurance risk: in order to face this question, different approaches have been followed in the actuarial literature. Sundt [5] based the computation on the Panjer recursion numerical method and Hürlimann [3] provided distribution-free approximations to pure premiums.

In a situation of incomplete information in which only some characteristics of the involved elements are known, it appears to be particularly interesting to set this problem in the framework of risk-adjusted premiums.

We start from the methodology developed by Sundt [5] to price excess of loss reinsurance with reinstatements for pure premiums and, with the aim of relaxing the basic hypothesis made by Walhin and Paris [6], who calculated the initial premium P under the Proportional Hazard transform premium principle, we address our analysis to the study of the role played by risk-adjusted premium principles. The particular choice in the proposal of Walhin and Paris of the PH-transform risk measure strengthens our interest in the study of risk-adjusted premiums that belong to the class of distortion risk measures defined by Wang [7].

In the mathematical model we studied (for more details see Campana [1]), when the reinstatements are paid ($0 \leq c_i \leq 1$ is the ith percentage of reinstatement) the total premium income $\delta(P)$ becomes a random variable which is correlated to the aggregate claims S. Since risk measures satisfy the properties of linearity and additivity for comonotonic risks (see [2]) and layers are comonotonic risks, we can define the function

$$F(P, c_1, c_2, \ldots, c_K) = P \left[1 + \frac{1}{m} \sum_{i=0}^{K-1} c_{i+1} \, W_{g_1}(L_X(im, (i+1)m)) \right] - \sum_{i=0}^{K} W_{g_2}(L_X(im, (i+1)m)) \tag{1}$$

where g_1 and g_2 are distortion functions and $W_g(X)$ denotes the distortion risk measure of X. This function gives a measure of the distance between two distortion risk measures: that of the total premium income $\delta(P)$ and that of the aggregate claims S. The choice of risk-adjusted premiums satisfying the expected value equation ensures that the previous distance is null: in this way, it is possible to study the initial premium P as a function of the percentages of reinstatement.

The paper is organised as follows. In Section 2 we first review some basic settings for describing the excess of loss reinsurance model and we review some definitions and preliminary results in the field of non-proportional reinsurance covers. Section 3 is devoted to the problem of detecting the total initial premium: we present the study of the case in which the reinstatements are paid in order to consider the total premium income as a random variable which is correlated to the aggregate claims. The analysis is set in the framework of distortion risk measures: some basic definitions and results in this field are recalled. Section 4 presents the main results related to the problem of measuring the total initial premium as a function of the percentages of reinstatement, dependence that it is generally neglected in the literature. Some concluding remarks in Section 5 end the paper.

2 Excess of loss reinsurance with reinstatements: problem setting

The excess of loss reinsurance model we study in this paper is related to the model that has been proposed and analysed by Sundt [5]. Some notations, abbreviations and conventions used throughout the paper are the following.

An excess of loss reinsurance for the layer m in excess of d, written m xs d, is a reinsurance which covers the part of each claim that exceeds the deductible d but with a limit on the payment of each claim, which is set equal to m; in other words, the reinsurer covers for each claim of size Y the amount

$$L_Y(d, d+m) = \min\{(Y-d)_+, m\}$$

where $(a)_+ = a$ if $a > 0$, otherwise $(a)_+ = 0$.

We consider an insurance portfolio: N is the number of claims that occurred in the portfolio during the reference year and Y_i is the ith claim size ($i = 1, 2, \ldots, N$). The aggregate claims to the layer is the random sum given by

$$X = \sum_{i=1}^{N} L_{Y_i}(d, d + m).$$

It is assumed that $X = 0$ when $N = 0$. An excess of loss reinsurance, or for short an XL reinsurance, for the layer m xs d with aggregate deductible D and aggregate limit M covers only the part of X that exceeds D but with a limit M:

$$L_X(D, D + M) = \min\{(X - D)_+, M\}.$$

This cover is called an XL reinsurance for the layer m xs d with aggregate layer M xs D.

Generally it is assumed that the aggregate limit M is given as a whole multiple of the limit m, i.e., $M = (K + 1)m$: in this case we say that there is a limit to the number of losses covered by the reinsurer. This reinsurance cover is called an XL reinsurance for the layer m xs d with aggregate deductible D and K reinstatements and provides total cover for the following amount

$$L_X(D, D + (K + 1)m) = \min\{(X - D)_+, (K + 1)m\}. \tag{2}$$

Let P be the initial premium: it covers the original layer, that is

$$L_X(D, D + m) = \min\{(X - D)_+, m\}. \tag{3}$$

It can be considered as the 0-th reinstatement.

The condition that the reinstatement is paid pro rata means that the premium for the ith reinstatement is a random variable given by

$$\frac{c_i P}{m} L_X(D + (i - 1)m, D + im) \tag{4}$$

where $0 \le c_i \le 1$ is the ith percentage of reinstatement. If $c_i = 0$ the reinstatement is free, otherwise it is paid.

The related total premium income is a random variable, say $\delta(P)$, which is defined as

$$\delta(P) = P\left(1 + \frac{1}{m}\sum_{i=0}^{K-1} c_{i+1} L_X(D + im, D + (i + 1)m)\right). \tag{5}$$

From the point of view of the reinsurer, the aggregate claims S paid by the reinsurer for this XL reinsurance treaty, namely

$$S = L_X(D, D + (K + 1)m) \tag{6}$$

satisfy the relation

$$S = \sum_{i=0}^{K} L_X(D + im, D + (i + 1)m). \tag{7}$$

3 Initial premium, aggregate claims and distortion risk measures

The total premium income $\delta(P)$ is a random variable which is correlated to the aggregate claims S in the case in which the reinstatements are paid. Then it follows that it is not obvious how to calculate the initial premium P.

Despite its importance in practice, only recently have some Authors moved their attention to the study of techniques to calculate the initial premium. More precisely, Sundt [5] proposed the methodology to calculate the initial premium P under pure premiums and premiums loaded by the standard deviation principle.

Looking at the pure premium principle for which the expected total premium income should be equal to the expected aggregate claims payments

$$E[\delta(P)] = E[S], \tag{8}$$

it is quite natural to consider the case in which premium principles belong on more general classes: with the aim of plugging this gap, we focus our attention on the class of distortion risk measures. Our interest is supported by Walhin and Paris [6], who calculated the initial premium P under the Proportional Hazard transform premium principle. Even if their analysis is conducted by a numerical recursion, the choice of the PH-transform risk measure as a particular concave distortion risk measure strengthens our interest.

Furthermore, in an excess of loss reinsurance with reinstatements the computation of premiums requires the knowledge of the claim size distribution of the insurance risk: with reference to the expected value equation of the XL reinsurance with reinstatements (8), Sundt [5] based the computation on the Panjer recursion numerical method and Hürlimann [3] provided distribution-free approximations to pure premiums.

Note that both Authors assumed only the case of equal reinstatements, a particular hypothesis on basic elements characterising the model.

In this paper we set our analysis in the framework of distortion risk measures: the core of our proposal is represented by the choice of a more general equation characterising the excess of loss reinsurance with reinstatements, in such a way that it is possible to obtain some general properties satisfied by the initial premium as a function of the percentages of reinstatement. In order to present the main results, we recall some basic definitions and results.

3.1 Distortion risk measures

A risk measure is defined as a mapping from the set of random variables, namely losses or payments, to the set of real numbers. In actuarial science common risk measures are premium principles; other risk measures are used for determining provisions and capital requirements of an insurer in order to avoid insolvency (see e.g., Dhaene et al. [2]).

In this paper we consider the distortion risk measure introduced by Wang [7]:

$$W_g(X) = \int_0^\infty g(H_X(x))dx \tag{9}$$

where the distortion function g is defined as a non-decreasing function $g : [0, 1] \rightarrow [0, 1]$ such that $g(0) = 0$ and $g(1) = 1$. As is well known, the *quantile risk measure* and the *Tail Value-at-Risk* are examples of risk measures belonging to this class. In the particular case of a power g function, i.e., $g(x) = x^{1/\rho}$, $\rho \geq 1$, the corresponding risk measure is the *PH-transform risk measure*, which is the choice made by Walhin and Paris [6].

Distortion risk measures satisfy the following properties (see Wang [7] and Dhaene et al. [2]):

P1. Additivity for comonotonic risks

$$W_g(S^c) = \sum_{i=1}^{n} W_g(X_i) \tag{10}$$

where S^c is the sum of the components of the random vector $\mathbf{X}^c$ with the same marginal distributions of $\mathbf{X}$ and with the comonotonic dependence structure.

P2. Positive homogeneity

$$W_g(a\,X) = a\,W_g(X) \quad \text{for any non-negative constant } a; \tag{11}$$

P3. Translation invariance

$$W_g(X + b) = W_g(X) + b \quad \text{for any constant } b; \tag{12}$$

P4. Monotonicity

$$W_g(X) \leq W_g(Y) \tag{13}$$

for any two random variables X and Y where $X \leq Y$ with probability 1.

In the particular case of a concave distortion measure, the related distortion risk measure satisfying properties *P1-P4* is also sub-additive and it preserves stop-loss order. It is well known that examples of concave distortion risk measures are the *Tail Value-at-Risk* and the *PH-transform risk measure*, whereas *quantile risk measure* is not a concave risk measure.

4 Risk-adjusted premiums

In equation (8) the expected total premium income is set equal to the expected aggregate claims payments: in order to refer to a class of premium principles that is more general than the pure premium principle, we consider a new expected value condition with reference to the class of distortion risk measures.

We impose that the distorted expected value of the total premium income $\delta(P)$ equals the distorted expected value of the aggregate claims S, given two distortion functions g_1 and g_2. Note that in our proposal it is possible to consider distortion functions that are not necessarily the same.

The equilibrium condition may be studied as an equation on the initial premium P: if it admits a solution which is unique, then we call *initial risk-adjusted premium* the corresponding premium P. This is formalised in the following definition.

Definition 1. *Let g_1 and g_2 be distortion functions. The initial risk-adjusted premium P is the unique initial premium, if it does exist, for which the distorted expected value of the total premium income $\delta(P)$ equals the distorted expected value of the aggregate claims S, that is*

$$W_{g_1}(\delta(P)) = W_{g_2}(S). \tag{14}$$

Equation (14) may be studied from several different perspectives, mostly concerned with the existence and uniqueness of the solutions. The next result presents a set of conditions ensuring a positive answer to both these questions: the choice of an excess of loss reinsurance for the layer m xs d with no aggregate deductible D and K reinstatements plays the leading role.

Proposition 1. *Given an XL reinsurance with K reinstatements and no aggregate deductible and given two distortion functions g_1 and g_2, the initial risk-adjusted premium P results to be a function of the percentages of reinstatement $c_1, c_2, \ldots, c_K$. Moreover, it satisfies the following properties:*

i) P is a decreasing function of each percentage of reinstatement c_i ($i = 1, \ldots, K$);

ii) P is a convex, supermodular, quasiconcave and quasiconvex function of the percentages of reinstatement $c_1, c_2, \ldots, c_K$.

Proof. Given the equilibrium condition between the distorted expected premium income and the distorted expected claim payments (14), the initial risk-adjusted premium P is well defined: in fact equation (14) admits a solution which is unique.

Since the layers $L_X(im, (i + 1)m)$, $i = 1, 2, \ldots, K + 1$, are comonotonic risks from (7) we find

$$W_{g_2}(S) = \sum_{i=0}^{K} W_{g_2}(L_X(im, (i + 1)m)). \tag{15}$$

From (5), by assuming the absence of an aggregate deductible (i.e. ,$D = 0$), we have

$$W_{g_1}(\delta(P)) = P\left(1 + \frac{1}{m} \sum_{i=0}^{K-1} c_{i+1} W_{g_1}(L_X(im, (i + 1)m))\right). \tag{16}$$

Therefore, the initial premium P is well defined and it is given by

$$P = \frac{\sum_{i=0}^{K} W_{g_2}(L_X(im, (i + 1)m))}{1 + \frac{1}{m} \sum_{i=0}^{K-1} c_{i+1} W_{g_1}(L_X(im, (i + 1)m))}. \tag{17}$$

The initial risk-adjusted premium P may be considered a function of the percentages of reinstatement $c_1, c_2, \ldots, c_K$. Let $P = f(c_1, c_2, \cdots, c_K)$.

Clearly the function f is a decreasing function of any percentage of reinstatement c_i (where $i = 1, \ldots, K$).

Moreover, if we set

$$A = \sum_{i=0}^{K} W_{g_2}(L_X(im, (i + 1)m)),$$

the gradient vector $\nabla f(\underline{c})$ is

$$\nabla f(\underline{c}) = \left(\frac{\partial f}{\partial c_l}(\underline{c}) \right) = \left(\frac{-A \, W_{g_1}(L_X((l-1)m, lm))}{m \left[1 + \frac{1}{m} \sum_{i=0}^{K-1} c_{i+1} W_{g_1}(L_X(im, (i+1)m)) \right]^2} \right)$$

for each $l = 1, \ldots, K$.

Convexity follows by the strict positivity and concavity of the function

$$1 + \frac{1}{m} \sum_{i=0}^{K-1} c_{i+1} W_{g_1}(L_X(im, (i+1)m)).$$

Moreover, the Hessian matrix $H_f(\underline{c})$ of the function f is given by

$$H_f(\underline{c}) = \left(\frac{\partial^2 f}{\partial c_l \partial c_n}(\underline{c}) \right) = \left(\frac{2A \, W_{g_1}(L_X((l-1)m, lm)) \, W_{g_1}(L_X((n-1)m, nm))}{m^2 \left[1 + \frac{1}{m} \sum_{i=0}^{K-1} c_{i+1} W_{g_1}(L_X(im, (i+1)m)) \right]^3} \right)$$

for each $l, n = 1, \ldots, K$. More compactly it can be expressed as

$$H_f(\underline{c}) = \left(W_{g_1}(L_X((l-1)m, lm)) \, W_{g_1}(L_X((n-1)m, nm)) \right) B$$

for each $l, n = 1, \ldots, K$, where

$$B = \frac{2A}{m^2 \left[1 + \frac{1}{m} \sum_{i=0}^{K-1} c_{i+1} W_{g_1}(L_X(im, (i+1)m)) \right]^3}.$$

Clearly, $H_f(\underline{c})$ is non-negative definite.

Given that any cross-partial derivative of the matrix $H_f(\underline{c})$ is non-negative, the function g is supermodular.

Finally, the initial risk-adjusted premium P is a quasiconcave and quasiconvex function of the percentages of reinstatement $c_1, c_2, \ldots, c_K$ because it is a ratio of affine functions. $\square$

Remark 1. Note that the regularity properties exhibited by the initial risk-adjusted premium P are not influenced by functional relations between the two distortion functions g_1 and g_2. Moreover, any hypothesis on concavity/convexity of distortion risk measures may be omitted because they are unnecessary to prove the smooth shape of the initial premium P as a function of $c_1, c_2, \ldots, c_K$.

Remark 2. The reinsurance companies often assess treaties under the assumption that there are only total losses. This happens, for example, when they use the rate on line method to price catastrophe reinsurance. Then it follows that the aggregate claims are generated by a discrete distribution and we have (for more details see Campana [1])

$$P = f(c_1, c_2, \cdots, c_K) = \frac{m \sum_{i=0}^{K} g_2(p_{i+1})}{1 + \sum_{i=0}^{K-1} c_{i+1} g_1(p_{i+1})} \tag{18}$$

where the premium for the ith reinstatement (4) is a two-point random variable distributed as $c_i \, P \, B_{p_i}$ and B_{p_i} denotes a Bernoulli random variable such that

$$Pr[B_{p_i} = 1] = p_i = 1 - Pr[B_{p_i} = 0].$$

Then

$$\nabla f(\underline{c}) = \left(\frac{\partial f}{\partial c_l}(\underline{c}) \right) = \left(\frac{-m \sum_{i=0}^{K} g_2(p_{i+1})}{\left[1 + \sum_{i=0}^{K-1} c_{i+1} g_1(p_{i+1}) \right]^2} \, g_1(p_l) \right)$$

and

$$H_f(\underline{c}) = \left(\frac{\partial^2 f}{\partial c_l \partial c_n}(\underline{c}) \right) = \left(\frac{2m \sum_{i=0}^{K} g_2(p_{i+1})}{\left[1 + \sum_{i=0}^{K-1} c_{i+1} g_1(p_{i+1}) \right]^3} \, g_1(p_l) g_1(p_n) \right)$$

for each $l, n = 1, \ldots, K$.

5 Conclusions

In actuarial literature excess of loss reinsurance with reinstatement has been essentially studied in the framework of collective model of risk theory for which the classical evaluation of pure premiums requires knowledge of the claim size distribution. Generally, in practice, there is incomplete information: few characteristics of the aggregate claims can be computed. In this situation, interest in general properties characterising the involved premiums is flourishing.

Setting this problem in the framework of risk-adjusted premiums, it is shown that if risk-adjusted premiums satisfy a generalised expected value equation, then the initial premium exhibits some regularity properties as a function of the percentages of reinstatement. In this way it is possible to relax the particular choice made by Walhin and Paris [6] of the *PH-transform risk measure* and to extend the analysis of excess of loss reinsurance with reinstatements to cover the case of not necessarily equal reinstatements.

The obtained results suggest that further research may be addressed to the analysis of optimal premium plans.

Acknowledgement. We are grateful to some *MAF2008* conference members for valuable comments and suggestions on an earlier version of the paper.

References

1. Campana, A.: Reinstatement Premiums for an Excess of Loss Reinsurance Contract. Giornale Istituto Italiano degli Attuari LXIX, Rome (2006)

2. Dhaene, J., Vanduffel, S., Tang, Q.H., Goovaerts, M.J., Kaas, R., Vyncke, D.: Solvency capital, risk measures and comonotonicity: a review. Research Report OR 0416, Department of Applied Economics, K.U. Leuven (2004)
3. Hürlimann, W.: Excess of loss reinsurance with reinstatements revisited. ASTIN Bull. 35, 211–238 (2005)
4. Mata, A.J.: Pricing excess of loss reinsurance with reinstatements. ASTIN Bull. 30, 349–368 (2000)
5. Sundt, B.: On excess of loss reinsurance with reinstatements. Bull. Swiss Assoc. Actuaries, Heft 1, 51–65 (1991)
6. Walhin, J.F., Paris, J.: Excess of loss reinsurance with reinstatements: premium calculation and ruin probability of the cedent. Proceedings of the third IME Congress, London (1999)
7. Wang, S.: Premium calculation by transforming the layer premium density. ASTIN Bull. 26, 71–92 (1996)

Some classes of multivariate risk measures

Marta Cardin and Elisa Pagani

Abstract. In actuarial literature the properties of risk measures or insurance premium principles have been extensively studied. We propose a new kind of stop-loss transform and a related order in the multivariate setting and some equivalent conditions. In our work there is a characterisation of some particular classes of multivariate and bivariate risk measures and a new representation result in a multivariate framework.

Key words: risk measures, distortion function, concordance measures, stochastic orders

1 Introduction

In actuarial sciences it is fairly common to compare two random variables that are risks by stochastic orderings defined using inequalities on expectations of the random variables transformed by measurable functions. By characterising the considered set of functions some particular stochastic orderings may be obtained such as stochastic dominance or stop-loss order. These stochastic order relations of integral form may be extended to cover also the case of random vectors.

The main contribution of this paper concerns the construction of a mathematical framework for the representation of some classes of multivariate risk measures; in particular we study the extension to the multivariate case of distorted risk measures and we propose a new kind of vector risk measure. Moreover, we introduce the product stop-loss transform of a random vector to derive a multivariate product stop-loss order.

2 Multivariate case

We consider only non-negative random vectors. Let Ω be the space of the states of nature, $\mathcal{F}$ be the σ-field and P be the probability measure on $\mathcal{F}$. Our random vector is the function $\mathbf{X} : \Omega \to \mathrm{R}_+^n$ such that $\mathbf{X}(\omega)$ represents the payoff obtained if state ω occurs. We also specify some notations: $F^{\mathbf{X}}(\mathbf{x}) : \mathrm{R}^n \to [0, 1]$ is the distribution function of $\mathbf{X}$, $S^{\mathbf{X}}(\mathbf{x}) : \mathrm{R}^n \to [0, 1]$ is its survival or tail function, and $(\mathbf{X}(\omega) - a)_+ = \max(\mathbf{X}(\omega) - a, 0)$ componentwise.

A risk measure, or a premium principle, is the functional $R : \mathcal{X} \to \tilde{R}$, where $\mathcal{X}$ is a set of non-negative random vectors and $\tilde{R}$ is the extended real line.

In what follows we present some desirable properties P for risk measures, that are our proposal to generalise the well known properties for the scalar case:

1. **Expectation boundedness**: $R[X] \geq E[X_1 \ldots X_n] \ \forall X$.
2. **Non-excessive loading**: $R[X] \leq \sup_{\omega \in \Omega} \{|X_1(\omega)|, \ldots, |X_n(\omega)|\}$.
3. **Translation invariance**: $R[X + a] = R[X] + \bar{a} \quad \forall X, \forall a \in R^n$, where a is a vector of sure initial amounts and $\bar{a}$ is the componentwise product of the elements of a.
4. **Positive homogeneity of order n**: $R[cX] = c^n R[X] \quad \forall X, \forall c \geq 0$.
5. **Monotonicity**: $R[X] \leq R[Y] \quad \forall X, Y$ such that $X \preceq Y$ in some stochastic sense.
6. **Constancy**: $R[b] = \bar{b} \quad \forall b \in R^n$. A special case is $R[0] = 0$, which is called normalisation property.
7. **Subadditivity**: $R[X + Y] \leq R[X] + R[Y] \quad \forall X, Y$, which reflects the idea that risk can be reduced by diversification.
8. **Convexity**: $R[\lambda X + (1 - \lambda)Y] \leq \lambda R[X] + (1 - \lambda)R[Y], \quad \forall X, Y$ and $\lambda \in [0, 1]$; this property implies diversification effects as subadditivity does.

We recall here also some notations about stochastic orderings for multivariate random variables: $X \preceq_{SD} Y$ indicates the usual stochastic dominance, $X \preceq_{UO} Y$ indicates the upper orthant order, $X \preceq_{LO} Y$ indicates the lower orthant order, $X \preceq_C Y$ indicates the concordance order and $X \preceq_{SM} Y$ indicates the supermodular order. For the definitions, look them up in, for instance, [5].

Let us now characterise another formulation for stop-loss transform in the multi-variate setting.

Definition 1. *The product stop-loss transform of a random vector* $X \in \mathcal{X}$ *is defined by* $\pi_X(t) = E\left[(X_1 - t_1)_+ \ldots (X_n - t_n)_+\right] \quad \forall t \in R^n$.

As in the univariate case, we can use this instrument to derive a multivariate stochastic order:

Definition 2. *Let* $X, Y \in \mathcal{X}$ *be two random vectors. We say that* X *precedes* Y *in the multivariate product stop-loss order* $\left(X \preceq_{SL_n} Y\right)$ *if it holds:*

$$\pi_X(t) \leq \pi_Y(t) \quad \forall t \in R^n.$$

It could be interesting to give some extensions to the theory of risk in the multivariate case, but sometimes it is not possible and we will be satisfied if the generalisation works at least in two dimensions. As is well known, different notions are equivalent in the bivariate case for risks with the same univariate marginal distribution [11], but this is no longer true for n-variate risks with $n \geq 3$ [8].

We now introduce the concept of *Fréchet space*: $\mathcal{R}$ denotes the *Fréchet space* given the margins, that is $\mathcal{R}(F_1, F_2)$ is the class of all bivariate distributions with given margins F_1, F_2. The lower Fréchet bound $\underline{X}$ of X is defined by $F^{\underline{X}}(t) :=$ $\max\{F_1(t_1) + F_2(t_2) - 1, 0\}$ and the upper Fréchet bound of X, $\overline{X}$, is defined by $F^{\overline{X}}(t) := \min_i\{F_i(t_i)\}$, where $t = (t_1, t_2) \in \mathbb{R}^2$ and $i = 1, 2$. The following theorems summon up some known results about stochastic orders. For a more interested reader, we cite [5].

Theorem 1. *Let* X, Y *be bivariate random variables, where* $X, Y \in \mathcal{R}(F_1, F_2)$. *Then:*
$$X \preceq_{UO} Y \quad \Leftrightarrow \quad Y \preceq_{LO} X \quad \Leftrightarrow \quad X \preceq_{SM} Y \quad \Leftrightarrow \quad X \preceq_C Y.$$

This result is no longer true when multivariate random variables are considered with $n \geq 3$.

Theorem 2. *Let* X, Y *be bivariate random variables in* $\mathcal{R}(F_1, F_2)$. *The following conditions are equivalent:*

i) $X \preceq_{SM} Y$;
ii) $E[f(X)] \leq E[f(Y)]$ *for every increasing supermodular function* f;
iii) $E[f_1(X_1)f_2(X_2)] \leq E[f_1(Y_1)f_2(Y_2)]$ *for all increasing functions* f_1, f_2;
iv) $\pi_X(t) \leq \pi_Y(t) \quad \forall t \in \mathbb{R}^2$.

3 Multivariate distorted risk measures

Distorted probabilities have been developed in the theory of risk to consider the hypothesis that the original probability is not adequate to describe the distribution (for example to protect us against some events). These probabilities generate new risk measures, called distorted risk measures, see for instance [4, 12, 13].

In this section we try to deepen our knowledge about distorted risk measures in the multidimensional case. Something about this topic is discussed in [9], but here there is not a representation through complete mathematical results.

We can define the distortion risk measure in the multivariate case as:

Definition 3. *Given a distortion* g, *which is a non-decreasing function such that* g : $[0, 1] \rightarrow [0, 1]$, *with* $g(0) = 0$ *and* $g(1) = 1$, *a vector distorted risk measure is the functional:* $R_g[X] = \int_0^{+\infty} \ldots \int_0^{+\infty} g(S^X(x)) dx_1 \ldots dx_n$.

We note that the function $g(S^X(x)) : \mathbb{R}_+^n \rightarrow [0, 1]$ is non-increasing in each component.

Proposition 1. *The properties of the multivariate distorted risk measures are the following: P1-P6, and P7, P8 if* g *is concave.*

Proof. P1 and P2 follow immediately from Definition 3, P3 follows recalling that $S^{X+a}(t) = S^X(t - a)$, P4 is a consequence of the fact that $S^{cX}(t) = S^X\left(\frac{t}{c}\right)$, P5 follows from the relationship between multivariate stochastic orders and P6 is given by $\int_0^{b_1} \ldots \int_0^{b_n} g(1) dt_n \ldots dt_1 = b_n \ldots b_1 = \bar{b}$. P7 follows from this definition of concavity: if g is a concave function, then we have that $g\,(a+c)-g\,(a) \geq g\,(b+c)-g\,(b)$ with $a \leq b$ and $c \geq 0$. We apply this definition pointwise to $S^{\mathbf{X}} \leq S^{\mathbf{Y}}$ with $S^{\mathbf{X}+\mathbf{Y}} \geq 0$. P8 is obvious from properties P4, P5 and P7. □

In the multivariate case the equality $F^{\mathbf{X}} = 1 - S^{\mathbf{X}}$ does not hold and thus, the relation $\int_{\mathbf{R}^n_+} g\left(S^{\mathbf{X}}(\mathbf{x})\right) d\mathbf{x} = \int_{\mathbf{R}^n_+} [1 - f\left(F^{\mathbf{X}}(\mathbf{x})\right)] d\mathbf{x}$ is not in general true with $f : [0,1] \to [0,1]$, increasing function.

Moreover, the duality relationship between the functions f and g does not hold, thus, in general, the equation $g\,(x) = 1 - f\,(1-x)$ is not true. Applying the concept of distortion of either the survival function or the distribution function, the relationship between f and g no longer holds.

Therefore we can observe the differences in the two different approaches.

Definition 4. *Given a distortion function $f : [0,1] \to [0,1]$, increasing and such that $f(0) = 0$ and $f(1) = 1$, a vector distorted risk measure is the functional:*

$$R_f\,[\mathbf{X}] = \int_{\mathbf{R}^n_+} [1 - f\left(F^{\mathbf{X}}(\mathbf{x})\right)] d\mathbf{x}.$$

Now we have subadditivity with a convex function f and this leads to the convexity of the measure R_f.

Remembering that a distortion is a univariate function even when we deal with random vectors and multivariate distributions, we can also define vector Values at Risk (VaR) and vector Conditional Values at Risk (CVaR), using slight alterations of the usual distortions for VaR and CVaR respectively, and composing these with the multivariate tail distributions or the distribution functions.

Definition 5. *Let $\mathbf{X}$ be a random vector that takes on values in $\mathbf{R}^n_+$. Vector VaR is the distorted measure $VaR\,[\mathbf{X}; p] = \int_0^{+\infty} \ldots \int_0^{+\infty} g\left(S^{\mathbf{X}}(\mathbf{x})\right) dx_1 \ldots dx_n$, expressed using the distortion*

$$g\left(S^{\mathbf{X}}(\mathbf{x})\right) = \begin{cases} 0 & 0 \leq S^{X_i}(x_i) \leq 1 - p_i \\ 1 & 1 - p_i < S^{X_i}(x_i) \leq 1 \end{cases}.$$

If we want to give to this formulation a more explicit form we can consider the componentwise order for which $\mathbf{x} > VaR\,[\mathbf{X}; p]$ stands for $x_i > VaR\,[X_i; p]$ $\forall i = 1, \ldots, n$ or more lightly $x_i > VaR_{X_i}$ and we can rewrite the distortion as:

$$g\left(S^{\mathbf{X}}(\mathbf{x})\right) = \begin{cases} 0 & x_i \geq VaR_{X_i} \\ 1 & 0 \leq x_i < VaR_{X_i} \end{cases}.$$

to obtain $VaR\,[\mathbf{X}; p] = \int_0^{VaR_{X_n}} \ldots \int_0^{VaR_{X_1}} 1\,dx_1 \ldots dx_n = VaR_{X_1} \ldots VaR_{X_n}$. Obviously this result suggests that considering a componentwise order is similar to considering an independency between the components of the random vector. Actually we are considering only the case in which the components are concordant.

In the same way we can define the vector Conditional Value at Risk:

Definition 6. *Let* $\mathbf{X}$ *be a random vector with values in* $\mathbb{R}^n_+$. *Vector CVaR is the distorted measure* $CVaR\,[\mathbf{X};\,p] = \int_0^{+\infty}\ldots\int_0^{+\infty} g\left(S^{\mathbf{X}}(\mathbf{x})\right)dx_1\ldots dx_n$, *expressed using the distortion:*

$$g\left(S^{\mathbf{X}}(\mathbf{x})\right) = \begin{cases} \dfrac{S^{\mathbf{X}}(\mathbf{x})}{\prod_{i=1}^n (1-p_i)} & 0 \le S^{X_i}(x_i) \le 1-p_i \\ 1 & 1-p_i < S^{X_i}(x_i) \le 1 \end{cases}.$$

A more tractable form is given by:

$$g\left(S^{\mathbf{X}}(\mathbf{x})\right) = \begin{cases} \dfrac{S^{\mathbf{X}}(\mathbf{x})}{\prod_{i=1}^n (1-p_i)} & x_i \ge VaR_{X_i} \\ 1 & 0 \le x_i < VaR_{X_i} \end{cases},$$

which allows this formula:

$$CVaR\,[\mathbf{X};\,p] =$$

$$\int_0^{VaR_{X_n}}\ldots\int_0^{VaR_{X_1}} 1\,dx_1\ldots dx_n + \int_{VaR_{X_n}}^{+\infty}\ldots\int_{VaR_{X_1}}^{+\infty} \frac{S^{\mathbf{X}}(\mathbf{x})}{\prod_{i=1}^n (1-p_i)}dx_1\ldots dx_n =$$

$$VaR\,[\mathbf{X};\,p] + \int_{VaR_{X_n}}^{+\infty}\ldots\int_{VaR_{X_1}}^{+\infty} \frac{S^{\mathbf{X}}(\mathbf{x})}{\prod_{i=1}^n (1-p_i)}dx_1\ldots dx_n.$$

The second part of the formula is not easy to render explicitly if we do not introduce an independence hypothesis.

If we follow Definition 4 instead of 3 we can introduce a different formulation for CVaR, very useful in proving a good result proposed later on.

The increasing convex function f used in the definition of CVaR is the following:

$$f\left(F^{\mathbf{X}}(\mathbf{x})\right) = \begin{cases} 0 & F^{X_i}(x_i) < p_i \quad 0 \le x_i < VaR_{X_i} \\ \dfrac{F^{\mathbf{X}}(\mathbf{x}) - 1 + \prod_{i=1}^n (1-p_i)}{\prod_{i=1}^n (1-p_i)} & F^{X_i}(x_i) \ge p_i \quad x_i \ge VaR_{X_i} \end{cases}$$

Definition 7. *Let* $\mathbf{X}$ *be a random vector that takes on values in* $\mathbb{R}^n_+$ *and* f *be an increasing function* $f : [0,1] \to [0,1]$, *such that* $f(0) = 0$ *and* $f(1) = 1$ *and defined as above. The Conditional Value at Risk distorted by such an* f *is the following:*

$$CVaR\,[\mathbf{X};\,p] = \int_0^{VaR_{X_n}}\ldots\int_0^{VaR_{X_1}} 1\,dx_1\ldots dx_n + \int_{VaR_{X_n}}^{+\infty}\ldots\int_{VaR_{X_1}}^{+\infty}$$

$$\left[1 - \frac{F^{\mathbf{X}}(\mathbf{x}) - 1 + \prod_{i=1}^n (1-p_i)}{\prod_{i=1}^n (1-p_i)}\right]dx_1\ldots dx_n =$$

$$\int_0^{+\infty}\ldots\int_0^{+\infty} 1 - \frac{[F^{\mathbf{X}}(\mathbf{x}) - 1 + \prod_{i=1}^n (1-p_i)]_+}{\prod_{i=1}^n (1-p_i)}dx_1\ldots dx_n.$$

We recall here that if $0 \leq x_i < VaR_{X_i}$, or $F^{X_i} < p_i$, then $F^{\mathbf{X}} < \min_i \{p_i\}$, while if $x_i \geq VaR_{X_i}$ or $F^{X_i} \geq p_i$ then $F^{\mathbf{X}} \geq \max_i \{\sum_{i=1}^{n} p_i - (n-1), 0\}$. Only in the bivariate case do we then know that $S^{\mathbf{X}} = F^{\mathbf{X}} - 1 + S^{X_1} + S^{X_2}$. Therefore if $F^{\mathbf{X}} \leq \min_i \{p_i\}$, also $F^{\mathbf{X}} \leq 1 - \prod_{i=1}^{n}(1 - p_i)$. This lets us consider the bounds for the joint distribution, not just for the marginals. Finally we can present an interesting result regarding the representation of subadditive distorted risk measures through convex combinations of Conditional Values at Risk.

Theorem 3. *Let* $\mathbf{X} \in \mathcal{X}$. *Consider a subadditive multivariate distortion in the form* $R_f[\mathbf{X}] = \int_{\mathbb{R}_+^n} [1 - f(F^{\mathbf{X}}(\mathbf{x}))]d\mathbf{x}$. *Then there exists a probability measure* μ *on* $[0, 1]$ *such that:* $R_f[\mathbf{X}] = \int_0^1 CVaR[\mathbf{X}; p]\,d\mu(p)$.

Proof. The multivariate distorted measure $R_f[\mathbf{X}] = \int_{\mathbb{R}_+^n} [1 - f(F^{\mathbf{X}}(\mathbf{x}))]d\mathbf{x}$ is subadditive if f is a convex, increasing function such that: $f : [0, 1] \to [0, 1]$ with $f(0) = 0$ and $f(1) = 1$. Let $p = 1 - \prod_{i=1}^{n}(1 - p_i)$, then a probability measure $\mu(p)$ exists such that this function f can be represented as: $f(u) = \int_0^1 \frac{(u-p)_+}{(1-p)}d\mu(p)$ with $p \in [0, 1]$. Then, $\forall \mathbf{X} \in \mathcal{X}$, we can write

$$R_f[\mathbf{X}] = \int_{\mathbb{R}_+^n} [1 - f(F^{\mathbf{X}}(\mathbf{x}))]d\mathbf{x}$$

$$= \int_{\mathbb{R}_+^n} [1 - \int_0^1 \frac{(F^{\mathbf{X}}(\mathbf{x}) - 1 + \prod_{i=1}^{n}(1 - p_i))_+}{\prod_{i=1}^{n}(1 - p_i)}]d\mu(p)\,d\mathbf{x}$$

$$= \int_{\mathbb{R}_+^n} d\mathbf{x} \int_0^1 [1 - \frac{(F^{\mathbf{X}}(\mathbf{x}) - 1 + \prod_{i=1}^{n}(1 - p_i))_+}{\prod_{i=1}^{n}(1 - p_i)}]d\mu(p)$$

$$= \int_0^1 d\mu(p) \int_{\mathbb{R}_+^n} [1 - \frac{(F^{\mathbf{X}}(\mathbf{x}) - 1 + \prod_{i=1}^{n}(1 - p_i))_+}{\prod_{i=1}^{n}(1 - p_i)}]d\mathbf{x}$$

$$= \int_0^1 CVaR[\mathbf{X}; p]d\mu(p). \qquad \square$$

Since not every result about stochastic dominance works in a multivariate setting, we restrict our attention to the bivariate one. However, this is interesting because it takes into consideration the riskiness not only of the marginal distributions, but also of the joint distribution, tracing out a course of action to multivariate generalisations. It is worth noting that this procedure has something to do with concordance measures (or measures of dependence), which we will describe later on.

We propose some observations about VaR and CVaR formulated through distortion functions when $\mathbf{X}$ is a random vector with values in $\mathbb{R}_+^2$; we have:

$$VaR[\mathbf{X}; p] = VaR_{X_1} VaR_{X_2}$$

and

$$CVaR[\mathbf{X}; p] = VaR_{X_1} VaR_{X_2} + \int_{VaR_{X_2}}^{+\infty} \int_{VaR_{X_1}}^{+\infty} \frac{S^{\mathbf{X}}(x_1, x_2)}{(1 - p_1)(1 - p_2)}dx_1 dx_2.$$

Under independence hypothesis we can rewrite CVaR in this manner:

$$CVaR[\mathbf{X}; p] = VaR_{X_1} VaR_{X_2}$$
$$+ \frac{1}{(1-p_1)(1-p_2)} \int_{VaR_{X_2}}^{+\infty} \int_{VaR_{X_1}}^{+\infty} S^{X_1}(x_1) S^{X_2}(x_2) dx_1 dx_2.$$

Then we consider

$$\int_{VaR_{X_2}}^{+\infty} \int_{VaR_{X_1}}^{+\infty} S^{X_1}(x_1) S^{X_2}(x_2) dx_1 dx_2 =$$

$$(1-p_1)(1-p_2) VaR_{X_1} VaR_{X_2} + (1-p_1) VaR_{X_1} \int_{VaR_{X_2}}^{+\infty} x_2 dS^{X_2}(x_2) +$$

$$(1-p_2) VaR_{X_2} \int_{VaR_{X_1}}^{+\infty} x_1 dS^{X_1}(x_1) + \int_{VaR_{X_1}}^{+\infty} x_1 dS^{X_1}(x_1) \int_{VaR_{X_2}}^{+\infty} x_2 dS^{X_2}(x_2),$$

which leads, with the first part, to:

$$CVaR[\mathbf{X}; p] = 2 VaR_{X_1} VaR_{X_2} - VaR_{X_1} E[X_2|X_2 > VaR_{X_2}] -$$
$$VaR_{X_2} E[X_1|X_1 > VaR_{X_1}] + E[X_2|X_2 > VaR_{X_2}] E[X_1|X_1 > VaR_{X_1}].$$

4 Measures of concordance

Concordance between two random variables arises if large values tend to occur with large values of the other and small values occur with small values of the other. So concordance considers nonlinear associations between random variables that correlation might miss. Now, we want to consider the main characteristics a measure of concordance should have. We restrict our attention to the bivariate case.

In 1984 Scarsini ([10]) defined a set of axioms that a bivariate dependence ordering of distributions should have in order that higher ordering means more positive concordance.

By a *measure of concordance* we mean a function that attaches to every continuous bivariate random vector a real number $\alpha(X_1, X_2)$ satisfying the following properties:

1. $-1 \leq \alpha(X_1, X_2) \leq 1$;
2. $\alpha(X_1, X_1) = 1$;
3. $\alpha(X_1, -X_1) = -1$;
4. $\alpha(-X_1, X_2) = \alpha(X_1, -X_2) = -\alpha(X_1, X_2)$;
5. $\alpha(X_1, X_2) = \alpha(X_2, X_1)$;
6. if X_1 and X_2 are independent, then $\alpha(X_1, X_2) = 0$;
7. if $(X_1, X_2) \preceq_C (Y_1, Y_2)$ then $\alpha(X_1, X_2) \leq \alpha(Y_1, Y_2)$
8. if $\{\mathbf{X}\}_n$ is a sequence of bivariate random vectors converging in distribution to $\mathbf{X}$, then $\lim_{n \to \infty} \alpha(\mathbf{X}_n) = \alpha(\mathbf{X})$.

Now we consider the dihedral group D_4 of the symmetries on the square $[0, 1]^2$. We have $D_4 = \{e, r, r^2, r^3, h, hr, hr^2, hr^3\}$ where e is the identity, h is the reflection about $x = \frac{1}{2}$ and r is a $90°$ counterclockwise rotation.

A measure μ on $[0, 1]^2$ is said to be D_4-invariant if its value for any Borel set A of $[0, 1]^2$ is invariant with respect to the symmetries of the unit square that is $\mu(A) = \mu(d(A))$.

Proposition 2. *If μ is a bounded D_4-invariant measure on $[0, 1]^2$, there exist $\alpha, \beta \in \mathbb{R}$ such that the function defined by*

$$\rho((X_1, X_2)) = \alpha \int_{[0,1]^2} F^{(X_1,X_2)}(x_1, x_2) d\mu(F^{X_1}(x_1), F^{X_2}(x_2)) - \beta$$

is a concordance measure.

Proof. A measure of concordance associated to a continuous bivariate random vector depends only on the copula associated to the vector since a measure of concordance is invariant under invariant increasing transformation of the random variables. So the result follows from Theorem 3.1 of [6]. □

5 A vector-valued measure

In Definition 1 we have introduced the concept of product stop-loss transform for random vectors. We use this approach to give a definition for a new measure that we call Product Stop-loss Premium.

Definition 8. *Consider a non-negative bivariate random vector $\mathbf{X}$ and calculate the Value at Risk of its single components. Product Stop-loss Premium (PSP) is defined as follows: $PSP[\mathbf{X}; p] = E\left[(X_1 - VaR_{X_1})_+ (X_2 - VaR_{X_2})_+\right]$.*

Of course this definition could be extended also in a general case, writing:

$$PSP[\mathbf{X}; p] = E\left[(X_1 - VaR_{X_1})_+ \cdots (X_n - VaR_{X_n})_+\right],$$

but some properties will be different, because not everything stated for the bivariate case works in the multivariate one.

Our aim is to give a multivariate measure that can detect the joint tail risk of the distribution. In doing this we also have a representation of the marginal risks and thus the result is a measure that describes the joint and marginal risk in a simple and intuitive manner.

We examine in particular the case $X_1 > VaR_{X_1}$ and $X_2 > VaR_{X_2}$ simultaneously, since large and small values will tend to be more often associated under the distribution that dominates the other one.

Random variables are concordant if they tend to be all large together or small together and in this case we have a measure with non-trivial values when the variables exceed given thresholds together and are not constant, otherwise we have $PSP[\mathbf{X}; p] = 0$.

It is clear that concordance affects this measure, and in general we know that concordance behaviour influences risk management of large portfolios of insurance contracts or financial assets. In these portfolios the main risk is the occurrence of many joint default events or simultaneous downside evolutions of prices.

PSP for multivariate distributions is interpreted as a measure that can keep the dependence structure of the components of the random vector considered, when specified thresholds are exceeded by each component with probability p_i; but indeed it is also a measure that can evaluate the joint as well as the marginal risk. In fact, we have:

$$PSP\left[\mathbf{X}; p\right] = E\left[\left(X_1 - VaR_{X_1}\right)_+ \left(X_2 - VaR_{X_2}\right)_+\right] =$$

$$\int_{VaR_{X_2}}^{+\infty} \int_{VaR_{X_1}}^{+\infty} S^{\mathbf{X}}(\mathbf{x}) \, dx_1 dx_2 - VaR_{X_2} E\left[X_1 | X_1 > VaR_{X_1}\right]$$

$$-VaR_{X_1} E\left[X_2 | X_2 > VaR_{X_2}\right] + VaR_{X_1} VaR_{X_2}.$$

Let us denote with $\overline{CVaR}[\mathbf{X}; p]$ the CVaR restricted to the bivariate independent case, with $X_1 > VaR_{X_1}$ and $X_2 > VaR_{X_2}$, then we have:

$$\overline{CVaR}\left[\mathbf{X}; p\right] = E\left[X_1 | X_1 > VaR_{X_1}\right] E\left[X_2 | X_2 > VaR_{X_2}\right] -$$

$$VaR_{X_1} E\left[X_2 | X_2 > VaR_{X_2}\right] - VaR_{X_2} E\left[X_1 | X_1 > VaR_{X_1}\right] + VaR_{X_1} VaR_{X_2}.$$

We can conclude that these risk measures are the same for bivariate vectors with independent components, on the condition of these restrictions.

We propose here a way to compare dependence, introducing a stochastic order based on our PSP measure.

Proposition 3. *If* $\mathbf{X}, \mathbf{Y} \in \mathcal{R}_2$, *then* $\mathbf{X} \preceq_{SM} \mathbf{Y} \iff PSP[\mathbf{X}; p] \leq PSP[\mathbf{Y}; p] \; \forall p$ *holds.*

Proof. If $\mathbf{X} \preceq_{SM} \mathbf{Y}$ then $E[f(\mathbf{X})] \leq E[f(\mathbf{Y})]$ for every supermodular function f, therefore also for the specific supermodular function that defines our PSP and then follows $PSP[\mathbf{X}; p] \leq PSP[\mathbf{Y}; p]$. Conversely if

$$PSP[\mathbf{X}; p] \leq PSP[\mathbf{Y}; p] \quad \text{and} \quad \mathbf{X}, \mathbf{Y} \in \mathcal{R}_2,$$

we have

$$\int_{VaR_{X_2}}^{+\infty} \int_{VaR_{X_1}}^{+\infty} S^{\mathbf{X}}(\mathbf{t}) \, d\mathbf{t} \leq \int_{VaR_{Y_2}}^{+\infty} \int_{VaR_{Y_1}}^{+\infty} S^{\mathbf{Y}}(\mathbf{t}) \, d\mathbf{t}$$

with $VaR_{X_2} = VaR_{Y_2}$ and $VaR_{X_1} = VaR_{Y_1}$. It follows that $S^{\mathbf{X}}(\mathbf{t}) \leq S^{\mathbf{Y}}(\mathbf{t})$, which leads to $\mathbf{X} \preceq_C \mathbf{Y}$. From Theorem 1 follows $\mathbf{X} \preceq_{SM} \mathbf{Y}$. $\square$

Obviously PSP is also consistent with the concordance order.

Another discussed property for risk measures is subadditivity; risk measures that are subadditive for all possible dependence structures of the vectors do not reflect the dependence between $(X_1 - \alpha_1)_+$ and $(X_2 - \alpha_2)_+$.

We can note that our PSP is not always subadditive; in fact, if we take the non-negative vectors $\mathbf{X}, \mathbf{Y} \in \mathcal{X}$, the following relation is not always satisfied:

$$E\left[\left(X_1 + Y_1 - VaR_{X_1} - VaR_{Y_1}\right)_+ \left(X_2 + Y_2 - VaR_{X_2} - VaR_{Y_2}\right)_+\right] \leq$$
$$E\left[\left(X_1 - VaR_{X_1}\right)_+ \left(X_2 - VaR_{X_2}\right)_+\right] + E\left[\left(Y_1 - VaR_{Y_1}\right)_+ \left(Y_2 - VaR_{Y_2}\right)_+\right].$$

After verifying all the possible combinations among scenarios

$$X_1 > VaR_{X_1}, \quad X_1 < VaR_{X_1}, \quad X_2 > VaR_{X_2}, \quad X_2 < VaR_{X_2},$$
$$Y_1 > VaR_{Y_1}, \quad Y_1 < VaR_{Y_1}, \quad Y_2 > VaR_{Y_2}, \quad Y_2 < VaR_{Y_2},$$

we can conclude that the measure is not subadditive when:

- the sum of the components is concordant and such that:

$$X_i + Y_i > VaR_{X_i} + VaR_{Y_i} \ \forall i = 1, 2,$$

  with discordant components of at most one vector;
- the sum of the components is concordant and such that:

$$X_i + Y_i > VaR_{X_i} + VaR_{Y_i} \ \forall i = 1, 2,$$

  with both vectors that have concordant components, but with a different sign: i.e.,

$$X_i > (<)VaR_{X_i} \quad \text{and} \quad Y_i < (>)VaR_{Y_i} \quad \forall i;$$

- $X_i > VaR_{X_i}$ and $Y_i > VaR_{Y_i} \ \forall i$ simultaneously.

Hence, in these cases, the measure reflects the dependence structure of the vectors involved.

6 Conclusions

In this paper we have proposed a mathematical framework for the introduction of multivariate measures of risk. After considering the main properties a vector measure should have, and recalling some stochastic orders, we have outlined our results on multivariate risk measures. First of all, we have generalised the theory about distorted risk measures for the multivariate case, giving a representation result for those measures that are subadditive and defining the vector VaR and CVaR. Then, we have introduced a new risk measure, called Product Stop-Loss Premium, through its definition, its main properties and its relationships with CVaR and measures of concordance. This measure lets us also propose a new stochastic order. We can observe that, in the literature, there are other attempts to study multivariate risk measures, we cite for example [1–3,7] and [9], but they all approach the argument from different points of view. Indeed, [9] is the first work that deals with multivariate distorted risk measures, but it represents only an outline for further developments, as we have done in the present work.

More recently, the study of risk measures has focused on weakening the definition of convenient properties for risk measures, in order to represent the markets in a more faithful manner, or on the generalisation of the space that collects the random vectors.

References

1. Balbás, A., Guerra, P.J.: Generalized vector risk functions. Working Paper no. 06–67, http://docubib.uc3m.es (2006)
2. Bentahar, I.: Tail conditional expectation for vector valued risks. Discussion paper, http://sfb649.wiwi.hu-berlin.de (2006)
3. Burgert, C., Rüschendorf, L.: Consistent risk measures for portfolio vectors. Ins. Math. Econ. 38, 289–297 (2006)
4. Denneberg, D.: Distorted probabilities and insurance premiums. Meth. Oper. Res. 63, 3–5 (1990)
5. Denuit, M., Dhaene, J., Goovaerts, M., Kaas, R.: Actuarial Theory for Dependent Risks. Measures, Orders and Models. John Wiley (2005)
6. Edwards, H., Mikusinski, P., Taylor, M.D.: Measures of concordance determined by D_4-invariant copulas. J. Math. Math. Sci. 70, 3867–3875 (2004)
7. Jouini, E., Meddeb, M., Touzi, N.: Vector valued coherent risk measures. Finan. Stoch. 4, 531–552 (2004)
8. Müller, A.: Stop-loss order for portfolios of dependent risks. Ins. Math. Econ. 21, 219–223 (1997)
9. Rüschendorf, L.: Law invariant convex risk measures for portfolio vectors. Stat. Decisions 24, 97–108 (2006)
10. Scarsini, M.: On measures of concordance. Stochastica 8, 201–218 (1984)
11. Tchen, A.H.: Inequalities for distributions with given marginales. Ann. Prob. 8, 814–827 (1980)
12. Wang, S.: Premium calculation by transforming the layer premium density. Astin Bull. 26, 71–92 (1996)
13. Wang, S., Young, V.R., Panjer, H.H.: Axiomatic characterization of insurance pricing. Ins. Math. Econ. 21, 173–183 (1997)

Assessing risk perception by means of ordinal models

Paola Cerchiello, Maria Iannario, and Domenico Piccolo

Abstract. This paper presents a discrete mixture model as a suitable approach for the analysis of data concerning risk perception, when they are expressed by means of ordered scores (ratings). The model, which is the result of a personal *feeling* (*risk perception*) towards the object and an inherent *uncertainty* in the choice of the ordinal value of responses, reduces the collective information, synthesising different risk dimensions related to a preselected domain. After a brief introduction to risk management, the presentation of the *CUB* model and related inferential issues, we illustrate a case study concerning risk perception for the workers of a printing press factory.

Key words: risk perception, *CUB* models, ordinal data

1 Introduction

During the past quarter-century, researchers have been intensively studying risk from many perspectives. The field of risk analysis has rapidly grown, focusing on issues of risk assessment and risk management. The former involves the identification, quantification and characterisation of threats faced in fields ranging from human health to the environment through a variety of daily-life activities (i.e., bank, insurance, IT-intensive society, etc.). Meanwhile, risk management focuses on processes of communication, mitigation and decision making. In normal usage, the notion of "risk" has negative connotations and involves involuntary and random aspects. Moreover, the conceptual analysis of the risk concept wavers from a purely statistical definition (objective) to a notion based on the mind's representation (subjective). In this context, perception of risk plays a prominent role in people's decision processes, in the sense that different behaviours depend on distinct risk perception evaluation. Both individual and group differences have been shown to be associated with differences in perceptions of the relative risk of choice options, rather than with differences in attitude towards (perceived) risk, i.e., a tendency to approach or to avoid options perceived as riskier [23, 24]. Risk is subjectively defined by individuals and is influenced by a wide array of psychological, social, institutional and cultural factors

[22]. However, there is no consensus on the relationship between personality and risk perception [5].

Another fundamental dimension related to the concept of risk deals with the dichotomy between experts' perceptions and those of the common people. The role of experts is central in several fields, especially when quantitative data are not sufficient for the risk assessment phase (i.e., in operational risk). Typically, experts' opinions are collected via questionnaires on ordinal scales; thereby several models have been proposed to elaborate and exploit results: linear aggregation [9], fuzzy methods [2, 25] and Bayesian approaches [4].

Our contribution follows this research path, proposing a class of statistical model able to measure the perceptions expressed either by experts or common people. In particular we focus on the problem of risk perception related to the workplace with regards to injury. Thus, some studies focusing on the relationship between organisational factors and risk behaviour in the workplace [21] suggest that the likelihood of injuries is affected especially by the following variables: working conditions, occupational safety training programmes and safety compliance. Rundmo [20] pointed out how the possibility of workplace injuries is linked to the perception of risk frequency and exposure.

2 *CUB* models: description and inference

A researcher faced with a large amount of raw data wants to synthesise it in a way that preserves essential information without too much distortion. The primary goal of statistical modelling is to summarise massive amounts of data within simple structures and with few parameters. Thus, it is important to keep in mind the trade-off between accuracy and parsimony. In this context we present an innovative data-reduction technique by means of statistical models (*CUB*) able to map different results into a parametric space and to model distinct and weighted choices/perceptions of each decision-maker.

CUB models, in fact, are devoted to generate probability structures adequate to interpret, fit and forecast the subject's perceived level of a given "stimulus" (risk, sensation, opinion, perception, awareness, appreciation, feeling, taste, etc.). All current theories of choice under risk or uncertainty assume that people assess the desirability and likelihood of possible outcomes of choice alternatives and integrate this information through some type of expectation-based calculus to reach a decision. Instead, the approach of *CUB* models is motivated by a direct investigation of the psychological process that generates the human choice [15].

Generally, the choices – derived by the perception of risk – are of a qualitative (categorical) nature and classical statistical models introduced for continuous phenomena are neither suitable nor effective. Thus, qualitative and ordinal data require specific methods to avoid incongruities and/or loss of efficiency in the analysis of real data. With this structure we investigate a probability model that produces interpretable results and a good fit. It decodes a discrete random variable (MUB, introduced by

D'Elia and Piccolo [8]) and we use *CUB* models when we relate the responses to subjects' covariates. The presence of Uniform and shifted Binomial distributions and the introduction of Covariates justify the acronym *CUB*. This model combines a personal feeling (risk awareness) towards the object and an inherent uncertainty in the choice of the ordinal value of responses when people are faced with discrete choices.

The result for interpreting the responses of the raters is a mixture model for ordered data in which we assume that the rank r is the realisation of a random variable R that is a mixture of Uniform and shifted Binomial random variables (both defined on the support $r = 1, 2, \ldots, m$), with a probability distribution:

$$Pr(R = r) = \pi \, \binom{m-1}{r-1}(1 - \xi)^{r-1}\xi^{m-r} + (1 - \pi) \, \frac{1}{m}, \quad r = 1, 2, \ldots, m. \quad (1)$$

The parameters $\pi \in (0, 1]$ and $\xi \in [0, 1]$, and the model is well defined for a given $m > 3$.

The risk-as-feelings hypothesis postulates that responses to risky situations (including decision making) result in part from direct (i.e., not correctly mediated) emotional influences, including feelings such as worry, fear, dread or anxiety. Thus, the first component, *feeling-risk awareness*, is generated by a continuous random variable whose discretisation is expressed by a *shifted Binomial* distribution. This choice is motivated by the ability of this discrete distribution to cope with several different shapes (skewness, flatness, symmetry, intermediate modes, etc.). Moreover, since risk is a continuous latent variable summarised well by a Gaussian distribution, the shifted Binomial is a convenient unimodal discrete random variable on the support $\{1, 2, \ldots, m\}$.

At the same time, feeling states are postulated to respond to factors, such as the immediacy of a risk, that do not enter into cognitive evaluations of the risk and also respond to probabilities and outcome values in a fashion that is different from the way in which these variables enter into cognitive evaluations. Thus, the second component, *uncertainty*, depends on the specific components/values (knowledge, ignorance, personal interest, engagement, time spent to decide) concerning people. As a consequence, it seems sensible to express it by a discrete *Uniform* random variable. Of course, the mixture (1) allows the perception of any people to be weighted with respect to this extreme distribution. Indeed, only if $\pi = 0$ does a person act as motivated by a total uncertainty; instead, in real situation, the quantity $(1 - \pi)$ measures the propensity of each respondent towards the maximal uncertainty.

An important characterisation of this approach is that we can map a set of expressed ratings into an estimated model via (π, ξ) parameters. Thus, an observed complex situation of preferences/choices may be simply related to a single point in the parametric space.

In this context, it is reasonable to assume that the main components of the choice mechanism change with the subjects' characteristics (covariates). Thus, *CUB* models are able to include explanatory variables that are characteristics of subjects and which influence the position of different response choices. It is interesting to analyse the values of the corresponding parameters conditioned to covariate values.

In fact, better solutions are obtained when we introduce *covariates* for relating both feeling and uncertainty to the subject's characteristic. Generally, covariates improve the model fitting, discriminate among different sub-populations and are able to make more accurate predictions. Moreover, this circumstance should enhance the interpretation of parameters' estimates and the discussion of possible scenarios.

Following a general paradigm [14, 18], we relate π and ξ parameters to the subjects' covariates through a logistic function. The chosen mapping is the simplest one among the many transformations of real variables into the unit interval and *a posteriori* it provides evidence of ease of interpretation for the problems we will be discussing.

When we introduce covariates into a *MUB* random variable, we define these structures as $CUB(p, q)$ models characterised by a general parameter vector $\underline{\theta} = (\pi, \xi)'$ via the logistic mappings:

$$(\pi \mid \underline{y}_i) = \frac{1}{1 + e^{-\underline{y}_i \beta}}; \quad (\xi \mid \underline{w}_i) = \frac{1}{1 + e^{-\underline{w}_i \gamma}}; \quad i = 1, 2, \ldots, n. \qquad (2)$$

Here, we denote by $\underline{y}_i$ and $\underline{w}_i$ the subject's covariates for explaining π_i and ξ_i, respectively. Notice that (2) allows the consideration of models without covariates ($p = q = 0$); moreover, the significant set of covariates may or may not present some overlapping [11, 13, 19].

Finally, inferential issues for *CUB* models are tackled by maximum likelihood (ML) methods, exploiting the E-M algorithm [16, 17]. The related asymptotic inference may be applied using the approximate variance and covariance matrix of the ML estimators [14]. This approach has been successfully applied in several fields, especially in relation to evaluations of goods and services [6] and other fields of analysis such as social analysis [10, 11], medicine [7], sensometric studies [19] and linguistics [1].

The models we have introduced are able to fit and explain the behaviour of a univariate rating variable while we realise that the expression of a complete ranking list of m objects/items/services by n subjects should require a multivariate setting. Thus, the analysis that will be pursued in this paper should be interpreted as a marginal if we studied the rank distributions of a single item without reference to the ranks expressed towards the remaining ones.

Then in the following section, we analyse both the different items and injuries; afterwards we propose a complex map that summarises the essential information without distortion or inaccuracy.

3 Assessing risk perception: some empirical evidence

3.1 Data analysis

A cross-sectional study was performed in a printing press factory in Northern Italy that manufactures catalogues, books and reproductions of artworks. The staff of the factory consists of 700 employees (300 office workers and 400 blue-collar workers).

The study focused on the blue-collar population of six different departments, each dealing with a specific industrial process. The subjects in the cohort are distributed among the following six units, whose main activities are also described. In the Plates department, workers must set plates and cylinders used during the printing operations and then carried out in the Rotogravure and Offset departments. The Packaging department is responsible for the bookbinding and packaging operations, while the Plants department operates several systems (e.g., electrical and hydraulic) and provides services (e.g., storage and waste disposal) that support the production side of the company. Lastly, the Maintenance department workers perform a series of operations connected with the monitoring and correct functioning of the different equipment of the plant.

With the purpose of studying injury risk perception among company workers, a structured "Workplace Risk Perception Questionnaire" was developed. The questionnaire asked the respondents to express their opinions on a series of risk factors present in their workplace. A 7-point Likert scale was used to elicit the workers' answers whose ranges are interpreted as: 1 = "low perceived risk"; 7 = "high perceived risk". Moreover, we pay particular attention to socio-demographic characteristics like 'gender' (dichotomous variable '0'=men and '1'=women) , 'number of working years' within the company (continuous variable ranging form 1 to 30) and 'type of injury' (dichotomous variable '0'= not severe injury and '1'= severe injury). Finally, $n = 348$ validated questionnaires were collected.

3.2 Control and measure risk perception: a map

As already discussed, we built a class of model to evaluate, control and measure the risk perception and, means of monitoring activity, to inform the stakeholders of the direction of new policies. In this case we show a map of synthesis which contains whole information related to different risk dimensions.

In Figure 1 we plot for each item the reactions of feeling and uncertainty expressed by people. We can observe that the uncertainty is concentrated between 0 and 0.6, a range indicating a high level of indecision. The characteristic of feeling, however, is extended over the whole parametric space. Both aspects illustrate how the responses interact to determine behaviour. Moreover, we deepen some specific aspects of risk-related phenomena that are regarded as more interesting.

In the case of *control*, for example, we can observe a dichotomous behaviour: less sensitivity for injuries such as eye-wound, hit, moving machinery clash (in these cases people do not seem to ask for more control), and more for other injuries where people, on their scale of risk, consider the aspect of control as a sensible variable for improving the conditions of their job.

Instead, an interesting evaluation is referred to as *training*, as it is considered an important variable of the survey. Less evidence appears for other items shared among different levels and whose estimates are spread over the parametric space.

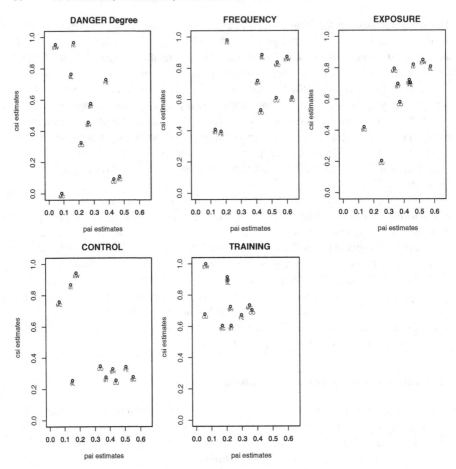

Fig. 1. Assessing risk perception: a map of items. 1=Structural Collapse (SC), 2=Short Circuit (SH), 3=Moving Machinery Clash (MC), 4=Eye Wound (EW), 5=Collision (CO), 6=Fire/Explosion (FE), 7=Slipping (SL), 8=Strain (ST), 9=Cut (CU), 10=Hit (HI)

3.3 Perception of fire/explosion risk

In this context we analyse the *degree of danger*, a principal item in measuring risk perception and we focus on the responses of samples with respect to *fire/explosion risk*. More specifically, we consider the degree of danger that people perceive with respect to *fire risk* and we connect it with some covariates.

In this kind of analysis, sensible covariates have to be introduced in the model by means of a stepwise strategy where a significant increase in the log-likelihoods (difference of deviances) is the criterion to compare different models. In order to simplify the discussion, we present only the full model and check it with respect to a model without covariates.

In Table 1 we list the estimation of parameters of a $CUB(0,3)$ with gender (Gen), working years (Year) and serious injury (Serinj) as sensible covariates.

Table 1. Parameters

Covariates	Parameters	Estimates	(Standard errors)
Uncertainty	π	0.440	(0.060)
Constant	γ_0	1.515	(0.319)
Gender	γ_1	1.029	(0.573)
Working years	γ_2	−0.032	(0.015)
Serious injury	γ_3	−0.928	(0.323)

Log-likelihood functions of $CUB(0,0)$ and $CUB(0,3)$ estimated models are $\ell_{00} = -660.07$ and $\ell_{03} = -651.81$, respectively. As a consequence, the model with covariates adds remarkable information to the generating mechanism of the data since $2 * (\ell_{03} - \ell_{00}) = 16.52$ is highly significant when compared to the $\chi^2_{0.05} = 7.815$ percentile with $g = 3$ degrees of freedom.

We may express the feeling parameters as a function of covariates in the following way:

$$\xi_i = \frac{1}{1 + e^{-1.515 - 1.029 \, Gen_i + 0.032 \, Year_i + 0.928 \, Serinj_i}}, \quad i = 1, 2, \ldots, n, \quad (3)$$

which synthesises the perception of danger of fire/explosion risk with respect to the chosen covariates. More specifically, this perception increases for men and for those who have worked for many years (a proxy of experience) and decreases for the part of sample that had not suffered from a serious accident. For correct interpretation, it must be remembered that if items are scored (as a vote, increasing from 1 to m as liking increases) then $(1 - \xi)$ must be considered as the actual measure of preference [10]. Although the value of the response is not metric (as it stems from a qualitative judgement), it may be useful for comparative purposes to compute the expected value of R, since it is related to the continuous proxy that generates the risk perception.

More specifically, Figure 2 shows the expectation and its relation to the varying working years and for all the profiles of gender and serious injury.

4 Conclusions

In this paper, we obtained some results about direct inference on the feeling/risk awareness and uncertainty parameters by means of CUB models with and without covariates. The experiments confirmed that this new statistical approach gives a different perspective on the evaluation of psychological processes and mechanisms that generate/influence risk perception in people. The results show that CUB models are a

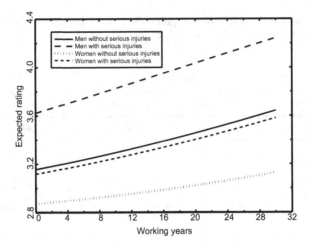

Fig. 2. Expected score as a function of working years, given gender and serious injury

suitable and flexible tool for examining and quantifying the change of response over one or more categorical and/or continuous covariates, and they provide a deeper insight into these kinds of dataset. They also allowed the summary of much information and some interesting evaluation of specific points investigated when covariates are both absent and present.

Moreover, we stress that the proposed model is a manifold target approach: in fact, it can be profitably applied to a variety of fields, ranging from credit and operational risk [3] to reputational and churn risk. Finally, it represents a convincing tool to exploit opinions expressed by field experts.

Acknowledgement. Methodological issues and related software are partially supported by the 2006 PRIN-MIUR research project: "Stima e verifica di modelli statistici per l'analisi della soddisfazione degli studenti universitari".

References

1. Balirano, G., Corduas, M.: Detecting semiotically expressed humor in diasporic tv productions. Humor: Int. J. Humor Res. 3, 227–251 (2008)
2. Bonafede, E.C., Cerchiello, P., Giudici, P.: A proposal for the fuzzification of causal variables, in operational risk management. Quaderni di Dipartimento di Statistica 44 (2007)
3. Bonafede, E.C., Cerchiello, P., Giudici, P.: Statistical models for Business Continuity Management. J. Oper. Risk Man. 2, 79–96 (2007)
4. Bonafede, E.C., Giudici, P.: Bayesian networks for enterprise risk assessment. Phys. A Stat. Mech. Appl. 382, 22–28 (2007)
5. Bromiley, P., Curley, S.P.: Individual differences in risk taking. In: Yates, F. (ed.) Risk-Taking Behavior. pp. 87–132. Wiley, New York (1992)

6. Corduas, M.: A study on university students' opinions about teaching quality: a model based approach for clustering data. In Proceedings of DIVAGO Meeting, University of Palermo (2008)

7. D'Elia, A.: A statistical modelling approach for the analysis of TMD chronic pain data. Stat. Meth. Med. Res. 17, 389–403 (2008)

8. D'Elia, A., Piccolo, D.: A mixture model for preference data analysis. Comput. Stat. Data Anal. 49, 917–934 (2005)

9. Godo, L., Torra, V.: On aggregation operators for ordinal qualitative information. Fuzzy Syst. 8, 143–154 (2000)

10. Iannario, M.: A statistical approach for modelling Urban Audit Perception Surveys. Quad. Stat. 9, 149–172 (2007)

11. Iannario, M.: A class of models for ordinal variables with covariates effects. Quad. Stat. 10, 53–72 (2008)

12. Iannario, M.: Dummy variables in *CUB* models. Statistica 2 (2008) in press

13. Iannario, M., Piccolo, D.: A new statistical model for the analysis of customer satisfaction. Quality Technology and Quantitative Management (2009) in press

14. King, G., Tomz, M., Wittenberg, J.: Making the most of statistical analyses: improving interpretation and presentation. Am. J. Pol. Sci. 44, 341–355 (2000)

15. Loewenstein, F., Hsee, C.K., Weber, U., Welch, N., Risk as feelings. Psychol. Bull. 127, 267–286 (2001)

16. McLachlan, G., Krishnan, G.J.: The EM Algorithm and Extensions. Wiley, New York (1997)

17. McLachlan, G., Peel, G.J.: Finite Mixture Models. Wiley, New York (2000)

18. Piccolo, D.: Observed information matrix for MUB models. Quad. Stat. 8, 33–78 (2006)

19. Piccolo, D., D'Elia, A.: A new approach for modelling consumers' preferences. Food Qual. Pref. 19, 247–259 (2008)

20. Rundmo T.: Risk perception and safety on offshore petroleum platforms Part 1: perception of risk. Saf. Sci. 15, 39–52 (1992)

21. Rundmo, T.: Associations between risk perception and safety. Saf. Sci. 24, 197–209 (1996)

22. Slimak, M.W., Dietz, T.: Personal values, beliefs, and ecological risk perception. Risk Anal. 26, 1689–1705 (2006)

23. Weber, E.U.: Personality and risk taking. In: Smelser, N.J., Baltes, P.B. (eds.): International Encyclopedia of the Social Behavioral Sciences. Oxford, UK: Elsevier Science Limited 11274–11276 (2001)

24. Weber, E.U., Millman, R.: Perceived risk attitudes: relating risk perception to risky choice. Management Science 43, 122–143 (1997)

25. Yu D., Park, W.S.: Combination and evaluation of expert opinions characterized in terms of fuzzy probabilities. Ann. Nuclear Energ. 27, 713–726 (2000)

A financial analysis of surplus dynamics for deferred life schemes[*]

Rosa Cocozza, Emilia Di Lorenzo, Albina Orlando, and Marilena Sibillo

Abstract. The paper investigates the financial dynamics of the surplus evolution in the case of deferred life schemes, in order to evaluate both the distributable earnings and the expected worst occurence for the portfolio surplus. The evaluation is based on a compact formulation of the insurance surplus defined as the difference between accrued assets and present value of relevant liabilities. The dynamic analysis is performed by means of Monte Carlo simulations in order to provide a year-by-year valuation. The analysis is applied to a deferred life scheme exemplar, considering that the selected contract constitutes the basis for many life insurance policies and pension plans. The evaluation is put into an asset and liability management decision-making context, where the relationships between profits and risks are compared in order to evaluate the main features of the whole portfolio.

Key words: financial risk, solvency, life insurance

1 Introduction

The paper investigates the financial dynamics of surplus analysis with the final aim of performing a breakdown of the distributable earnings. The question, put into an asset and liability management context, is aimed at evaluating and constructing a sort of budget of the distributable earnings, given the current information. To this aim, a general reconstruction of the whole surplus is performed by means of an analytical breakdown already fully developed elsewhere [1], and whose main characteristic is the computation of a result of the portfolio, that actuaries would qualify as surplus, accountants as income and economists as profit.

The analysis is developed with the aim of evaluating what share of each year's earnings can be distributed without compromising future results. This share is only a sort of minimum level of distributed earnings which can serve as a basis for business decisions and that can be easily updated year-by-year as market conditions modify. Then the formal model is applied to a life annuity cohort in a stochastic context in

---

[*] Although the paper is the result of a joint study, Sections 1, 2 and 4 are by R. Cocozza, whilst Section 3 is by E. Di Lorenzo, A. Orlando and M. Sibillo.

order to exemplify the potential of the model. In this paper deferred schemes are selected considering that they can be regarded as the basis for many life insurance policies and pension plans. Nevertheless, the model can be applied, given the necessary adjustments, to any kind of contract as well as to non-homogeneous portfolios.

The rest of the paper is organised as follows. Section 2 introduces the logical background of the model itself, while Section 3 detaches the mathematical framework and the computational application. Section 4 comments on the numerical results obtained and Section 5 concludes.

2 The model

As stated [1], the surplus of the policy is identified by the difference between the present value of the future net outcomes of the insurer and the (capitalised) flows paid by the insureds. This breakdown is evaluated year by year with the intent to compile a full prospective account of the surplus dynamics. In the case of plain portfolio analysis, the initial surplus is given by the loadings applied to pure premiums; in the case of a business line analysis, the initial surplus, set as stated, is boosted by the initial capital allocated to the business line or the product portfolio.

The initial surplus value, in both cases, can be regarded as the proper initial capital whose dynamic has to be explored with the aim of setting a general scheme of distributable and undistributable earnings. More specifically, given that at the beginning of the affair the initial surplus is set as S_0, the prospective future t-outcomes, defined as S_t, can be evaluated by means of simulated results to assess worst cases given a certain level of probability or a confidence interval.

The build up of these results, by means of the selected model and of Monte Carlo simulations (see Section 3), provides us with a complete set of future outcomes at the end of each period t. These values do not depend on the amount of the previous distributed earnings. Those results with an occurrence probability lower than the threshold value (linked to the selected confidence interval) play the role of worst cases scenarios and their average can be regarded as the *expected worst occurrence* corresponding to a certain level of confidence when it is treated as a Conditional Value-at-Risk (CVaR). Ultimately, for each period of time, we end up with a complete depiction of the surplus by means of a full set of outcomes, defined by both expected values and corresponding CVaR.

The results we obtain for each period can therefore be used as a basis for the evaluation of the distributable earnings, with the final aim of assessing the distributable surplus share. If the CVaR holds for the expected worst occurrence given a level of confidence, its interpretation is pragmatically straightforward: it is the expected worst value of the surplus for the selected confidence level. So for any t-period, the CVaR estimates the threshold surplus at the confidence level selected and automatically sets the maximum distributable earnings of the preceding period. In other words, the CVaR of S_t can be regarded as the maximum distributable amount of S_{t-1}; at the same time:

- the ratio of the CVaR of S_t to S_{t-1} can be regarded as the distributable share (DS) of the $t-1$ result; and

- the ratio of the CVaR of S_t minus S_{t-1} to the $t-1$ result can be regarded as the worst expected Return on Surplus (RoS) for the selected level of confidence.

Analogous conclusions can be inferred when the analysis is referred to a business line and the surplus is enhanced by the allocated capital: the interpretation of the result is similar and even clearer, as the last ratio is a proper worst expected return on equity.

From a methodological point of view, we would like to stress that the analysis and, therefore, the simulation procedure could be performed with reference to all the risk factors relevant to the time evolution of the portfolio. Many dynamics can simultaneously contribute to the differentials that depend on risk factors linked to both the assets in which premiums are invested and the value of liabilities for which capitalised premiums are deferred. Together with the demographic dynamic, the most important factor is the nature of the assets: if these are financial, the risks faced will be mainly financial, they will depend directly on the asset type and will not have any autonomous relevance. Besides, the crux of the problem is the difference between the total rate of return on assets and the rate of interest originally applied in premium calculation, so that it can be precisely addressed as *investment risk*, in order to highlight the composite nature of relevant risk drivers. At the same time, other factors can contribute to the difference such as the quality of the risk management process, with reference to both diversification and risk pooling. This implies that the level of the result and its variability is strictly dependent on individual company elements that involve both exogenous and endogenous factors.

Since our focus is on the financial aspect of the analysis, we concentrate in the following of the paper only on the question of the investment rate, excluding any demographic component and risk evaluation from our analysis. Bearing in mind this perspective, the rate actually used as a basis for the simulation procedure has to be consistent with the underlying investment and the parameters used to describe the rate process have to be consistent with the features of the backing asset portfolio. Therefore, once we decide the strategy, the evaluation is calibrated to the expected value and estimated variance of the proper return on asset as set by the investment portfolio. In other words, if we adopt, for instance, a bond strategy the relevant parameters will be estimate from the bond market, while if we adopt an equity investment, the relevant values will derive from the equity market, and so on, once we have defined the composition of the asset portfolio.

3 Surplus analysis

3.1 The mathematical framework

In the following we take into account a stochastic scenario involving the financial and the demographic risk components affecting a portfolio of identical policies issued to a cohort of c insureds aged x at issue.

We denote by X_s the stochastic cash flow referred to each contract at time s and by N_s the number of claims at time s, $\{N_s\}$ being i.i.d. and multinomial $(c, \mathbb{E}[\mathbf{1}_s])$, where the random variable $\mathbf{1}_s$ takes the value 1 if the insured event occurs, 0 otherwise.

The value of the business at time t is expressed by the portfolio surplus S_t at that time, that is the stochastic difference of the value of the assets and the liabilities assessed at time t. In general we can write:

$$S_t = \sum_s N_s X_s e^{\int_s^t \delta_u du}, \tag{1}$$

where δ_u is the stochastic force of interest and X_s is the difference between premiums and benefits at time s.

Assuming that the random variables N_s are mutually independent on the random interest δ_s and denoting by F_t the information flow at time t,

$$\mathfrak{S}_t = \mathbb{E}[S_t|F_t] = \sum_s c X_s \, \mathbb{E}[1_s]\mathbb{E}[e^{\int_s^t \delta_u du}]. \tag{2}$$

Formula (2) can be easily specialised in the case of a portfolio of m-deferred life annuities, with annual level premiums P payable at the beginning of each year for a period of n years ($n \leq m$) and constant annual instalments, R, paid at the end of each year, payable if the insured is surviving at the corresponding payment date. It holds:

$$\mathfrak{S}_t = \mathbb{E}[S_t] = \sum_s c X_s \, {}_s p_x \, \mathbb{E}[e^{\int_s^t \delta_u du}] \tag{3}$$

where ${}_s p_x$ denotes the probability that the individual aged x survives at the age $x + s$ and

$$X_s = \begin{cases} P & \text{if } s < n \\ -R & \text{if } s > m \\ 0 & \text{if } n < s < m. \end{cases} \tag{4}$$

As widely explained in the previous section, the surplus analysis provides useful tools for the equilibrium appraisal, which can be synthesised by the following rough but meaningful and simple relationship:

$$Prob(S_t > 0) = \varepsilon. \tag{5}$$

For a deeper understanding of the choice of ε, refer to [1]. From a more general perspective, we can estimate the maximun loss $\overline{S}_\alpha$ of the surplus at a certain valuation time t with a fixed confidence level α, defined as

$$Prob(S_t > \overline{S}_\alpha) = \alpha, \tag{6}$$

that is:

$$\overline{S}_\alpha = F^{-1}(1 - \alpha), \tag{7}$$

F being the cumulative distribution function of S_t.

In the following we will take advantage of a simulative procedure to calculate the quantile surplus involved in (6), basing our analysis on the portfolio *mean surplus* at time t.

We focus on capturing the impact on the financial position at time t – numerically represented by the surplus on that date – of the financial uncertainty, which constitutes a systematic risk source, and is thus independent of the portfolio size. In fact in this case the pooling effect does not have any consequences, in contrast to the effect of specific risk sources, as the accidental deviations of mortality.

Formally the valuation of the mean surplus can be obtained observing that it is possible to construct a proxy of the cumulative distribution function of $\mathbf{S_t}$ since (cf. [4])

$$\lim_{c \to \infty} \mathcal{P} \left(\left| \frac{N_s}{c} - \mathbb{E}[\mathbf{1_s}] \right| \geq \epsilon \right) = 0,$$

hence, when the number of policies tends to infinity, $\mathbf{S_t}/c$ converges in distribution to the random variable

$$\Gamma_t = \sum_s X_s \mathbb{E}[\mathbf{1_s}] e^{\int_s^t \delta_u du}. \tag{8}$$

In the case of the portfolio of m-deferred life annuities described above, we set:

$$x_s = \begin{cases} P & \text{if } s < n \\ -R & \text{if } s > m \end{cases}, \qquad y_s = \begin{cases} R & \text{if } s > m \\ -P & \text{if } s < n \end{cases},$$

so, making explicit the surplus' formalisation, we can write

$$\mathbf{S_t} = \sum_{i=1}^{c} \left(\sum_{s=0}^{min(K_{x_i},t)} x_s e^{\int_s^t \delta_u du} - \sum_{s=t+1}^{min(K_{x_i},T)} y_s e^{-\int_t^s \delta_u du} \right)$$

where K_{x_i} denotes the curtate future lifetime of the ith insured aged x at issue and T is the contract maturity ($T \leq \omega - x$, ω being the ultimate age). We can point out that the second term on the right-hand side represents the mathematical provision at time t.

So, remembering the homogeneity assumptions about the portfolio components, formula 8 can be specialised as follows:

$$\Gamma_t = \mathbb{E}\left[\left(\sum_{s=0}^{min(K_x,t)} x_s e^{\int_s^t \delta_u du} - \sum_{s=t+1}^{min(K_x,T)} y_s e^{-\int_t^s \delta_u du} \right) |\{\delta_u\}_{u \geq 0}\right] = \tag{9}$$

$$= \sum_{s \leq t} x_s \, {}_s p_x \mathbb{E}[e^{\int_s^t \delta_u du}] - \sum_{s > t} y_s \, {}_s p_x \mathbb{E}[e^{-\int_t^s \delta_u du}].$$

3.2 The computational application

As the computational application of the preceding model we consider a portfolio of unitary 20-year life annuities with a deferment period of 10 years issued to a group of 1000 male policyholders aged 40. The portfolio is homogeneous, since it is assumed that policyholders have the same risk characteristics and that the future lifetimes are independent and identically distributed. As far as the premiums are concerned, we build up the cash flow mapping considering that premiums are paid periodically at the beginning of each year of the deferment period. The market premium has a global loading percentage of 7% compensating for expenses, safety and profit. Pure premiums are computed by applying 2% as the policy rate and by using as lifetables the Italian IPS55.

Since our analysis is focused on the financial aspect, the single local source of uncertainty is the spot rate, which is a diffusion process described by a Vasicek model

$$dr(t) = k(\mu - r(t))dt + \sigma dW(t), r(0) = r_0, \tag{10}$$

where k, μ, σ and r_0 are positive constants and μ is the long-term rate. As informative filtration, we use the information set available at time 0. As a consequence, for instance, in calculating the flows accrued up to time t, the starting value r_0 for the simulated trajectories is the value known at time 0. Analogously, in discounting the flows of the period subsequent to t, the starting value of the simulated trajectories is $E[r_t|F_0]$. The parameter estimation is based on Euribor-Eonia data with calibration set on 11/04/2007 (cf. [2]), since we make the hypothesis that the investment strategy is based on a roll-over investment in short-term bonds, as we face an upward term structure. The estimated values are $\mu = 4.10\%$, $\sigma = 0.5\%$ and $r_0 = 3.78\%$.

In order to evaluate the Expected Surplus and the CVaR in a simulation framework, we consider the Vasicek model to describe the evolution in time of the global rate of return on investments earned by the asset portfolio. The α-quantile, q_α, of the surplus distribution is defined as:

$$Prob\{S(t) < q_\alpha\} = 1 - \alpha. \tag{11}$$

In the simulation procedure we set $\alpha = 99\%$. The expected $(1 - \alpha)$ worst case is given by the following:

$$E[worst\ cases\ (1 - \alpha)] = (1 - \alpha)^{-1} \int_\alpha^1 q_p dp, \tag{12}$$

q_p being the p-quantile of the surplus distribution. The last equation is then the average of the surplus value lower than the α-quantile, q_α.

4 Results

Recalling Section 2, the simulation results provide us with the expected value of the surplus for each period, the first value at time 0 being the portfolio difference between the pure premium and the market premium. Therefore, scrolling down the table we can very easily see the evolution of the surplus over time together with the corresponding CVaR.

As far as the time evolution is concerned, the surplus shows an increasing trend, which is consistent with the positive effect of the financial leverage, since we invest

Table 1. Surplus behaviour and related parameters

Time	$E(S_t)$	CvaR	DS (99%)	RoS (99%)
0	1649,899			
1	1732,851	179,9519	10.91%	−89.09%
2	1820,197	362,4452	20.92%	−79.08%
3	1912,03	556,0326	30.55%	−69.45%
4	2008,6	727,887	38.07%	−61.93%
5	2110,173	890,2828	44.32%	−55.68%
6	2217,03	1051,625	49.84%	−50.16%
7	2329,47	1211,922	55.66%	−45.34%
8	2447,811	1333,802	57,26%	−42,74%
9	2572,389	1409,132	57.57%	−42.43%
10	2703,561	1470,693	55.17%	−42.83%
11	2841,708	1478,472	54.69%	−45.31%
12	2987,235	1416,793	49.86%	−50.14%
13	3140,569	1332,953	44.62%	−55.38%
14	3302,167	1228,486	39.12%	−60.88%
15	3472,516	1109,346	33.59%	−66.41%
16	3652,13	982,2259	28.29%	−71.71%
17	3841,561	840,4843	23.01%	−76.99%
18	4041,392	660,9939	17.21%	−82.79%
19	4252,246	443,7866	10.98%	−89.02%

for the whole period of time at a rate which is systematically higher ($\mu = 4.10\%$ and $r_0 = 3.78\%$) than the premium rate (2%), thus giving rise to a return on assets always higher than the average rate of financing. As far as the CVaR is concerned, it shows a dynamic which is totally consistent with the mathematical provision time evolution, as one can expect as has already been shown elsewhere [3], the financial risk dynamic is mainly driven by the mathematical provision time progression. Accordingly, the time evolution of the RoS, as defined in Section 2, is directly influenced by the mathematical provision and its absolute value, as can easily be seen, is dependent on the confidence level chosen. As far as the connection with the reserve dynamic is concerned, we can state that both DS and worst expected RoE prove to be fully consistent with the traditional and pragmatic idea that the lower the reserve the higher the risk of the business. Therefore, distributable earnings can be quantified and managed, through this approach, in order to minimise the ruin probability on the basis of both the general investment strategy and the specific market condition available at time of issue.

5 Conclusions and future research prospect

The sketched model proves to be a way to quantify the amount of distributable earnings year-by-year with reference to a specific portfolio of policies as it gives the opportunity to build upon a complete distribution budget. Since we concentrate solely on the financial dynamics, the first extension could be the inclusion of a demographic component and the modelling of the surplus dynamic by means also of stochastic demographic rates, in order to incorporate, where appropriate, the systematic and unsystematic components. Another extension could be the evaluation at a whole series of critical confidence intervals in order to end up with a double-entry DS table where the definition of its levels can be graduated by means of different levels of probability, in order to control how the actual distributed amount can influence the future performance of the portfolio.

References

1. Cocozza, R., Di Lorenzo, E.: Solvency of life insurance companies: methodological issues. J. Actuarial Practice 13, 81–101 (2006)
2. Cocozza, R., Orlando, A.: Decision making in structured finance: a case of risk adjusted performance evaluation. In: Proceedings of the IV International Conference on Computational and Management Sciences, 20–22 April 2007, http://www.cms2007.unige.ch/
3. Cocozza, R., Di Lorenzo, E., Sibillo, S.: The current value of the mathematical provision: a financial risk prospect. Prob. Persp. Man. 5 (2007)
4. Lisenko, N., Parker, G.: Stochastic analysis of life insurance surplus. Proceedings of "AFIR Colloquium", Stockholm 2007

Checking financial markets via Benford's law: the S&P 500 case

Marco Corazza, Andrea Ellero and Alberto Zorzi

Abstract. In general, in a given financial market, the probability distribution of the first significant digit of the prices/returns of the assets listed therein follows Benford's law, but does not necessarily follow this distribution in case of anomalous events. In this paper we investigate the empirical probability distribution of the first significant digit of S&P 500's stock quotations. The analysis proceeds along three steps. First, we consider the overall probability distribution during the investigation period, obtaining as result that it essentially follows Benford's law, i.e., that the market has ordinarily worked. Second, we study the day-by-day probability distributions. We observe that the majority of such distributions follow Benford's law and that the non-Benford days are generally associated to events such as the Wall Street crash on February 27, 2007. Finally, we take into account the sequences of consecutive non-Benford days, and find that, generally, they are rather short.

Key words: Benford's law, S&P 500 stock market, overall analysis, day-by-day analysis, consecutive rejection days analysis

1 Introduction

It is an established fact that some events, not necessarily of an economic nature, have a strong influence on the financial markets in the sense that such events can induce anomalous behaviours in the quotations of the majority of the listed assets. For instance, this is the case of the Twin Towers attack on September 11, 2001. Of course, not all such events are so (tragically) evident. In fact, several times the financial markets have been passed through by a mass of anomalous movements which are individually not perceptible and whose causes are generally unobservable.

In this paper we investigate this phenomenon of "anomalous movements in financial markets" in a real stock market, namely the S&P 500, by using the so-called Benford's law. In short (see the next section for more details), Benford's law is the probability distribution associated with the first significant digit[1] of numbers belonging to a certain typology of sets. As will be made clear in section 2, it is reasonable to

---

[1] Here significant digit is meant as not the digit zero.

guess that the first significant digit of financial prices/returns follows Benford's law in the case of ordinary working of the considered financial markets, and that it does not follow such a distribution in anomalous situations.

The remainder of this paper is organised as follows. In the next section we provide a brief introduction to Benford's law and the intuitions underlying our approach. In section 3 we present a short review of its main financial applications. In section 4 we detail our methodology of investigation and give the results coming from its application to the S&P 500 stock market. In the last section we provide some final remarks and some cues for future researches.

2 Benford's law: an introduction

Originally, Benford's law was detected as empirical evidence. In fact, some scientists noticed that, for extensive collections of heterogeneous numerical data expressed in decimal form, the frequency of numbers which have d as the first significant digit, with $d = 1, 2, \ldots, 9$, was not $1/9$ as one would expect, but strictly decreases as d increases; it was about 0.301 if $d = 1$, about 0.176 if $d = 2, \ldots$, about 0.051 if $d = 8$ and about 0.046 if $d = 9$. As a consequence, the frequency of numerical data with the first significant digit equal to 1, 2 or 3 appeared to be about 60%. The first observation of this phenomenon traces back to Newcomb in 1881 (see [9]), but a more precise description of it was given by Benford in 1938 (see [2]). After the investigation of a huge quantity of heterogeneous numerical data,[2] Benford guessed the following general formula for the probability that the first significant digit equals d:

$$\Pr(\text{first significant digit} = d) = \log_{10}\left(1 + \frac{1}{d}\right), \quad d = 1, \ldots, 9.$$

This formula is now called Benford's law.

Only in more recent times the Benford's law obtained well posed theoretical foundations. Likely, the two most common explanations for the emergence of probability distributions which follow Benford's law are linked to scale invariance and multiplicative processes (see [11] and [6]).[3] With attention to the latter explanation – which is of interest for our approach – and without going into technical details, Hill proved, under fairly general conditions, using random probability measures, that ≪*if* [probability] *distributions are selected at random and random samples are taken from each of these distributions, the significant digits of the combined sample will converge to Benford distribution*≫ (see [6]). This statement offers the basis for the main intuition underlying our paper. In fact, we consider the stocks of the S&P 500 market as the randomly selected probability distributions, and the prices/returns of each of these different assets as the generated random samples. The first significant

---

[2] For instance, lake surface areas, river lengths, compounds molecular weights, street address numbers and so on.

[3] In other studies it has been proved that also powers of [0, 1]-uniform probability distribution asymptotically satisfy Benford's law (see [1] and [7]).

digit of such prices/returns should follow the Benford distribution. But, if some exceptional event affects these stocks in ways similar among them, the corresponding asset prices/returns could be considered "less random" than that stated in [6], and the probability distribution of their first significant digit should depart from the Benford one. In this sense the fitting, or not, to Benford's law provides an indication of the ordinary working, or not, of the corresponding financial market.

3 Benford's law: financial applications

Investigations similar to ours have been sketched in a short paper by Ley (see [8]), which studied daily returns of the Dow Jones Industrial Average (DJIA) Index from 1900 to 1993 and of the Standard and Poor's (S&P) Index from 1926 to 1993. The author found that the distribution of the first significant digit of the returns roughly follows Benford's law. Similar results have been obtained for stock prices on single trading days by Zhipeng et al. (see [12]).

An idea analogous to the one traced in the previous section, namely that the detection of a shunt from Benford's law might be a symptom of data manipulation, has been used in tax-fraud detection by Nigrini (see [10]), and in fraudulent economic and scientific data by Günnel et al. and by Diekmann, respectively (see [5] and [4]).

Benford's law has been used also to discuss tests concerning the presence of "psychological barriers" and of "resistance levels" in stock markets. In particular De Ceuster et al. (see [3]) claimed that differences of the distribution of digits from uniformity are a natural phenomenon; as a consequence they found no support for the psychological barriers hypothesis.

All these different financial applications support the idea that in financial markets that are not "altered", Benford's law holds.

4 Do the S&P 500's stocks satisfy Benford's law?

The data set we consider consists of 3067 daily close prices and 3067 daily close logarithmic returns for 361 stocks belonging to the S&P 500 market,[4] from August 14, 1995 to October 17, 2007. The analysis we perform proceeds along three steps:

- in the first one we investigate the overall probability distribution of the first significant digit both on the whole data set of prices and on the whole data set of returns;
- in the second step we study the day-by-day distribution of the first significant digit of returns;
- finally, in the third step we analyse the sequences of consecutive days in which the distribution of the first significant digit of returns does not follow Benford's law, i.e., the consecutive days in which anomalous behaviours happen.

---

[4] In this analysis we take into account only the S&P 500 stocks that are listed for each of the days belonging to the investigation period.

4.1 Overall analysis

Here we compare the overall probability distributions of the first significant digit of the considered prices and returns against Benford's law and the uniform probability distribution (see Fig. 1) by means of the chi-square goodness-of-fit test. Uniform probability distribution is used as the (intuitive) benchmark alternative to the (counterintuitive) Benford's law.

At a visual inspection, both the empirical probability distributions seem to be rather Benford-like (in particular, the one associated to returns). Nevertheless, in both the comparisons the null is rejected. In Table 1 we report the values of the associated chi-square goodness-of-fit tests with 8 degrees of freedom (we recall that $\chi^2_{8,0.95} = 15.51$).

From a qualitative point of view, our results are analogous to the ones obtained by Ley (see [8]). In particular, that author observed that, despite the fact that the chi-square goodness-of-fit tests on DJIA and S&P Indexes suggest rejection of the null, this was due to the large number of observations considered. In fact, the same kind of analysis performed only on 1983–1993 data suggested acceptance of the null. Moreover, the rejection with respect to the uniform probability distribution is stronger and stronger than the rejection with respect to Benford's law. In other words,

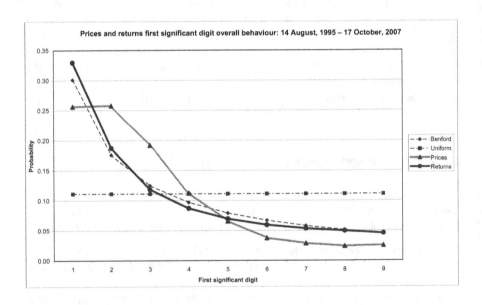

Table 1. Overall calculated chi-square

Reference probability distribution	χ^2 w.r.t. prices	χ^2 w.r.t. returns
Benford	151527.74	7664.84
Uniform	780562.24	673479.62

looking at chi-square as a distance, the empirical probability distributions are closer to Benford's law than to the uniform probability distribution. In this sense we agree with Ley (see [8]) claiming that the distributions of the first significant digit of prices and returns essentially follow Benford's law. In other terms, the S&P 500 stock market behaviour as a whole in the period August 14, 1995 to October 17, 2007 can be considered as "ordinary".

Finally, we observe that the empirical probability distribution related to returns is significantly closer to Benford's law than the empirical probability distribution related to prices. In particular, the latter is 19.77 times further away from Benford's law than the former. This evidence is theoretically coherent with that stated in the paper of Pietronero *et al.* (see [11]), since logarithmic returns are obtained from prices by a multiplicative process.

4.2 Day-by-day analysis

Here, we address our attention to returns since their empirical probability distribution is closer to Benford's law than that of prices. We day-by-day perform the same kind of analysis considered in the previous subsection, but only with respect to Benford's law.

Over the investigated 3067 days, the null is rejected 1371 times, i.e., in about 44.70% of cases. In Figure 2 we represent the values of the day-by-day calculated chi-square goodness-of-fit tests (the horizontal white line indicates the value of $\chi^2_{8,0.05}$).

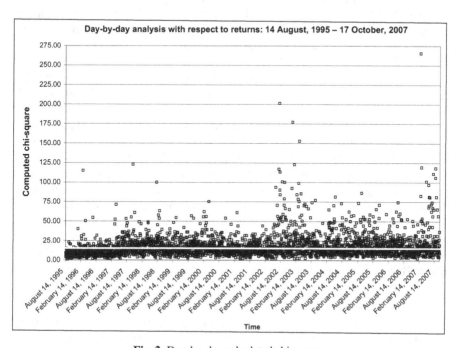

Fig. 2. Day-by-day calculated chi-square

We deepen our analysis taking into account also a confidence level α equal to 1% (we recall that $\chi^2_{8,0.99} = 20.09$); it results that the null is rejected 890 times, i.e., in about 29.02% of cases.

In order to check if such rejection percentages are reasonable, we perform the following computational experiment:

- First, for each of the considered 361 stocks we generate a simulated time series of its logarithmic returns which has the same length as the original time series, and whose probability distribution follows a Gaussian one with mean and variance equal to the real ones estimated for the stock[5] (the Gaussian probability distribution is chosen for coherence with the classical theory of financial markets);
- Second, we perform a day-by-day analysis on the generated financial market in the same way as for the true financial market.

Repeating the experiment 50 times, we obtain the following mean values of the rejection percentages: 57.92% if $\alpha = 5\%$ (about 1776 cases) and 33.50% if $\alpha = 1\%$ (about 1027 cases). This results have not to be considered particularly surprising. In fact, to each of the considered stocks we associate always the same kind of probability distribution, the Gaussian one, instead of selecting it at random as would be required to obtain a Benford distribution (see section 2).

The fact that the rejection percentages in a classical-like market are greater than the corresponding percentages in the true one denotes that a certain number of deviations from Benford's law, i.e., a certain number of days in which the financial market is not ordinary working, is physiological. Moreover, the significant differences between rejection percentages concerning the classical-like market and the true one can be interpreted as a symptom of the fact that, at least from a distributional point of view, the true financial market does not always follow what is prescribed by the classical theory.

In Table 2 we report the 45 most rejected days at a 5% significance level with the corresponding values of the chi-square goodness-fit-of tests with 8 degrees of freedom.

We notice that some of the days and periods reported in Table 2 are characterised by well known events. For instance, the Wall Street crash on February, 2007 (the most rejected day) and the troubles of important hedge funds since 2003 (24.44% of the first 45 most rejected days falls in 2003). Nevertheless, in other rejection days/periods the link with analogous events cannot generally be observed. In such cases the day-by-day analysis can be profitaby used to detect hidden anomalous behaviours in financial markets. On the other hand, the most accepted day is September 5, 1995, whose value of the chi-square goodness-fit-of test is 0.91. In Figure 3 we graphically compare the empirical probability distributions of the most rejected and of the most accepted days against Benford's law and the uniform probability distribution.

---

[5] In generating this simulated financial market, we do not consider the correlation structure existing among the returns of the various stocks because, during the investigation period, such a structure does not appear particularly relevant. So, the simulated financial market can be reckoned as a reasonable approximation of the true one.

Table 2. Most rejected days and related calculated chi-square

Rank	Day	χ^2	Rank	Day	χ^2	Rank	Day	χ^2
1	02.27.2007	265.55	16	10.01.2002	100.19	31	06.22.2007	80.86
2	07.29.2002	201.43	17	08.04.1998	99.96	32	09.26.2002	78.90
3	01.02.2003	177.33	18	03.17.2003	98.81	33	03.10.2003	78.44
4	03.24.2003	152.84	19	06.07.2007	96.93	34	10.01.2003	77.21
5	01.24.2003	122.93	20	06.17.2002	93.67	35	03.21.2003	75.63
6	10.27.1997	122.89	21	12.27.2002	92.29	36	04.14.2000	75.22
7	03.13.2007	119.38	22	08.05.2002	90.05	37	06.05.2006	74.57
8	08.29.2007	118.19	23	03.30.2005	86.39	38	07.20.2007	74.27
9	07.24.2002	117.10	24	04.14.2003	85.39	39	08.22.2003	74.10
10	03.08.1996	115.01	25	02.24.2003	84.14	40	02.22.2005	73.40
11	08.06.2002	113.42	26	08.08.2002	84.05	41	08.05.2004	72.96
12	08.03.2007	111.14	27	03.02.2007	82.86	42	05.30.2003	72.24
13	08.28.2007	106.23	28	05.25.2004	81.47	43	07.10.2007	71.97
14	05.10.2007	101.04	29	06.13.2007	81.47	44	04.11.1997	71.44
15	09.03.2002	100.76	30	09.07.2007	81.29	45	10.28.2005	71.23

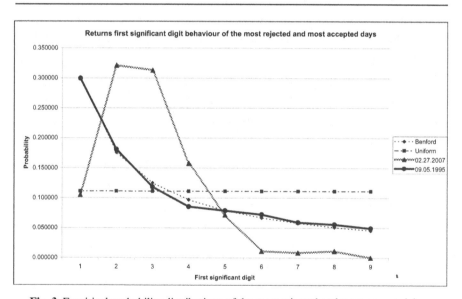

Fig. 3. Empirical probability distributions of the most rejected and most accepted days

Finally, we spend a few words on the day of the Twin Towers attack, which has been chosen as the central one of the data set. We remark that 42.89% of all the rejected days falls before this day and 57.11% of them after. Moreover, if we limit our attention to the first 45 most rejected days, the difference between such percentages considerably increases to 11.11% before and 88.89% after, respectively.

These results show that the S&P 500 stock market was subject to anomalous activities after September 11, 2001 rather than before.

4.3 Consecutive rejection days analysis

Here, addressing our attention to returns once more, we analyse the sequences of consecutive rejection days detected using $\alpha = 5\%$. In Table 3 we report the number of such sequences having lengths from 1 day to 12 days, respectively (12 days is the *maximum* length detected in the investigation period). For deepening the investigation, we also report the results obtained using $\alpha = 1\%$.

We observe that the length of the large majority of the sequences of consecutive rejection days is rather low. This fact can be interpreted as the capability of the S&P 500 stock market to "absorb" anomalous events in short time periods.

On the contrary, given such a capability, the presence of long sequences of consecutive rejection days is an indicator of malaise of the market. For instance, this is the case of a 9-day sequence (September 20, 2001 to October 2, 2001) that started immediately after the Twin Towers attack and of a 6-day sequence (February 27, 2007 to March 6, 2007) that started on the day of the Wall Street crash. Moreover, analogously to what we already observed in the previous subsection, since the events/causes associated to such sequences are not always observable, the consecutive rejection days analysis might be profitably used for detecting continued anomalous behaviours in financial markets.

Table 3. Sequences of consecutive rejection days

Sequence length	# with $\alpha = 5\%$	# with $\alpha = 1\%$
1	412	366
2	148	105
3	71	52
4	48	12
5	24	7
6	13	2
7	7	1
8	1	0
9	3	0
10	1	1
11	0	0
12	1	0

5 Conclusions

To the best of our knowledge, several aspects concerning the use of Benford's law-based analyses in financial markets have not yet been investigated. Among the various ones, we consider the following:

- Given the few studies on this topic, the actual capability of this kind of approach to detect anomalous behaviours in financial markets has to be carefully checked and measured. To this end, the systematic applications of these approaches to a large number of different financial markets is needed;
- From a methodological point of view, we guess that restricting the analysis we performed in this paper to the different sectors compounding the financial market could be useful for detecting, in the case of anomalous behaviours of the market as a whole, which sectors are the most reliable;
- We guess also that, in order to make this analysis more careful, we should at least take into account the probability distribution of the second significant digit (see [6]), i.e.,

$$
\Pr(\text{second significant digit} = d) = \log_{10} \sum_{k=1}^{9} \left(1 + \frac{1}{10k + d}\right), \quad d = 0, \ldots, 9; \,^6
$$

- Finally, the results we presented in this paper are *ex post*. Currently, we are beginning to develop and apply a new Benford's law-based approach in order to check some predictive capabilities. The first very preliminary results seem to be encouraging.

References

1. Adhikari, A., Sarkar, B.: Distribution of most significant digit in certain functions whose arguments are random variables. Sankhya, Series B 30, 47–58 (1968)
2. Benford, F.: The law of anomalous numbers. Proc. Am. Phil. Soc. 78, 551–572 (1938)
3. De Ceuster, M.J.K., Dhaene, G., Schatteman T.: On the hypothesis of psychological barriers in stock markets and Benford's law. J. Emp. Finan. 5, 263–279 (1998)
4. Diekmann, A.: Note the first digit! Using Benford's law to detect fraudulent scientific data. J. Appl. Stat. 34, 321–329 (2007)
5. Günnel, S., Tödter, K-H.: Does Benford's law hold in economic research and forecasting? Discussion Paper, Series 1: Economic Studies, Deutsche Bundesbank 32/2007, (2007)
6. Hill, T.P.: A statistical derivation of the significant-digit law. Stat. Sci. 10, 354–363 (1995)
7. Janvresse, É., de la Rue, T.: From uniform distributions to Benford's law. J. Appl. Prob. 41, 1203–1210 (2004)
8. Ley, E.: On the peculiar distribution of the U.S. stock indexes' digits. Am. Stat. 50, 311–313 (1996)

---

[6] We remark that zero is significant as a second digit.

9. Newcomb, S.: Note on the frequency of use of the different digits in natural numbers. Am. J. Math. 4, 39–40 (1881)
10. Nigrini, M.: A taxpayer compliance application of Benford's law. J. Am. Tax. Assoc. 18, 72–91 (1996)
11. Pietronero, L., Tossati, E., Tossati, V., Vespignani, A.: Explaining the uneven distribution of numbers in nature: the laws of Benford and Zipf. Physica A. 293, 297–304 (2001)
12. Zhipeng L., Lin C., Huajia W.: Discussion on Benford's law and its applications. In: Cornell University Library (2004) Available via arXiv.org. http://arxiv.org/abs/math/0408057v2.Cited4Oct2004

Empirical likelihood based nonparametric testing for CAPM

Pietro Coretto and Maria Lucia Parrella

Abstract. The Capital Asset Pricing Model (CAPM) predicts a linear relation between assets' return and their betas. However, there is empirical evidence that such a relationship does not necessarily occur, and in some cases it might even be nonlinear. In this paper we explore a nonparametric approach where the linear specification is tested against a nonparametric alternative. This methodology is implemented on S&P500 data.

Key words: CAPM, goodness-of-fit test, empirical likelihood

1 Introduction

An asset pricing model provides a method for assessing the riskiness of cash flows from a project. The model provides an estimate of the relationship between that riskiness and the cost of capital. According to the "capital asset pricing model" (CAPM), the only relevant measure of a project's risk is a variable unique to this model, known as the project's beta. In the CAPM, the cost of capital, i.e., the return, is a linear function of the the beta of the project being evaluated. A manager who has an estimate of the beta of a potential project can use the CAPM to estimate the cost of capital for the project. If the CAPM captures investors' behaviour adequately, then the historical data should reveal a positive linear relation between return on financial assets and their betas. Also, no other measure of risk should be able to explain the differences in average returns across financial assets that are not explained by CAPM betas. The fact that CAPM theory predicts the existence of a cross-section linear relation between returns and betas can be empirically tested. To this end we propose a nonparametric testing methodology (see [10] and [3] among others).

The first test of the CAPM was run by Fama and MacBeth [7] and their study validated the theory. The authors tested the linearity against some parametric nonlinear alternatives. However subsequent empirical analysis highlighted that the validity of the CAPM could depend on the testing period. There is a huge amount of literature on this topic (for a comprehensive review see [11]), however, final conclusions have not been made.

The famous Fama-MacBeth contribution (and the following) tests the linear spec-ification against a number of nonlinear parametric specifications. The main contri-bution of this paper is that we test the linear specification of the CAPM against a nonlinear nonparametric specification. And by this we do not confine the test to a specific (restricting) nonlinear alternative. Our testing method is based on kernel smoothing to form a nonparametric specification for the null hypothesis that the re-lation between returns and betas is linear against the alternative hypothesis that there is a deviation from the linearity predicted by the CAPM. We apply our methodology to the S&P 500 market.

The paper is organised as follows: we introduce the theoretical model, we intro-duce the Fama and MacBeth two-stage parametric estimation procedure, we outline the nonparametric testing methodology and finally we discuss some empirical findings based on the analysis of the S&P 500 market.

2 The CAPM in a nutshell

CAPM was first developed by Sharpe and Treynor; Mossin, Lintner and Black brought the analysis further. For a comprehensive review see [5] and [1]. We will refer to SLB as the Sharpe-Lintner-Black version of the model. The SLB model is based on the assumption that there is a positive trade-off between any asset's risk and its expected return. In this model, the expected return on an asset is determined by three variables: the risk-free rate of return, the expected return on the market portfolio and the asset's beta. The last one is a parameter that measures each asset's systematic risk, i.e., the share of the market portfolio's variance determined by each asset.

2.1 Theoretical model

The CAPM equation is derived by imposing a number of assumptions that we discuss briefly. An important building-block of the CAPM theory is the so-called *perfect market hypothesis*. This is about assuming away any kind of frictions in trading and holding assets. Under the perfect market hypothesis, unlimited short-sales of risky assets and risk-free assets are possible.

The second assumption is that all investors choose their portfolios based on mean (which they like) and variance (which they do not like). This assumption means that people's choices are consistent with Von-Neumann- Morgenstern's axiomatisation.

All investors make the same assessment of the return distribution. This is referred to as "homogenous expectations". The implication of this hypothesis is that we can draw the same minimum-variance frontier for every investor.

Next is the "market equilibrium" hypothesis (i.e., supply of assets equals demand). The market portfolio is defined as the portfolio of assets that are in positive net supply, weighted by their market capitalisations. Usually it is assumed that the risk-free instrument is in zero net supply. On the demand side, the net holdings of all investors equal aggregate net demand. The last assumption states that all assets are marketable, i.e., there is a market for each asset.

On the basis of these assumptions we can derive a model that relates the expected return of a risky asset with the risk-free rate and the return of market portfolio; in the latter, all assets are held according to their value weights. We will denote $\tilde{R}_j$ a random variable that describes the return of risky asset j. Let $\tilde{R}_f$ be the risk-free rate and $\tilde{R}_M$ the return on market portfolio. Under the assumptions above and assuming that the following expectations exist, the theory of CAPM states that there exists the following relation:

$$E[\tilde{R}_j] = E[\tilde{R}_f] + \beta_j \left(E[\tilde{R}_M] - E[\tilde{R}_f] \right). \tag{1}$$

The term β_j in the CAPM equation (1) is the key to the whole model's implications. β_j represents the risk asset j contributes in the market portfolio, measured relative to the market portfolio's variance:

$$\beta_j = \frac{\text{Cov}[\tilde{R}_j, \tilde{R}_M]}{\text{Var}[\tilde{R}_M]}. \tag{2}$$

β is a measure of systematic risk: since it is correlated with the market portfolio's variance and the market portfolio is efficient, an investor cannot possibly diversify away from it. The theory predicts that each asset's return depends linearly on its beta. Notice that the CAPM equation is a one-period model; this means that this equation should hold period by period. In order to estimate and test the CAPM equation date by date, we need to make further assumptions in order to estimate the betas first.

2.2 Testing strategy

The beauty of the CAPM theory is that in order to predict assets' return we only need information about prices and no further expensive information is needed. The tests conducted over the last 45 years have brought up different issues and contrasting views and results. Whereas the first test found no empirical evidence for the theory of equilibrium asset prices, a very famous test, conducted in 1973 by Fama and MacBeth (see [7]), provided evidence in favour of the validity of the SLB-CAPM model. However, later studies (e.g., Fama and French [6]) have challenged the positive and linear relationship between betas and returns (i.e., CAPM's theory's main conclusion) by introducing other variables which proved to have a much greater explanatory power but at some costs.

The main contribution of this paper is a nonparametric test about the linear specification of the CAPM. Our nonparametric testing approach is based on the comparison between the predicted returns obtained via the parametric linear model implied by the CAPM and the returns predicted by a kernel estimator. This testing strategy implies two steps: the first step (or "parametric step") is to estimate the predicted returns based on the CAPM equation; the second step (or nonparametric step) is to predict the returns on the basis of a kernel regression. We describe the two steps in detail.

2.3 The parametric step

We complete the first step by using the same methodology developed by Fama and
MacBeth [7]. In order to apply this methodology we need to make some further assumptions. The CAPM is a model of expected returns in a one-period economy. What
we actually observe, though, is a time series of asset prices and other variables from
which we can compute the realised returns over various holding periods. We need
to assume that investors know the return distribution over one particular investment
period. In order to estimate the parameters of that distribution it is convenient to
assume that the latter is stationary. In addition, we assume that returns are drawn
independently over time. Although the last assumption appears to be too strong, several empirical studies proved that this cannot seriously affect the first-step estimation
(see [7]). The latter comment applies in particular when short sequences of daily
returns are used to estimate the betas (see below).

What about the "market portfolio"? Can the market portfolio be easily identified?
It is worth remembering that the CAPM covers all marketable assets and it does not
distinguish between different types of financial instruments. This is the focus of Roll's
Critique [14]. As a market proxy we will use the S&P500 index. The CAPM provides
us with no information about the length of the time period over which investors choose
their portfolios. It could be a day, a month, a year or a decade.

Now we describe the Fama-MacBeth estimation methodology. We have a time
series of assets' prices recorded in some financial market. Let us assume that $R_{j,t}$ is
the log-return at time t for the asset j, where $j = 1, 2, \ldots, S$ and $t = 1, 2, \ldots, T$.
Let $R_{M,t}$ be the market log-return at time t. The relation (1) has to hold at each t for
each asset. We have to estimate the CAPM for each t. To do the latter we need a time
series of βs.

The first stage is to obtain a time series of estimated betas based on a rolling
scheme. For each asset $j = 1, 2, \ldots, S$, and for fixed w and $p = 1, 2, \ldots, T-w+1$,
we take the pairs $\{R_{j,t}, R_{M,t}\}_{t=p,p+1,\ldots,p+w-1}$ and we estimate the market equation

$$R_{j,t} = \alpha_{j,p} + \beta_{j,p} R_{M,t} + \varepsilon_{j,t}, \tag{3}$$

where $\{\varepsilon_{j,t}\}_{t=p,p+1,\ldots,p+w-1}$ is an i.i.d. sequence of random variables with zero mean
and finite variance. The (3) is estimated for each $j = 1, 2, \ldots, S$, to obtain β_j for
periods $p = 1, 2, \ldots, T-w+1$. The estimated $\hat{\beta}_{j,p}$ is the estimate of the systematic
risk of the jth asset in period p. From this first regression we also store the estimated
standard deviation of the error term, say $\hat{\sigma}_{j,p} = \sqrt{\hat{\text{Var}}(\hat{\varepsilon}_{j,t})}$. The latter is a measure
of the unsystematic risk connected to the jth asset in period p. The use of $\hat{\sigma}_{j,p}$ will
be clear afterwards.

In the second stage for each period $p = 1, 2, \ldots, T-w+1$ we estimate the
linear model implied by the CAPM applying a cross-section (across $j = 1, 2, \ldots, S$)
linear regression of assets' returns on their estimated betas. For each period $p =
1, 2, \ldots, T-w+1$ the second-stage estimation is:

$$R_{j,p+1} = \gamma_1^A + \gamma_2^A \hat{\beta}_{j,p} + \xi_{j,p+1}^A, \tag{4}$$

where $\{\xi_{j,p}^A\}_{j=1,2,\ldots,S}$ is an i.i.d. sequence of random variables having zero mean and finite variance. Notice that we regress $R_{j,p+1}$ on $\hat{\beta} j, p$; this is because it is assumed that investors base current investment decisions on the most recent available β. The Fama-MacBeth testing procedure consisted in testing the linear relation (4) for each period p. If the linear model (4) holds in period p, that means that the model (1) statistically holds in period p. For a long time it has been thought that when the CAPM fails this is due to the fact that unsystematic risk affects returns as well as possible nonlinearities in betas. Two further second-stage equations have been considered to check for the aforementioned effects. The first alternative to (4) is:

$$R_{j,p+1} = \gamma_1^B + \gamma_2^B \hat{\beta}_{j,p} + \gamma_3^B \hat{\sigma}_{j,p} + \xi_{j,p+1}^B, \tag{5}$$

where we add a further regressor which is the unsystematic risk measure. The second alternative is:

$$R_{j,p+1} = \gamma_1^C + \gamma_2^C \hat{\beta}_{j,p}^2 + \gamma_3^C \hat{\sigma}_{j,p} + \xi_{j,p+1}^C, \tag{6}$$

where the betas enter in the regression squared. In (5) and (6) we assume that the errors $\{\xi_{j,p}^B\}_{j=1,2,\ldots,S}$ and $\{\xi_{j,p}^C\}_{j=1,2,\ldots,S}$ are two i.i.d. sequence of random variables with zero mean and finite variance. As for (4), (5) and (6) are also estimated for each period $p = 1, 2, \ldots, T - w + 1$. The models A, B and C represented by equations (4),(5) and (6) are estimated and tested in the famous paper by Fama and MacBeth [7].

3 The nonparametric goodness-of-fit test

In this section we apply the method proposed by Härdle et al. [8] for testing the linear specification of the CAPM model, that is, models A, B and C defined in the second stage of the previous step. This goodness-of-fit test is based on the combination of two nonparametric tools: the Nadaraya-Watson kernel estimator and the Empirical Likelihood of Owen [13]. Here we briefly describe the testing approach, then, in the next section, we apply it to the CAMP model estimated on the S&P500 stock market.

Let us consider the following nonparametric model

$$R_{j,p+1} = m\left(\mathbf{X}_{j,p}\right) + e_{j,p+1}, \qquad \begin{array}{l} j = 1, 2, \ldots, S \\ p = 1, 2, \ldots, T - w + 1 \end{array} \tag{7}$$

where $R_{j,p+1}$ is the log-return of period p for the asset j, $\mathbf{X}_{j,p} \in R^d$ is the vector of d regressors observed in period p for the asset j and $e_{j,p+1}$ is the error, for which we assume that $E(e_{j,p+1}|\mathbf{X}_{j,p}) = 0$ for all j. We also assume that the regressors $\mathbf{X}_{j,p}$ and the errors $e_{j,p}$ are independent for different js, but we allow some conditional heteroscedasticity in the model. The main interest lies in testing the following hypothesis

$$H_0 : m(\mathbf{x}) = m_\gamma(\mathbf{x}) = \gamma^T \mathbf{x} \quad \text{versus} \quad H_1 : m(\mathbf{x}) \neq m_\gamma(\mathbf{x}), \tag{8}$$

where $m_\gamma(\mathbf{x}) = \gamma^T \mathbf{x}$ is the linear parametric model and γ is the vector of unknown parameters belonging to a parameter space $\Gamma \in R^{d+1}$. Let us denote with $m_{\hat{\gamma}}(\mathbf{x})$ the estimate of $m_\gamma(\mathbf{x})$ given by a parametric method consistent under H_0.

The Nadaraya-Watson estimator of the regression function $m(\mathbf{x})$ is given by

$$\hat{m}_h(\mathbf{x}) = \frac{\sum_{j=1}^{S} R_{j,p+1} K_h(\mathbf{x} - \mathbf{X}_{j,p})}{\sum_{j=1}^{S} K_h(\mathbf{x} - \mathbf{X}_{j,p})}, \tag{9}$$

where $K_h(\mathbf{u}) = h^{-d} K(h^{-1}\mathbf{u})$ and K is a d-dimensional product kernel, as defined in [8]. The parameter h is the bandwidth of the estimator, which regulates the smoothing of the estimated function with respect to all regressors. We use a common bandwidth because we assume that all the regressors have been standardised.

When applied to kernel estimators, empirical likelihood can be defined as follows. For a given $\mathbf{x}$, let $p_j(\mathbf{x})$ be nonnegative weights assigned to the pairs $(\mathbf{X}_{j,p}, R_{j,p+1})$, for $j = 1, \ldots, S$. The empirical likelihood for a smoothed version of $m_{\hat{\gamma}}(\mathbf{x})$ is defined as

$$L\{\tilde{m}_{\hat{\gamma}}(\mathbf{x})\} = \max \left\{ \prod_{j=1}^{S} p_j(\mathbf{x}) \right\}, \tag{10}$$

where the maximisation is subject to the following constraints

$$\sum_{j=1}^{S} p_j(\mathbf{x}) = 1; \quad \sum_{j=1}^{S} p_j(\mathbf{x}) K\left(\frac{\mathbf{x} - \mathbf{X}_{j,p}}{h}\right) [R_{j,p+1} - \tilde{m}_{\hat{\gamma}}(\mathbf{x})] = 0. \tag{11}$$

As is clear from equation (11), the comparison is based on a smoothed version of the estimated parametric function $m_{\hat{\gamma}}(\mathbf{x})$ (see [8] for a discussion), given by

$$\tilde{m}_{\hat{\gamma}}(\mathbf{x}) = \frac{\sum_{j=1}^{S} m_{\hat{\gamma}}(\mathbf{X}_{j,p}) K_h(\mathbf{x} - \mathbf{X}_{j,p})}{\sum_{j=1}^{S} K_h(\mathbf{x} - \mathbf{X}_{j,p})}. \tag{12}$$

By using Lagrange's method, the empirical log-likelihood ratio is given by

$$l\{\tilde{m}_{\hat{\gamma}}(\mathbf{x})\} = -2 \log[L\{\tilde{m}_{\hat{\gamma}}(\mathbf{x})\} S^S]. \tag{13}$$

Note that S^S comes from the maximisation in (10), since the maximum is achieved at $p_j(\mathbf{x}) = S^{-1}$.

Theorem 1. *Under H_0 and the assumptions A.1 in [8], we have*

$$l\{\tilde{m}_{\hat{\gamma}}(\mathbf{x})\} \xrightarrow{d} \chi_1^2. \tag{14}$$

Proof (sketch). The proof of the theorem is based on the following asymptotic equivalence (see [4] and [8])

$$l\{\tilde{m}_{\hat{\gamma}}(\mathbf{x})\} \approx \left[(Sh^d)^{1/2} \frac{\{\hat{m}_h(\mathbf{x}) - \tilde{m}_\gamma(\mathbf{x})\}}{V^{1/2}(\mathbf{x}; h)} \right]^2, \tag{15}$$

where $V(\mathbf{x}; h)$ is the conditional variance of $R_{j,p+1}$ given $\mathbf{X}_{j,p} = \mathbf{x}$. For theorem 3.4 of [2], the quantity in brackets is asymptotically $N(0, 1)$.   □

As shown in theorem 1, the empirical log-likelihood ratio is asymptotically equivalent to a Studentised L_2-distance between $\tilde{m}_\gamma(\mathbf{x})$ and $\hat{m}_h(\mathbf{x})$, so it may be compared to the statistic tests used in [9] and [12]. The main attraction of the test procedure described here is its ability to automatically studentising the statistic, so we do not have to estimate $V(\mathbf{x}; h)$, contrary to what happens with other nonparametric goodness-of-fit tests. Based on Theorem 1 and on the assumed independence of the regressors, we use the following goodness-of-fit test statistic

$$\sum_{r=1}^{S^*} l\left\{\tilde{m}_{\hat{\gamma}}(\mathbf{x}_{r,p})\right\}, \tag{16}$$

which is built on a set of $S^* < S$ points $\mathbf{x}_{r,p}$, selected equally spaced in the support of the regressors. The statistic in (16) is compared with the percentiles of a χ^2 distribution with S^* degrees of freedom.

4 Empirical results and conclusions

In this section we discuss some results obtained by estimating the models A, B and C presented in equations (4), (5) and (6) and we apply the nonparametric step to test the linearity of such models. Note that here we consider specifically the linear functions under the null, but the hypotheses stated in (8) might refer to other functional forms for $m_\gamma(x)$.

The market log-return is given by the S&P500 index, while the asset log-returns are the $S = 498$ assets included in the S&P stock index. The time series are observed from the 3rd of January 2000 to the 31st of December 2007, for a total of 1509 time observations. We consider three different rolling window lengths, that is $w = 22$, 66 and 264, which correspond roughly to one, three and twelve months of trading. The total number of periods in the cross-section analysis (second stage of the Fama and MacBeth method) is 1487 when $w = 22$, 1443 when $w = 66$ and 1245 when $w = 264$.

For each asset, we estimate the coefficients $\hat{\beta}_{j,p}$, $j = 1, \ldots, S$, $p = 1, \ldots, T - w + 1$ from equation (3). We obtain a matrix of estimated betas, of dimension $(1510 - w, 498)$. For each resulting period we estimate the cross-section models A, B and C and we apply the nonparametric testing scheme. The assumptions A.1 in [8] are clearly satisfied for the data at hand. The bandwidth used in the kernel smoothing in (9), (11) and (12) has been selected automatically for each period p, by considering optimality criteria based on a generalised cross-validation algorithm. In (16) we have considered $S^* = 30$ equally spaced points. It is well known that kernel estimations generally suffer from some form of instability in the tails of the estimated function, due to the local sparseness of the observations. To avoid such problems, we selected the S^* points in the internal side of the support of the regressors, ranging on the central 95% of the total observed support.

In Table 1 we summarise the results of the two testing procedures (parametric and nonparametric) described in previous sections.

Table 1. Percentage of testing periods for each specified model and window when cases 1–4 occur. Cases 1–4 are as follows. Case 1: the estimated linear coefficients in the parametric step are jointly equal to zero at level $\alpha = 5\%$, and we do not reject the H_0 hypothesis in the nonparametric stage at the same level; Case 2: the estimated linear coefficients in the parametric step are jointly equal to zero at level $\alpha = 5\%$, and we reject the H_0 hypothesis in the nonparametric stage at the same level; Case 3: the estimated linear coefficients in the parametric step are jointly different from zero at level $\alpha = 5\%$, and we do not reject the H_0 hypothesis in the nonparametric stage at the same level; Case 4: the estimated linear coefficients in the parametric step are jointly different from zero at level $\alpha = 5\%$, and we reject the H_0 hypothesis in the nonparametric stage at the same level

Window	Regressors (model)		
	β (Model A)	β, σ (Model B)	β^2, σ (Model C)
Case 1			
$w = 22$	49.42	45.595	57.78
$w = 66$	43.87	44.144	51.195
$w = 264$	43.449	43.213	52.099
Case 2			
$w = 22$	31.141	13.114	23.925
$w = 66$	38.41	16.078	28.373
$w = 264$	38.41	12.53	28.148
Case 3			
$w = 22$	10.155	26.564	9.695
$w = 66$	8.238	23.909	10.409
$w = 264$	9.183	25.141	9.465
Case 4			
$w = 22$	9.284	14.728	8.6
$w = 66$	9.483	15.87	10.023
$w = 264$	8.959	19.116	10.288

As Case 1 we label the percentage of testing periods where the estimated linear coefficients in the parametric step are jointly equal to zero at the testing level $\alpha = 5\%$, and we do not reject the H_0 hypothesis (e.g., the linear relation statistically holds) in the nonparametric stage at the same level. This case is of particular interest because the percentage of testing periods when it occurs is not smaller than 43% for all rolling windows and all sets of regressors. If we combine the results of the two stages (both the parametric and the nonparametric) this means that in almost half of the testing periods we validate the linearity of the regression function, but this is probably a constant.

As Case 2 we label the percentage of testing periods where the estimated linear coefficients in the parametric step are jointly equal to zero at level $\alpha = 5\%$, and we reject the H_0 hypothesis in the nonparametric stage at the same level. That is, no linear relationship can be detected but there is some evidence of nonlinear structures.

As Case 3 we report the percentage of testing periods where the estimated linear coefficients in the parametric step are jointly different from zero at level $\alpha = 5\%$, and we do not reject the H_0 hypothesis in the nonparametric stage at the same level. The case is in favour of the CAPM theory because here we are saying that the estimated linear coefficients in the parametric step are jointly different from zero at level $\alpha = 5\%$, and we do not reject the linearity hypothesis (H_0) in the nonparametric stage at the same level. This means that when this situation occurs we are validating the idea behind the CAPM, that is: historical information on prices can be useful to explain the cross-section variations of assets' returns. For this case the best occurs for model B, which is the one that uses the pair of regressors (β, σ). The statistical conclusion we draw from case 3 is that when $w = 22$ for 26.6% of the testing periods we validate the linear relation between assets' returns, βs and the nonsystematic risk (σ). Moreover this result does not depend on the rolling window (even though $w = 66$ produces slightly better results).

Finally, as Case 4 we label the percentage of testing periods where the estimated linear coefficients in the parametric step are jointly different from zero at level $\alpha = 5\%$, and we reject the H_0 hypothesis in the nonparametric stage at the same level. That is, a relationship is present but the nonparametric testing step supports the evidence for nonlinear effects. It is worthwhile to observe that for all the cases considered, conclusions do not seem to be related to the particular choice of the rolling window. They remain basically stable when moving across different choices. Despite the limitations of the present analysis the open question remains whether the betas are a determinant of the cross-section variations of assets' return at all. From this study we cannot conclude that model B is validated. But certainly, this occurs in approximately one quarter of the testing periods. This encourages us to investigate other possibilities that could reveal stronger paths in the data. There are several issues that would be worth investigating further: dependence structures, group structures in the risk behaviour of assets, robustness issues and nonparametric specifications of the first stage.

References

1. Ang, A., Chen, J.: Capital market equilibrium with restricted borrowing. J. Emp. Finan. 14, 1–40 (2007)
2. Bosq, D.: Nonparametric Statistics for Stochastic Processes, vol. 110, Lecture Notes in Statistics. Springer, New York (1998)
3. Chen, S.X., Gao, J.: An adaptive empirical likelihood test for time series models. J. Econometrics 141, 950–972 (2007)

4. Chen, S., Qin, Y.S.: Empirical likelihood confidence intervals for local linear smoothers. Biometrika 87, 946–953 (2000)
5. Elton, E.J., Gruber, M.J., Brown, S.J., Goetzmann, W.N.: Modern Portfolio Theory and Investment Analysis. John Wiley, New York, 6th edition (2003)
6. Fama, E., French, K.R.: The cross-section of expected stock returns. J. Finan. 47, 427–465 (1992)
7. Fama, E., MacBeth, J.: Risk, return and equilibrium: empirical tests. J. Pol. Econ. 81, 607–636 (1973)
8. Härdle, W., Chen, S., Li, M.: An empirical likelihood goodness-of-fit test for time series. J. R. Stat. Soc. B 65, 663–678 (2003)
9. Härdle, W., Mammen, E.: Comparing nonparametric versus parametric regression fits. Ann. Stat. 21, 1926–1947 (1993)
10. Horowitz, J.L., Spokoiny, V.G.: An adaptive, rate-optimal test of a parametric mean-regression model against a nonparametric alternative. Econometrica 69, 599–631 (2001)
11. Jagannathan, R., Wang, Z.: The CAPM is alive and well. J. Finan. 51, 3–53 (1996)
12. Kreiss, J., Neumann, M.H., Yao, Q.: Bootstrap tests for simple structures in nonparametric time series regression. Discussion paper. Humboldt-Universität zu Berlin, Berlin (1998)
13. Owen, A.B.: Empirical Likelihood. Chapman & Hall, London (2001)
14. Roll, R.W.: A critique of the asset pricing theory's tests. J. Finan. Econ. 4, 129–176 (1977)

Lee-Carter error matrix simulation: heteroschedasticity impact on actuarial valuations

Valeria D'Amato and Maria Russolillo

Abstract. Recently a number of approaches have been developed for forecasting mortality. In this paper, we consider the Lee-Carter model and we investigate in particular the hypothesis about the error structure implicitly assumed in the model specification, i.e., the errors are homoschedastic. The homoschedasticity assumption is quite unrealistic, because of the observed pattern of the mortality rates showing a different variability at old ages than younger ages. Therefore, the opportunity to analyse the robustness of estimated parameter is emerging. To this aim, we propose an experimental strategy in order to assess the robustness of the Lee-Carter model by inducing the errors to satisfy the homoschedasticity hypothesis. Moreover, we apply it to a matrix of Italian mortality rates. Finally, we highlight the results through an application to a pension annuity portfolio.

Key words: Lee-Carter model, mortality forecasting, SVD

1 Introduction

The background of the research is based on the bilinear mortality forecasting methods. These methods are taken into account to describe the improvements in the mortality trend and to project survival tables. We focus on the Lee-Carter (hereinafter LC) method for modelling and forecasting mortality, described in Section 2. In particular, we focus on a sensitivity issue of this model and in order to deal with it, in Section 3, we illustrate the implementation of an experimental strategy to assess the robustness of the LC model. In Section 4, we run the experiment and apply it to a matrix of Italian mortality rates. The results are applied to a pension annuity portfolio in Section 5. Finally, Section 6 concludes.

2 The Lee-Carter model: a sensitivity issue

The LC method is a powerful approach to mortality projections. The traditional LC model analytical expression [7] is the following:

$$ln\left(M_{x,t}\right) = \alpha_x + \beta_x \kappa_t + E_{x,t}, \tag{1}$$

describing the log of a time series of age-specific death rates $m_{x,t}$ as the sum of an age-specific parameter independent of time α_x and a component given by the product of a time-varying parameter κ_t, reflecting the general level of mortality and the parameter β_x, representing how rapidly or slowly mortality at each age varies when the general level of mortality changes. The final term $E_{x,t}$ is the error term, assumed to be homoschedastic (with mean 0 and variance σ_ϵ^2).

On the basis of equation (1), if $\tilde{M}_{x,t}$ is the matrix holding the mean centred log-mortality rates, the LC model can be expressed as:

$$\tilde{M}_{x,t} = ln\left(M_{x,t}\right) - \alpha_x = \beta_x \kappa_t + E_{x,t}. \tag{2}$$

Following LC [7], the parameters β_x and κ_t can be estimated according to the Singular Value Decomposition (SVD) with suitable normality constraints. The LC model incorporates different sources of uncertainty, as discussed in LC [8], Appendix B: uncertainty in the demographic model and uncertainty in forecasting. The former can be incorporated by considering errors in fitting the original matrix of mortality rates, while forecast uncertainty arises from the errors in the forecast of the mortality index. In our contribution, we deal with the demographic component in order to consider the sensitivity of the estimated mortality index. In particular, the research consists in defining an experimental strategy to force the fulfilment of the homoschedasticity hypothesis and evaluate its impact on the estimated κ_t.

3 The experiment

The experimental strategy introduced above, with the aim of inducing the errors to satisfy the homoschedasticity hypothesis, consists in the following phases [11]. The error term can be expressed as follows:

$$\hat{E}_{x,t} = \tilde{M}_{x,t} - \hat{\beta}_x \hat{\kappa}_t, \tag{3}$$

i.e., as the difference between the matrix $\tilde{M}_{x,t}$, referring to the mean centred log-mortality rates and the product between β_x and κ_t deriving from the estimation of the LC model. The successive step consists in exploring the residuals by means of statistical indicators such as: range, interquartile range, mean absolute deviation (MAD) of a sample of data, standard deviation, box-plot, etc. Afterward, we proceed in finding those age groups that show higher variability in the errors. Once we have explored the residuals $\hat{E}_{x,t}$, we may find some non-conforming age groups. We rank them according to decreasing non-conformity, i.e., from the more widespread to the more homogeneous one. For each selected age group, it is possible to reduce the variability by dividing the entire range into several quantiles, leaving aside each time the fixed $\alpha\%$ of the extreme values. We replicate each running under the same conditions a large number of times (i.e., 1000). For each age group and for each percentile, we define a new error matrix. The successive runnings give more and more homogeneous error terms. By way of this experiment, we investigate the residual's heteroschedasticity deriving from two factors: the age group effect and the number of altered values in each age group. In particular, we wish to determine the hypothetical pattern of κ_t by increasing the homogeneity in the residuals. Thus, under these assumptions, we analyse the changes in κ_t that can be derived from every simulated error matrix. In particular, at each running we obtain a different error matrix $\overline{E}_{x,t}$, which is used for computing a new data matrix $\overline{M}_x, t$, from which it is possible to derive the correspondent κ_t. To clarify the procedure analytically, let us introduce the following relation:

$$\left[\widetilde{M}_{x,t} - \overline{E}_{x,t}\right] = \overline{M}_{x,t} \rightarrow \beta_x \kappa_t, \tag{4}$$

where $\overline{M}_{x,t}$ is a new matrix of data obtained by the difference between $\widetilde{M}_x, t$ (the matrix holding the raw mean centred log mortality rates) and $\overline{E}_{x,t}$ (the matrix holding the mean of altered errors). From $\overline{M}_{x,t}$, if β_x is fixed, we obtain the κ_t as the ordinary least square (OLS) coefficients of a regression model. We replicate the procedure by considering further non-homogenous age groups with the result of obtaining at each step a new κ_t. We mean to carry on the analysis by running a graphical exploration of the different κ_t patterns. Thus, we plot the experimental results so that all the κ_t's are compared with the ordinary one. Moreover, we compare the slope effect of the experimental κ_t through a numerical analysis.

4 Running the experiment

The experiment is applied to a data matrix holding the Italian mean centred log-mortality rates for the male population from 1950 to 2000 [6]. In particular, the rows of the matrix represent the 21 age groups [0], [1–4], [5–9], ..., [95–99] and the columns refer to the years 1950–2000. Our procedure consists of an analysis of the residuals' variability through some dispersion indices which help us to determine the age groups in which the model hypothesis does not hold (see Table 1).

We can notice that the residuals in the age groups 1–4, 5–9, 15–19 and 25–29 (written in bold character) are far from being homogeneous. Thus the age groups 1–4, 15–19, 5–9, 25–29 will be sequentially, and according to this order, entered in the experiment. Alongside the dispersion indices, we provide a graphical analysis by displaying the boxplot for each age group (Fig. 1), where on the x-axis the age groups are reported and on the y-axis the residuals' variability. If we look at the age groups 1–4 and 15–19 we can notice that they show the widest spread compared to the others. In particular, we perceive that for those age groups the range goes from -2 to 2.

For this reason, we explore to what extent the estimated κ_t are affected by such a variability. A way of approaching this issue can be found by means of the following replicating procedure, implemented in a Matlab routine. For each of the four age groups we substitute the extreme residual values with the following six quantiles: 5%, 10%, 15%, 20%, 25%, 30%. Then we generate 1000 random replications (for each age group and each interval). From the replicated errors (1000 times × 4 age groups × 6 percentiles) we compute the estimated κ_t (6 × 4 × 1000 times) and then we work out the 24 averages of the 1000 simulated κ_t. In Figure 2 we show the 24,000 estimated κ_t through a Plot-Matrix, representing the successive age groups entered in

Table 1. Different dispersion indices to analyse the residuals' variability

Age	IQ Range	MAD	Range	STD
0	0.107	0.059	0.300	0.075
1–4	*2.046*	*0.990*	*4.039*	*1.139*
5–9	*1.200*	*0.565*	*2.318*	*0.653*
10–14	0.165	0.083	0.377	0.099
15–19	*1.913*	*0.872*	*3.615*	*1.007*
20–24	0.252	0.131	0.510	0.153
25–29	*0.856*	*0.433*	*1.587*	*0.498*
30–34	0.536	0.250	1.151	0.299
35–39	0.240	0.186	0.868	0.239
40–44	0.787	0.373	1.522	0.424
45–49	0.254	0.126	0.436	0.145
50–54	0.597	0.311	1.290	0.367
55–59	0.196	0.151	0.652	0.187
60–64	0.247	0.170	0.803	0.212
65–69	0.207	0.119	0.604	0.147
70–74	0.294	0.171	0.739	0.202
75–79	0.230	0.117	0.485	0.133
80–84	0.346	0.187	0.835	0.227
85–89	0.178	0.099	0.482	0.124
90–94	0.307	0.153	0.701	0.186
95–99	0.071	0.042	0.220	0.051

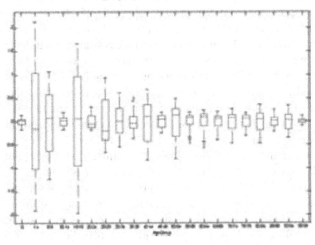

Fig. 1. Box-plot of the residuals'variability for each age group, starting from 0 up to 95–99

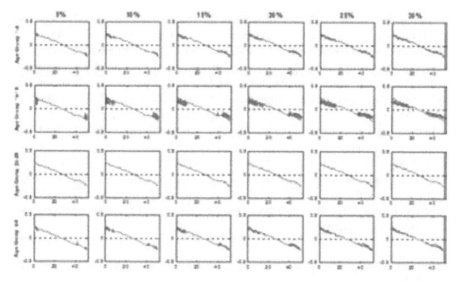

Fig. 2. κ_t resulting from different experimental conditions: the age group (on the rows) and the different percentiles (on the columns) effect

the experiment in the four rows and the successive increment in the percentage of outer values which have been transformed in the 6 columns. We can notice the different κ_t behaviour in the four rows as more age groups and percentiles are considered.

For better interpretation of these results, we have plotted a synthetic view of the resulting average of the 1000 κ_t under the 24 conditions (see Fig. 3) and compared them with the series derived by the traditional LC estimation.

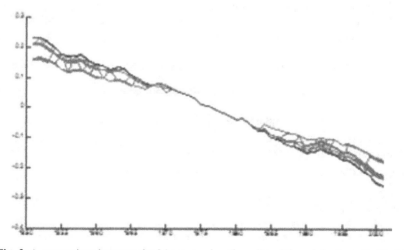

Fig. 3. A comparison between the 24 averaged κ_t (in red) and the original one (in black)

In Figure 3, where on the x-axis there are the years from 1950 to 2000 and on the y-axis there are the κ_t values, we represent the 24,000 κ_t grouped according to the 24 different experimental conditions. We can observe the impact on the κ_t series of the age groups change and of the increase of percentage of random values considered in the selected age groups. We can notice that the κ_t derived by the experiment (in red) tends to be flatter than the original one (drawn in black), i.e., there are changes in homogeneity on the κ_t for each of the four age groups. By comparing the ordinary κ_t to the simulated ones, we obtain information about the effect of the lack of homoschedasticity on the LC estimates. To what extent does it influence the sensitivity of the results? We note that the more homogenous the residuals are, the flatter the κ_t is. From an actuarial point of view, the κ_t series reveals an important result: when we use the new κ_t series to generate life tables, we find survival probabilities lower than the original ones. The effect of that on a pension annuity portfolio will be illustrated in the following application.

5 Numerical illustrations

In this section, we provide an application of the previous procedure for generating survival probabilities by applying them to a pension annuity portfolio in which beneficiaries enter the retirement state at the same time. In particular, having assessed the breaking of the homoschedasticity hypothesis in the Lee-Carter model, we intend to quantify its impact on given quantities of interest of the portfolio under consideration. The analysis concerns the dynamic behaviour of the financial fund checked year by year arising from the two flows in and out of the portfolio, the first consisting in the increasing effect due to the interest maturing on the accumulated fund and the second in the outflow represented by the benefit payments due in case the pensioners are still alive. Let us fix one of the future valuation dates, say corresponding to time κ, and consider what the portfolio fund is at this valuation date. As concerns the portfolio fund consistency at time κ, we can write [2]:

$$Z_\kappa = Z_{\kappa-1}\left(1 + i_\kappa^*\right) + N^\kappa P \quad \text{with} \quad \kappa = 1, 2, \cdots, n-1, \tag{5}$$

$$Z_\kappa = Z_{\kappa-1}\left(1 + i_\kappa^*\right) - N^\kappa R \quad \text{with} \quad \kappa = n, n+1, \cdots, \varpi - x, \tag{6}$$

where N^0 represents the number of persons of the same age x at contract issue $t = 0$ reaching the retirement state at the same time n, that is at the age $x + n$, and i_κ^* is a random financial interest rate in the time period $(k - 1, k)$. The formulas respectively refer to the accumulation phase and the annuitisation phase.

5.1 Financial hypotheses

Referring to the financial scenario, we refer to the interest rate as the rate of return on investments linked to the assets in which insurer invests. In order to compare, we consider both a deterministic interest rate and a stochastic interest rate framework.

As regards the former, we assume that the deposited portfolio funds earn at the financial interest rate fixed at a level of 3%. As regards the latter, we adopt the Vasicek model [12]. This stochastic interest rate environment seems to be particularly suitable for describing the instantaneous global rate of return on the assets linked to the portfolio under consideration, because of potential negative values. As is well known, this circumstance is not in contrast with the idea of taking into account a short rate reflecting the global investment strategy related to the portfolio [9].

5.2 Mortality hypotheses

As concerns the mortality model, we consider the survival probabilities generated by the above-described simulation procedure (hereinafter simulation method) and by the classical estimation of the Lee-Carter model (traditional method). In the former methodology we consider the κ_t series arising from the experiment. Following the Box-Jenkins procedure, we find that an ARIMA (0,1,0) model is more feasible for our time series. After obtaining the κ_t projected series, we construct the projected life table and then we extrapolate the probabilities referred to insured aged $x = 45$. In Figure 4 we report the survival probability distribution as a function of different LC estimation methods: the traditional and the simulation methods. We can notice that the pattern of simulated probabilities lies under the traditional probabilities. Moreover this difference increases as the projection time increases.

Thus, referring to the financial and the demographic stochastic environments described above, we evaluate the periodic portfolio funds. As regards the premium calculation hypotheses, we use two different assumptions (simulated LC, classical LC) and the fixed interest rate at 4%. We use the same mortality assumptions made in the premium calculation even for the portfolio fund dynamics from the retirement age on, i.e., which means to resort to a sort of homogeneity quality in the demographic

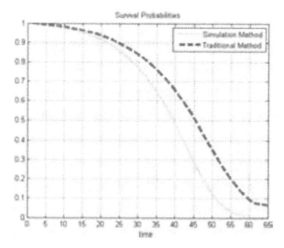

Fig. 4. Comparison between the two different methods for generating survival probabilities on the basis of the Lee-Carter model: traditional and simulation method

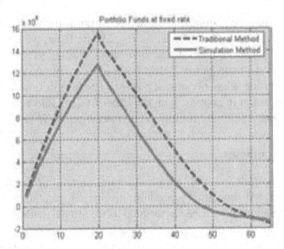

Fig. 5. Portfolio of 1000 pension annuities, $x = 45$, $t = 20$, $r = 100$. Fixed rate at 3%

description, in the light of the main results of [2]. In the following graphs (see Figs. 5 and 6) we represent the portfolio funds along with the potential whole contract life, i.e., both into the *accumulation phase* and into the *annuitisation phase*. The portfolio funds trend is calculated on a pension annuity portfolio referred to a cohort of $c = 1000$ beneficiaries aged $x = 45$ at time $t = 0$ and entering in the retirement state 20 years later, that is at age 65. The cash flows are represented by the constant premiums P, payable at the beginning of each year up to $t = 20$ in case the beneficiary is still alive at that moment (*accumulation phase*) and by the constant benefits $R = 100$ payable at the beginning of each year after $t = 20$ (*annuitisation phase*) in case the beneficiary is still alive at that moment. Figure 5 shows how the portfolio funds increase with

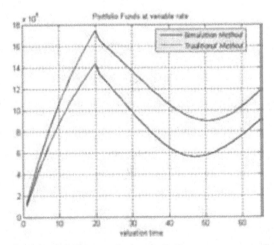

Fig. 6. Portfolio of 1000 pension annuities, $x = 45$, $t = 20$, $r = 100$. Stochastic rate of return

better survival probabilities. In particular, in this figure is represented the portfolio funds earning interest, term by term, at the fixed rate of return of 3%, from the time issue on. As a first result we find out that the portfolio fund amount is overestimated when the survival probabilities are calculated on the basis of the projection of the traditional LC estimation. On the basis of the results reported above, we can notice how the lack of homoschedasticity affects the portfolio risk assessment.

Finally, we evaluate the portfolio fund consistency from the contract issue on, adopting the Vasicek model for describing the instantaneous global rate of return on the assets linked to the portfolio under consideration. As in the previous case, Figure 6 shows that the traditional forecasting method blows up the portfolio funds amount both into the accumulation and into the annuitisation phases. Our findings are confirmed also in the case of the stochastic rate of return. For this reason, we provide evidence that the lack of homoschedasticity has a strong effect on the actuarial results.

6 Conclusions

The simulation procedure proposed in this paper is characterised by an experimental strategy to stress the fulfilment of the homoschedasticity hypothesis of the LC model. In particular, we simulate different experimental conditions to force the errors to satisfy the model hypothesis in a fitting manner. Besides, we develop the κ_t series for generating more realistic survival probabilities. Finally we measure the impact of the two different procedures for generating survival probabilities, using the traditional and simulation methods, on a portfolio of pension annuity. The applications, referred to the male population, show that the probabilities generated on the basis of the simulation procedure are lower than the probabilities obtained through the traditional methodology by the LC model. In particular, if we apply the simulated projections to a financial valuation of periodic portfolio funds of pension annuity portfolio, we can observe lower corresponding values than the traditional one, in both the so-called accumulation and annuitisation phases. Especially, we can notice more sizeable portfolio funds in the event of traditional methodology. In other words, the insurer's financial position would be overestimated by means of the traditional method in comparison with the simulation method. The results of the appraisal arise from the different behaviours of the residuals. In fact, in the traditional methodology, we get heteroschedasticity in the residuals for some age groups which can lead to more optimistic survival projections. On the other hand, on the basis of the simulation procedure, the final result shows how a more regular residual matrix leads to a flatter κ_t series according to the LC model hypothesis. This circumstance determines more pessimistic survival projections.

Acknowledgement. This paper is partially supported by MIUR grant "Problemi di proiezione del fenomeno sopravvivenza"(responsible: M. Russolillo).

References

1. Box, G.E.P., Draper, N.R.: Empirical Model Building and Response Surfaces. Wiley, New York (1987)
2. Coppola, M., D'Amato, V., Di Lorenzo, E., Sibillo, M.: Life office management perspectives by actuarial risk indexes. Int. Sci. Invest. Man. Finan. Innov. 2, 73–78 (2008)
3. Deaton, A., Paxson, C.: Mortality, Income, and Income Inequality Over Time in the Britain and the United States. Technical Report 8534, National Bureau of Economic Research, Cambridge, MA (2004): http://www.nber.org/papers/w8534
4. Eckart, C., Young, G.: The approximation of one matrix by another of lower rank. Psychometrika 1, 211–218 (1936)
5. Gabriel, K.R.: The biplot-graphic display of matrices with application to principal component analysis. Biometrika 58, 453–467 (1971)
6. Human Mortality Database. University of California, Berkeley (USA), and Max Planck Institute for Demographic Research (Germany). Available at www.mortality.org or www.humanmortality.de (referred to the period 1950 to 2000)
7. Lee, R.D., Carter, L.R.: Modelling and forecasting U.S. mortality. J. Am. Stat. Assoc. 87, 659–671 (1992)
8. Lee, R.D.: The Lee-Carter method for forecasting mortality, with various extensions and applications. N. Am. Actuarial J. 4, 80–91 (2000)
9. Olivieri, A., Pitacco, E.: Solvency requirements for pension annuities. J. Pension Econ. Finan. 2, 127–157 (2003)
10. Renshaw, A., Haberman, S.: Lee-Carter mortality forecasting: a parallel generalised linear modelling approach for England and Wales mortality projections. Appl. Stat. 52, 119–137 (2003)
11. Russolillo, M., Giordano, G.: A computational experiment to assess sensitivity in bilinear mortality forecasting. Proceedings of XXXI Convegno AMASES. Lecce, Italy (2007)
12. Vasicek, O.: An equilibrium characterisation of the term structure. J. Finan. Econ. 5, 177–188 (1977)

Estimating the volatility term structure

Antonio Díaz, Francisco Jareño, and Eliseo Navarro

Abstract. In this paper, we proceed to estimate term structure of interest rate volatilities, finding that these estimates depend significantly on the model used to estimate the term structure (Nelson and Siegel or Vasicek and Fong) and the heteroscedasticity structure of errors (OLS or GLS weighted by duration). We conclude in our empirical analysis that there are significant differences between these volatilities in the short (less than one year) and long term (more than ten years). Finally, we can detect that three principal components explain 90% of the changes in volatility term structure. These components are related to level, slope and curvature.

Key words: volatility term structure (VTS), term structure of interest rates (TSIR), GARCH, principal components (PCs)

1 Introduction

We define the term structure of volatilities as the relationship between the volatility of interest rates and their maturities. The importance of this concept has been growing over recent decades, particularly as interest rate derivatives have developed and interest rate volatility has become the key factor for the valuation of assets such as caplets, caps, floors, swaptions, etc. Moreover, interest rate volatility is one of the inputs needed to implement some term structure models such as those of Black, Derman and Toy [4] or Hull and White [12], which are particularly popular among practitioners.

However, one of the main problems concerning the estimation of the volatility term structure (VTS) arises from the fact that zero coupon rates are unobservable. So they must be previously estimated and this requires the adoption of a particular methodology. The problem of the term structure estimation is an old question widely analysed in the literature and several procedures have been suggested over the last thirty years.

Among the most popular methods are those developed by Nelson and Siegel [14] and Vasicek and Fong [17]. In Spain, these methods have been applied in Núñez [15] and Contreras et al. [7] respectively.

A large body of literature focuses on the bond valuation ability of these alternative models without analysing the impact of the term structure estimation method on second or higher moments of the zero coupon rates. Nevertheless, in this paper we focus on the second moment of interest rates derived from alternative term structure methods. So, the aim of this paper is to analyse if there are significant differences between the estimates of the VTS depending on the model used for estimating the term structure of interest rates (TSIR).

In this study we compare Nelson and Siegel [14], NS^O, Vasicek and Fong [17], VF^O, and both models using two alternative hypotheses about the error variance. First we assume homoscedasticity in the bond price errors and so does the term structure as estimated by OLS. Alternatively, a heteroscedastic error structure is employed estimating by GLS weighting pricing errors by the inverse of its duration, NS^G and VF^G.

In the literature, to minimise errors in prices is usual in order to optimise any model for estimating the TSIR. Nevertheless, this procedure tends to misestimate short-term interest rates. This is because an error in short-term bond prices induces an error in the estimation of short-term interest rates greater than the error in long-term interest rates produced by the same error in long-term bond prices. In order to solve this problem, it is usual to weight pricing errors by the reciprocal of bond Macaulay's duration.[1]

Once estimates of TSIR are obtained, we proceed to estimate interest rate volatilities using conditional volatility models (GARCH models).

In addition, we try to identify the three main components in the representation of the VTS for each model. Some researchers have studied this subject, finding that a small number of factors are able to represent the behaviour of the TSIR [3, 13, 15]. Nevertheless, this analysis has not been applied, to a large extent, to the VTS (except, e.g., [1]).

We apply our methodology to the VTS from estimates of the Spanish TSIR. The data used in this empirical analysis are the Spanish Treasury bill and bond prices of actual transactions from January 1994 to December 2006.

We show statistically significant differences between estimates of the term structure of interest rate volatilities depending on the model used to estimate the term structure and the heteroscedasticity structure of errors (NS^O, NS^G, VF^O and VF^G), mainly in the short-term (less than one year) and in the long-term (more than ten years) volatility. This inspection could have significant consequences for a lot of issues related to risk management in fixed income markets. On the other hand, we find three principal components (PCs) that can be interpreted as level, slope and curvature and they are not significantly different among our eight proposed models.

The rest of our paper is organised as follows. The next section describes the data used in this paper and the methodologies employed to estimate the TSIR: the Nelson and Siegel [14], NS, and Vasicek and Fong [17], VF, models. The third section describes the model used to estimate the term structure of volatilities. The fourth section analyses the differences in the VTS from our eight different models. Finally, the last two sections include a principal component analysis of VTS and, finally, summary and conclusions.

---

[1] This correction is usual in official estimations of the central banks [2].

2 Data

The database we use in this research contains daily volume-weighted averages of all the spot transaction prices and yields of all Spanish Treasury bills and bonds traded and registered in the dealer market or Bank of Spain's book entry system. They are obtained from annual files available at the "Banco de España" website.[2] We focus on 27 different maturities between 1 day and 15 years. Our sample runs from January 1994 to December 2006.

First of all, in order to refine our data, we have eliminated from the sample those assets with a trading volume less than 3 million euros (500 million pesetas) in a single day and bonds with term to maturity less than 15 days or larger than 15 years. Besides, in order to obtain a good adjustment in the short end of the yield curve, we always include in the sample the one-week interest rate from the repo market.

From the price (which must coincide with the quotient between effective volume and nominal volume of the transaction) provided by market, we obtain the yield to maturity on the settlement day. Sometimes this yield diverges from the yield reported by the market. Controlling for these conventions, we recalculate the yield using compound interest and the year basis ACT/ACT for both markets.[3]

We estimate the zero coupon bond yield curve using two alternative methods. The first one we use fits Nelson and Siegel's [14] exponential model for the estimation of the yield curve.[4] The second methodology is developed in Contreras et al. [7] where the Vasicek and Fong [17] term structure estimation method (VF^O) is adapted to the Spanish Treasury market. VF^O uses a non-parametric methodology based on exponential splines to estimate the discount function. A unique variable knot, which is located to minimise the sum of squared residuals, is used to adjust exponential splines.

With respect to the estimation methodology we apply both OLS and GLS. In the second case we adjust the bond price errors by the inverse of the bond Macaulay duration in order to avoid penalisation of more interest rate errors in the short end of the term structure.

In Figure 1 we illustrate the resulting estimations of the term structure in a single day depending on the weighting scheme applied to the error terms. It can be seen how assuming OLS or GLS affects mainly the estimates in the short and long ends of the TSIR even though in both cases we use the Nelson and Siegel model.[5]

---

[2] http://www.bde.es/banota/series.htm. Information reported is only about traded issues. It contains the following daily information for each reference: number of transactions, settlement day, nominal and effective trading volumes, maximum, minimum and average prices and yields.

[3] These divergences are due to simple or compound interest and a 360-day or 365-day year basis depending on the security term to maturity. http://www.bde.es/banota/actuesp.pdf

[4] See, for example, Díaz and Skinner [10], Díaz et al. [8] and Díaz and Navarro [9] for a more detailed explanation. Also, a number of authors have proposed extensions to the NS model that enhance flexibility [16].

[5] When using the Vasicek and Fong model, these differences are mainly shown in the short term. We observe differences depending on the model employed (VF or NS) even when the same error weighting scheme is used.

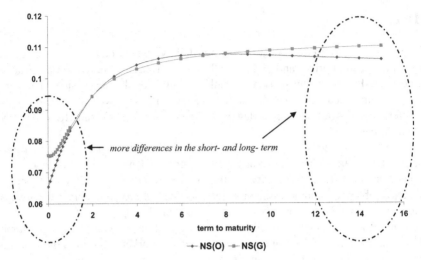

Fig. 1. TSIR estimated by NS^O and NS^G (01.07.1994)

In summary, we use four different estimation models: Nelson and Siegel [14], NS^G, and Vasicek and Fong [17], VF^G, which take into account residuals weighted by the reciprocal of maturity, and NS^O and VF^O, that is, with non-weighted residuals. These alternative estimation procedures provide the input of the subsequent functional principal component analysis.

3 GARCH models

VTS is an essential issue in finance, so it is important to have good volatility forecasts, which are based on the fact that volatility is time-varying in high-frequency data. In general, we can assume that there are several reasons to model and forecast volatility. First of all, it is necessary to analyse the risk of holding an asset[6] and the value of an option which depends crucially of the volatility of the underlying asset. Finally, more efficient estimators can be obtained if heteroscedasticity in the errors is handled properly.

In order to achieve these forecasts, extensive previous literature has used autoregressive conditional heteroscedasticity (ARCH) models, as introduced by Engle [11] and extended to generalized ARCH (GARCH) in Bollerslev [5]. These models normally improve the volatility estimates, to a large extent, compared with a constant variance model and they provide good volatility forecasts, so they are widely used in various branches of econometrics, especially in financial time series analysis. In fact, it is usually assumed that interest rate volatility can be accurately described by GARCH models.

---

[6] In fact, VaR estimates need as the main input the volatility of portfolio returns.

Taking into account a great variety of models (GARCH, ARCH-M, TGARCH, EGARCH ...), we identify the best one for each estimate of the TSIR: Nelson and Siegel (NS^O), Vasicek and Fong (VF^O) and both models weighted by duration (NS^G and VF^G), using Akaike Information Criterion (AIC). We select the ML aproach for estimating the GARCH parameters.[7] In particular, GARCH models fit very well when we use NS^O and VF^G. Nevertheless, T-GARCH and E-GARCH seem to be the best models for VF^O and NS^G estimations, respectively.

4 Differences in the volatility from different models

In this section we study the differences between the volatility term structure from different estimation models of the TSIR (NS^O, VF^O, NS^G and VF^G) and conditional volatility models (GARCH models in each previous case). In the first type of model, we obtain the historical volatility using 30-, 60- and 90-day moving windows and the standard deviation measure. We show the results with a 30-day moving window.

As a whole we can see a repeating pattern in the shape of the VTS: initially decreasing, then increasing until one to two years term and finally we can observe a constant or slightly decreasing interest rate volatility as we approach the long term of the curve. This is consistent with Campbell et al. [6], who argue that the hump of the VTS in the middle run can be explained by reduced forecast ability of interest rate movements at horizons around one year. They argue that there is some short-run forecastability arising from Federal Reserve operating procedures, and also some long-run forecastability from business-cycle effects on interest rates.

At first glance, volatility estimates for the different models used to estimate the interest rate term structure reveal how the methodology employed to estimate zero coupon bonds may have an important impact, both in level and shape, on the subsequent estimate of the VTS. This can be more clearly seen in Figure 2, where we show the VTS for our 8 cases on some particular days:

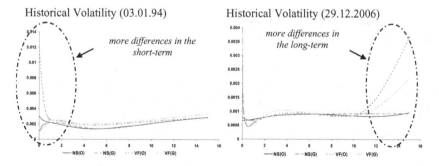

Fig. 2. Volatility Term Structure (VTS) among different models

---

[7] The selected model for each maturity and estimation model of the TSIR is available, but we do not exhibit these results so as to lighten the article.

In order to improve our analysis, we proceed to measure the average differences between volatility estimates using two alternative and different methods. We can detect that these differences seem to be higher in the short term (less than one year) and in the long term (more than ten years). Finally, we use some statistics to test whether volatility series have the same mean, median and variance (Table 1). In order to perform this analysis, we obtain an Anova-F test for the mean analysis, Kruskal-Wallis and van der Waerden test for the median analysis and, finally, a Levene and Brown-Forsythe test for analysing the significance of the VTS variance.

Table 1. Tests of equality of means, medians and variances among different models for each maturity

Test	Maturity (years)								
	0.25	0.5	0.75	1	3	5	10	12	15
F	477.8088^c	254.7131^c	97.67177^c	27.64625^c	0.653847	0.305357	2.175614^b	5.938809^c	175.7461^c
K-W	4349.893^c	2636.512^c	1151.682^c	433.3387^c	7.505526	3.589879	6.862232	44.18141^c	1141.098^c
vW	4454.727^c	2607.419^c	1100.201^c	379.1995^c	8.914100	4.463184	11.93060	55.15105^c	1170.865^c
L	194.7067^c	80.67102^c	20.38274^c	4.522192^c	0.106095	0.259973	4.682544^c	6.543890^c	165.7889^c
B-F	145.3684^c	58.94565^c	14.20114^c	2.158965^b	0.092483	0.217367	2.481528^b	3.134396^c	91.89411^c

$^a p < 0.10$, $^b p < 0.05$, $^c p < 0.01$
F: Anova-F Test, K-W: Kruskal-Wallis Test, vW: van der Waerden Test, L: Levene Test, B-F: Brown-Forsythe Test

On the one hand, statistics offer evidence against the null hypothesis of homogeneity for the shorter maturities (below to 1 year) and also for the longer maturities (more than 10 years), in mean and median. On the other hand, statistics to test for whether the volatility produced by the eight models has the same variance show the same results as mean and median analysis, that is, we find evidence against the null hypothesis for the shorter and longer maturities.

To summarise, this analysis shows that volatility estimates using different models and techniques display statistically significant differences, mainly in the shorter and longer maturities, as would be expected.

5 A principal component analysis of volatility term structure (VTS)

In this section, we try to reduce the dimensionality of the vector of 27 time series of historical/conditional volatilities,[8] working out their PCs, because this analysis is often used to identify the key uncorrelated sources of information.

This technique decomposes the sample covariance matrix or the correlation matrix computed for the series in the group. The row labelled "eigenvalue" in Table 2 reports the eigenvalues of the sample second moment matrix in descending order from left to right. We also show the variance proportion explained by each PC. Finally, we collect the cumulative sum of the variance proportion from left to right, that is, the variance proportion explained by PCs up to that order. The first PC is computed as a linear

---

[8] Note that we analyse volatility changes (see, for example, [3]).

Table 2. Main results of the principal component analysis

	NS^O	NS^G	VF^O	VF^G	GNS^O	GNS^G	GVF^O	GVF^G
	Historical Volatility				Conditional Volatility			
First Principal Component								
Eigenvalue	14.47963	14.42727	12.50790	14.95705	15.11595	14.52248	13.30651	15.23433
Var. prop.	0.536283	0.534343	0.463255	0.553965	0.559850	0.537870	0.492834	0.564234
Cum. prop.	0.536283	0.534343	0.463255	0.553965	0.559850	0.537870	0.492834	**0.564234**
Second Principal Component								
Eigenvalue	8.191949	7.305261	7.484512	6.623200	7.767501	7.520611	7.930251	6.769219
Var. prop.	0.303406	0.270565	0.277204	0.245304	0.287685	0.278541	0.293713	0.250712
Cum. prop.	0.839688	0.804909	0.740460	0.799268	**0.847535**	0.816411	0.786547	0.814946
Third Principal Component								
Eigenvalue	2.440719	2.549777	2.997161	2.321565	2.149763	2.366136	2.400861	2.120942
Var. prop.	0.090397	0.094436	0.111006	0.085984	0.079621	0.087635	0.088921	0.078553
Cum. prop.	**0.930085**	0.899345	0.851466	0.885252	0.927156	0.904045	0.875467	0.893499
Fourth Principal Component								
Eigenvalue	1.216678	1.318653	2.161253	1.388741	1.234067	1.270866	1.866241	1.237788
Var. prop.	0.045062	0.048839	0.080046	0.051435	0.045706	0.047069	0.069120	0.045844
Cum. prop.	**0.975147**	0.948184	0.931512	0.936687	0.972862	0.951114	0.944588	0.939343
Fifth Principal Component								
Eigenvalue	0.500027	0.711464	0.755749	0.812430	0.473576	0.690566	0.677145	0.788932
Var. prop.	0.018520	0.026351	0.027991	0.030090	0.017540	0.025577	0.025079	0.029220
Cum. prop.	**0.993667**	0.974534	0.959503	0.966777	0.990402	0.976691	0.969667	0.968563

G-before the name of the model indicates that we have used a GARCH model

combination of the series in the group with weights given by the first eigenvector. The second PC is the linear combination with weights given by the second eigenvector and so on.

We can emphasise the best values for the percentage of cumulative explained variance for each PC: 56% in case of GVF^G (first PC), 84% in case of GNS^O (second PC) and 93% (third PC), 97% (fourth PC) and 99% (fifth PC) in case of NS^O. Thus, the first five factors capture, at least, 97% of the variation in the volatility time series.

In this section, we can assert that the first three PCs are quite similar among different models. Particularly, the first PC keeps quasi constant over the whole volatility term structure (VTS) and the eight models. So, we can interpret it as the general level of the volatility (level or trend). With respect to the second PC, it presents coefficients of opposite sign in the short term and coefficients of the same sign in the long term, so this component can be interpreted as the difference between the levels of volatility between the two ends of the VTS (slope or tilt). Finally, the third PC shows changing signs of the coefficients, so this PC could be interpreted as changes in the curvature of the VTS (curvature). So, an important insight is that the three factors may be interpreted in terms of level, slope and curvature.

With regard to the fourth and fifth PC, they present some differences among each model; nevertheless, these PCs can be related with higher or lower hump of the VTS.

In order to finish this analysis, we want to test whether the first three PCs, which clearly reflect level, slope and curvature of the VTS, and the last two PCs are different among our eight models (historical and conditional volatilities).

Considering the results from Table 3, we can assert that statistics related to differences in mean evidence homogeneity in mean for our eight models as we cannot

Table 3. Tests of equality of means, medians and variances among different models

TEST	PC1	PC2	PC3	PC4	PC5
F	0.012749	0.056012	0.020015	0.179951	0.024021
K-W	1.016249	2.452214	3.810190	11.82140	55.13159 [c]
vW	0.518985	0.795634	2.032438	8.070648	45.21040 [c]
L	4.033919 [c]	23.92485 [c]	16.57419 [c]	66.74642 [c]	67.33491 [c]
B-F	4.064720 [c]	23.87826 [c]	16.51119 [c]	65.80991 [c]	67.06584 [c]

[a] $p < 0.10$, [b] $p < 0.05$, [c] $p < 0.01$
F: Anova-F Test, K-W: Kruskal-Wallis Test, vW: van der Waerden Test, L: Levene Test,
B-F: Brown-Forsythe Test

reject the null hypothesis. In case of differences in median, we find evidence against the null hypothesis of equal medians for the fifth PC. Nevertheless, the other PCs offer evidence in favour of the null hypothesis.

On the other hand, statistics to test whether the PC variance produced by our eight models is the same or not also appear in Table 3. For all the PCs, these statistics offer strong evidence against the null hypothesis.

Summarising, in this section we have concluded that the first three PCs can be related to level, slope and curvature of the VTS and, besides, these PCs are not significantly different in mean and median among our eight models. Nevertheless, PC4 and PC5 are significantly different between our models.

6 Conclusions

This paper aims to provide new insights into the behaviour of the VTS of interest rates by using historical volatility estimates from four different models of the term structure of interest rate (TSIR) and applying alternative conditional volatility specifications (using GARCH models) from 1994 to 2006. We have used the mentioned models, and we have worked out the volatility time series using 30-, 60- and 90-day moving windows in order to construct the VTS.

First of all, the results of our analysis show that there are statistically significant differences between estimates of the term structure of interest rate volatilities depending on the model used to estimate the term structure and the heteroscedasticity structure of errors (NS^O, NS^G, VF^O and VF^G), mainly in the short term (less than one year) and in the long term (more than ten years), but these differences do not depend on procedures to estimate the VTS. Secondly, the previous evidence suggests that the dynamics of term structures of volatilities can be well described by relatively few common components. The possible interpretation of these principal components in terms of level, slope and curvature can describe how the VTS shifts or changes shape in response to a shock on a PC.

We find that the first three PCs are quite similar among different models and they can be identified as trend, tilt and curvature. Regarding the fourth and fifth PCs, they can be related with higher or lower hump of the VTS. Also, the first three PCs are not significantly different in mean and median among our eight models. Nevertheless, PC4 and PC5 are significantly different between our models.

Acknowledgement. We acknowledge the financial support provided by *Junta de Comunidades de Castilla-La Mancha* grant PCI08-0089 and *Ministerio de Educación y Ciencia* grant ECO2008-05551/ECON which is partially supported by FEDER funds.

References

1. Abad, P., Novales, A.: Credit, liquidity and market risk in the term structure of swaps in pesetas. Rev. Econ. Finan. 5, 67–85 (2005)
2. Bank of International Settlements: Zero-coupon yield curves: technical documentation. BIS Papers 25, October (2005)
3. Benito, S., Novales, A.: A factor analysis of volatility across the term structure: the Spanish case. Rev. Econ. Finan. 13, 8–27 (2007).
4. Black, F., Dermand, E., Toy, W.: A one-factor model of interest rates and its application to treasury bond options. Finan. Anal. J. 46, 33–39 (1990)
5. Bollerslev, T.: Generalized autoregressive conditional heteroskedasticity. J. Econ. 31, 307–327 (1986)
6. Campbell, J.Y., Lo, A., Mackinlay, A.C.: The Econometrics of Financial Markets. Princeton University Press (1997)
7. Contreras, D., Ferrer, R., Navarro, E., Nave, J.M.: Análisis factorial de la estructura temporal de los tipos de interés en España. Revista Española de Financiación y Contabilidad 25, 139–160 (1996)
8. Díaz, A., Merrick, J.J. Jr., Navarro, E.: Spanish treasury bond market liquidity and volatility pre- and post-European Monetary Union. J. Banking Finan. 30, 1309–1332 (2006)
9. Díaz, A., Navarro, E.: Yield spread and term to maturity: default vs liquidity. Eur. Finan. Man. 8, 449–477 (2002)
10. Díaz, A., Skinner, F.: Estimating corporate yield curves. J. Fixed Income 11, 95–103 (2001)
11. Engle, R.F.: Autoregressive conditional heteroskedasticity with estimates of the variance of U.K. inflation. Econometrica 50, 987–1008 (1982)
12. Hull, J., White, A.: The pricing of options on assets with stochastic volatility. J. Finan. 42, 281–300 (1987)
13. Navarro, E., Nave, J.M.: A two-factor duration model for interest rate risk management. Investigaciones Económicas 21, 55–74 (1997)
14. Nelson, C.R., Siegel, A.F.: Parsimonious modeling of yield curves. J. Bus. 60, 473–489 (1987)
15. Núñez, S.: Estimación de la estructura temporal de los tipos de interés en España: elección entre métodos alternativos. Documentos de trabajo del Banco de España 22, 5–51 (1995)
16. Svensson, L.E.: Estimating and interpreting forward interest rates: Sweden 1992–1994. Centre for Economic Policy Research. Discussion Paper 1051 (1994)
17. Vasicek, O.A., Fong, H.G.: Term structure modeling using exponential splines. J. Finan. 37, 339–348 (1982)

Exact and approximated option pricing in a stochastic volatility jump-diffusion model

Fernanda D'Ippoliti, Enrico Moretto, Sara Pasquali, and Barbara Trivellato

Abstract. We propose a stochastic volatility jump-diffusion model for option pricing with contemporaneous jumps in both spot return and volatility dynamics. The model admits, in the spirit of Heston, a closed-form solution for European-style options. To evaluate more complex derivatives for which there is no explicit pricing expression, such as barrier options, a numerical methodology, based on an "exact algorithm" proposed by Broadie and Kaya, is applied. This technique is called exact as no discretisation of dynamics is required. We end up testing the goodness of our methodology using, as real data, prices and implied volatilities from the DJ Euro Stoxx 50 market and providing some numerical results for barrier options and their Greeks.

Key words: stochastic volatility jump-diffusion models, barrier option pricing, rejection sampling

1 Introduction

In recent years, many authors have tried to overcome the Heston setting [11]. This is due to the fact that the ability of stochastic volatility models to price short-time options is limited [1,14]. In [2], the author added (proportional) log-normal jumps to the dynamics of spot returns in the Heston model (see [10] for log-uniform jumps) and extended the Fourier inversion option pricing methodology of [11,15] for European and American options. This further improvement has not been sufficient to capture the rapid increase of volatility experienced in financial markets. One documented example of this feature is given by the market stress of Fall 1987, when the volatility jumped up from roughly 20% to over 50%. To fill this gap, the introduction of jumps in volatility has been considered the natural evolution of the existing diffusive stochastic volatility models with jumps in returns. In [9], the authors recognised that "although the motivation for jumps in volatility was to improve on the dynamics of volatility, the results indicate that jumps in volatility also have an important cross-sectional impact on option prices".

In this context, we formulate a stochastic volatility jump-diffusion model that, in the spirit of Heston, admits a closed-form solution for European-style options. The evolution of the underlying asset is driven by a stochastic differential equation with jumps that contains two diffusion terms: the first has constant volatility, as in the Black and Scholes (B&S) model [4], while the latter is of the Heston type. The dynamics of the volatility follow a square-root process with jumps. We suppose that the arrival times of both jumps are concurrent, hence we will refer to our model as a stochastic volatility with contemporaneous jumps (*SVCJ*) model. We claim that two diffusion terms in the dynamics of spot returns make our model more flexible than the Heston one.

Valuation of non-European options usually requires numerical techniques; in most cases some kind of discretisation is necessary so that a pricing bias is present. To avoid this flaw, we opt for the "exact simulation" approach developed by Broadie and Kaya (B&K) [5, 6] for stochastic volatility and other affine jump-diffusion models. This method is based on both a Fourier inversion technique and some conditioning arguments so to simulate the evolution of the value and the variance of an underlying asset. Unlike B&K's algorithm, to determine the integral of the variance, we replace the inverse transform method with a rejection sampling technique. We then compare the results of the closed-form expression for European-style option prices with their approximated counterparts using data from the DJ Euro Stoxx 50 derivative market. Having found that the modified algorithm returns reliable values, we determine prices and Greeks for barrier options for which no explicit formula exists.

2 Stochastic volatility jump-diffusion model

Let $(\Omega, \mathcal{F}, \mathbf{Q})$ be a complete probability space where $\mathbf{Q}$ is a risk-neutral probability measure and consider $t \in [0, T]$. We suppose that a bidimensional standard Wiener process $W = (W_1, W_2)$ and two compound Poisson processes Z_S and Z_v are defined. We assume that W_1, W_2, Z_S and Z_v are mutually independent. We suppose that

$$dS(t) = S(t^-)\left[(r - \lambda j_S)\, dt + \sigma_S\, dW_1(t) + \xi\sqrt{v(t^-)}\, dW_2(t) + dZ_S(t)\right], \quad (1)$$

$$dv(t) = k^*(\theta^* - v(t^-))\, dt + \sigma_v\sqrt{v(t^-)}\, dW_2(t) + dZ_v(t), \quad (2)$$

where $S(t)$ is the underlying asset, $\sqrt{v(t)}$ is the volatility process, and parameters r, σ_S, ξ, k^*, θ^* and σ_v are real constants (r is the riskless rate). The processes $Z_S(t)$ and $Z_v(t)$ have the same constant intensity $\lambda > 0$ (annual frequency of jumps). The process $Z_S(t)$ has log-normal distribution of jump sizes; if J_S is the relative jump size, then $\log(1+J_S)$ is distributed according to the $\mathcal{N}\left(\log(1 + j_S) - \frac{1}{2}\delta_S^2, \delta_S^2\right)$ law, where j_S is the unconditional mean of J_S. The process $Z_v(t)$ has an exponential distribution of jump sizes $J_v > 0$ with mean j_v. Note that $J_S \in (-1, +\infty)$ implies that the stock price remains positive for all $t \in [0, T]$. The variance $v(t)$ is a mean reverting process with jumps where k^*, θ^* and σ_v are, respectively, the speed of adjustment, the long-run mean and the variation coefficient. If $k^*, \theta^*, \sigma_v > 0$, $2k^*\theta^* \geq \sigma_v^2$, $v(0) \geq 0$ and $J_v > 0$, then the process $v(t)$ is positive for all $t \in [0, T]$ with probability 1 (see [12] in the no-jump case) and captures the large positive outliers in volatility documented in [3]. Jumps in both asset price and variance occur concurrently according to the counting process $N(t)$. The instantaneous correlation between S and v, when a jump does not occur, is $\rho(t) = \xi\sqrt{v(t)}/\left(\sigma_S^2 + \xi^2 v(t)\right)$, depends on two parameters and is stochastic because it contains the level of volatility in t. We claim that this improves the Heston model in which correlation between the underlying and volatility is constant. Further, ξ in $\rho(t)$ gauges the B&S constant volatility component $\left(\sigma_S^2\right)$ with the one driven by $v(t)$ (see [7]). Lastly, the instantaneous variance of returns $\sigma_S^2 + \xi^2 v(t)$ is uniformly bounded from below by a positive constant, and this fact proves to be useful in many control and filtering problems (see [13]).

3 Closed formula for European-style options

By analogy with B&S and Heston formulæ, the price of a call option with strike price K and maturity T written on the underlying asset S is

$$C(S, v, t) = S P_1(S, v, t) - K e^{-r(T-t)} P_2(S, v, t), \tag{3}$$

where $P_j(S, v, t)$, $j = 1, 2$, are cumulative distribution functions (cdf). In particular, $\tilde{P}_j(z) := P_j(e^z)$, $z \in \mathbf{R}$, $j = 1, 2$, are the conditional probabilities that the call option expires in-the-money, namely,

$$\tilde{P}_j(\log S, v, t; \log K) = \mathbf{Q}\{\log S(T) \geq \log K \,|\, \log S(t) = S, v(t) = v\}. \tag{4}$$

Using a Fourier transform method one gets

$$\tilde{P}_j(\log S, v, t; \log K) =$$
$$\frac{1}{2} + \frac{1}{\pi} \int_0^\infty \mathcal{R}\left(\frac{e^{-iu_1 \log K}\,\varphi_j(\log S, v, t; u_1, 0)}{iu_1}\right) du_1, \tag{5}$$

where $\mathcal{R}(z)$ denotes the real part of $z \in \mathbf{C}$, and $\varphi_j(\log S, v, t; u_1, u_2)$, $j = 1, 2$, are characteristic functions. Following [8] and [11], we guess

$$\varphi_j(Y, v, t; u_1, u_2) =$$
$$\exp\left[C_j(\tau; u_1, u_2) + J_j(\tau; u_1, u_2) + D_j(\tau; u_1, u_2)v + iu_1 Y\right], \tag{6}$$

where $Y = \log S$, $\tau = T - t$ and $j = 1, 2$. The explicit expressions of the characteristic functions are obtained to solutions to partial differential equations (PDEs) (see [7] for details); densities $\tilde{p}_j(Y, v, t; \log K)$ of the distribution functions $\tilde{F}_j(Y, v, t; \log K) = 1 - \tilde{P}_j(Y, v, t; \log K)$ are then

$$\tilde{p}_j(Y, v, t; \log K) =$$
$$-\frac{1}{\pi} \int_0^\infty \mathcal{R}\left(-e^{-iu_1 \log K}\varphi_j(Y, v, t; u_1, 0)\right) du_1, \quad j = 1, 2. \tag{7}$$

4 Generating sample paths

Following [5,6], we now give a Monte Carlo simulation estimator to compute option price derivatives without discretising processes S and v. The main idea is that, by appropriately conditioning on the paths generated by the variance and jump processes, the evolution of the asset price can be represented as a series of log-normal random variables. This method is called *Exact Simulation Algorithm (ESA) for the SVCJ Model*. In Step 3 of this method, the sampling from a cdf is done through an inverse transform method. Since the inverse function of the cdf is not available in closed form, the authors apply a Newton method to obtain a value of the distribution. To avoid the inversion of the cdf, we use a rejection sampling whose basic idea is to sample from a known distribution proportional to the real cdf (see [7]). This modification involves an improvement of efficiency of the algorithm, as the numerical results in Table 2 show.

To price a path-dependent option whose payoff is a function of the asset price vector $(S(t_0), \ldots, S(t_M))$ ($M = 1$ for a path-independent option), let $0 = t_0 < t_1 < \ldots < t_M = T$ be a partition of the interval $[0, T]$ into M possibly unequal segments of length $\Delta t_i := t_i - t_{i-1}$, for $i = 1, \ldots, M$. Now consider two consecutive time steps t_{i-1} and t_i on the time grid and assume $v(t_{i-1})$ is known. The algorithm can be summarised as follows:

Step 1. Generate a Poisson random variable with mean $\lambda \Delta t_i$ and simulate n_i, the number of jumps. Let $\tau_{i,1}$ be the time of the first jump after t_{i-1}. Set $u := t_{i-1}$ and $t := \tau_{i,1}$ ($u < t$). If $t > t_i$, skip Steps 5. and 6.

Step 2. Generate a sample from the distribution of $v(t)$ given $v(u)$, that is, a non-central chi-squared distribution.

Step 3. Generate a sample from the distribution of $\int_u^t v(q)dq$ given $v(u)$ and $v(t)$: this is done by writing the conditional characteristic function of the integral and then the density function. We simulate a value of the integral applying the rejection sampling.

Step 4. Recover $\int_u^t \sqrt{v(q)}dW_2(q)$ given $v(u)$, $v(t)$ and $\int_u^t v(q)dq$.

Step 5. If $t \leq t_i$, generate J_v by sampling from an exponential distribution with mean j_v. Update the variance value by setting $\tilde{v}(t) = v(t) + J_v^{(1)}$, where $J_v^{(1)}$ is the first jump size of the variance.

Step 6. If $t < t_i$, determine the time of the next jump $\tau_{i,2}$ after $\tau_{i,1}$. If $\tau_{i,2} \leq t_i$, set $u := \tau_{i,1}$ and $t := \tau_{i,2}$. Repeat the iteration Steps 2–5. up to t_i. If $\tau_{i,2} > t_i$, set $u := \tau_{i,1}$ and $t := t_i$. Repeat once the iteration Steps 2–4.

Step 7. Define the average variance between t_{i-1} and t_i as

$$\bar{\sigma}_i^2 = \frac{n_i \delta_S^2 + \sigma_S^2 \Delta t_i}{\Delta t_i}, \tag{8}$$

and an auxiliary variable

$$\beta_i = e^{n_i \log(1+js) - \lambda js \Delta t_i - \frac{\xi^2}{2} \int_{t_{i-1}}^{t_i} v(q)dq + \xi \int_{t_{i-1}}^{t_i} \sqrt{v(q)}dW_2(q)}. \tag{9}$$

Using (8) and (9), the value $S(t_i)$ given $S(t_{i-1})$ can be written as

$$S(t_i) = S(t_{i-1})\beta_i \exp\left\{\left(r - \frac{\overline{\sigma}_i^2}{2}\right)\Delta t_i + \overline{\sigma}_i\sqrt{\Delta t_i}\,R\right\},\qquad(10)$$

where $R \sim \mathcal{N}(0, 1)$, hence $S(t_i)$ is a lognormal random variable.

5 Barrier options and their Greeks

To price barrier options, we choose to apply the conditional Monte Carlo (CMC) technique, first used in finance in [16]. This method is applicable to path-dependent derivatives whose prices have a closed-form solution in the B&S setting. It exploits the following variance-reducing property of conditional expectation: for any random variables X and Y, $\mathbf{var}[\mathbf{E}[X|Y]] \leq \mathbf{var}[X]$, with strict inequality excepted in trivial cases.

Now, we illustrate the CMC method for discrete barrier options. Let $C(S(0), K, r, T, \sigma)$ denote the B&S price of a European call option with constant volatility σ, maturity T, strike K, written on an asset with initial price $S(0)$. The discounted payoff for a discrete knock-out option with barrier $H > S(0)$ is given by

$$f(X) = e^{-rT}(S(T) - K)^+ \mathbf{1}_{\{\max_{1\leq i\leq M} S(t_i) < H\}},\qquad(11)$$

where $S(t_i)$ is the asset price at time t_i for a time partition $0 = t_0 < t_1 < \ldots < t_M = T$. Using the law of iterated expectations, we obtain the following unconditional price of the option

$$\mathbf{E}\left[e^{-rT}(S(T) - K)^+ \mathbf{1}_{\{\max_{1\leq i\leq M} S(t_i) < H\}}\right]$$

$$= \mathbf{E}\left[\mathbf{E}\left[e^{-rT}(S(T) - K)^+ \mathbf{1}_{\{\max_{1\leq i\leq M} S(t_i) < H\}}\,\Big|\,\int_0^T v(q)dq, \int_0^T v(q)dW_2(q), J_S\right]\right]$$

$$= \mathbf{E}\left[C\left(S(0)\beta_M, K, r, T, \overline{\sigma}_M\right)\mathbf{1}_{\{\max_{1\leq i\leq M} S(t_i) < H\}}\right],\qquad(12)$$

where $\overline{\sigma}_M$ and β_M are defined in (8) and (9), respectively.

This approach can also be used to generate an unbiased estimator for delta, gamma and rho, exploiting the likelihood ratio (LR) method.

Suppose that $p \in \mathbf{R}^n$ is a vector of parameters with probability density $g_p(X)$, where X is a random vector that determines the discounted payoff function $f(X)$ defined in (11). The option price is given by

$$\alpha(p) = \mathbf{E}[f(X)],\qquad(13)$$

and we are interested in finding the derivative $\alpha'(p)$. From (13), one gets

$$\alpha'(p) = \frac{d}{dp}\mathbf{E}[f(X)] = \int_{R^n} f(x)\frac{d}{dp}g_p(x)dx = \mathbf{E}\left[f(X)\frac{g_p'(x)}{g_p(x)}\right].\qquad(14)$$

The expression $f(X)\frac{g_p'(x)}{g_p(x)}$ is an unbiased estimator of $\alpha'(p)$ and the quantity $\frac{g_p'(x)}{g_p(x)}$ is called score function. Note that this latter does not depend on $f(X)$ and that the Greek for each option is computed according to which quantity is considered a parameter in the expression of g.

Consider a discrete knock-out barrier option whose payoff is given by (11). From (14), it follows that the LR estimator for the option Greeks are given by the product of $f(X)$ and the score function. The score function is determined by using the key idea of CMC method: by appropriately conditioning on the paths generated by the variance and jump processes, the evolution of the asset price S is a log-normal random variable (see (10)), hence its conditional density is

$$g(x) = \frac{1}{x\overline{\sigma}_i\sqrt{\Delta t_i}}\phi(d_i(x)), \tag{15}$$

where $\overline{\sigma}_i$ is defined in (8), $\phi(\cdot)$ is the standard normal density function and

$$d_i(x) = \frac{\log\left(\frac{x}{S(t_{i-1})\beta_i}\right) - (r - \frac{1}{2}\overline{\sigma}_i^2)\Delta t_i}{\overline{\sigma}_i\sqrt{\Delta t_i}}. \tag{16}$$

Now, to find the estimator of delta and gamma, i.e., the first and the second derivative with respect to the price of the underlying asset, respectively, we let $p = S$ in (14) and compute the derivative of g in $S(0)$. After some algebra, we have

$$\left(\frac{\partial g(x)}{\partial S}\right)_{S=S(0)} = \frac{d_i(x)\phi(d_i(x))}{xS(0)\overline{\sigma}_i^2\Delta t_i}. \tag{17}$$

Dividing this latter by $g(x)$ and evaluating the expression at $x = S(t_1)$, we have the following score function for the LR delta estimator

$$\frac{d_1}{S(0)\overline{\sigma}_1\sqrt{\Delta t_1}}, \tag{18}$$

where d_i is defined in (16), and $\overline{\sigma}_i$ in (8). The case of the LR gamma estimator is analogous. The estimator of delta is given by

$$e^{-rT}(S(T) - K)^+\mathbf{1}_{\{\max_{1\leq i\leq M} S(t_i)<H\}}\left(\frac{d_1}{S(0)\overline{\sigma}_1\sqrt{\Delta t_1}}\right), \tag{19}$$

and the estimator of gamma is

$$e^{-rT}(S(T) - K)^+\mathbf{1}_{\{\max_{1\leq i\leq M} S(t_i)<H\}}\left(\frac{d_1^2 - d_1\overline{\sigma}_1\sqrt{\Delta t_1} - 1}{S^2(0)\overline{\sigma}_1^2\Delta t_1}\right). \tag{20}$$

To compute the estimator of rho, it is sufficient to compute the derivative of g with respect to r,

$$e^{-rT}(S(T) - K)^+\mathbf{1}_{\{\max_{1\leq i\leq M} S(t_i)<H\}}\left(-T + \sum_{i=1}^{M}\frac{d_i\sqrt{\Delta t_i}}{\overline{\sigma}_i}\right). \tag{21}$$

6 Numerical results

In this section, we apply our model to the DJ Euro Stoxx 50 market (data provided by Banca IMI, Milan), using the set of parameters reported in Table 1. These parameters have been chosen in order to test the efficiency of our algorithm and obtain a good approximation of market volatilities. Model calibration is beyond the scope of this paper and is left for further research.

Table 1. Values of parameters of the models (1) and (2)

θ^*	k^*	σ_S	σ_v	ξ	λ	j_S	δ_S	j_v
0.175	0.25	0.08	0.2	-0.4	0.05	0.025	0.02	0.03

The Dow Jones Euro Stoxx 50 (DJ50) 'blue-chip' index covers the fifty EuroZone largest sector leaders whose stocks belong to the Dow Jones Euro Stoxx Index. DJ50's option market is very liquid and ranges widely in both maturities (from one month to ten years) and strike prices (moneyness from 90% up to 115%). It is worth noting that indexes carry dividends paid by companies so that a dividend yield d has to be properly considered by subtracting it from the drift term in the dynamics of S. Volatilities in Table 2 (column 2) represent the term $\sqrt{\sigma_S^2 + \xi^2 v(0)}$ (the instantaneous variance of spot return at $t = 0$, and not simply σ_S as in the B&S model), where $v(0)$ is the initial value of the stochastic volatility dynamics. It follows that we can obtain $v(0)$ from $v(0) = \left(\sigma_{MKT}^2 - \sigma_S^2\right)/\xi^2$, where σ_{MKT} is the market volatility.

Exact versus approximated pricing

We present some numerical comparisons of the *ESA* described in Section 4 and other simulation methods. For this purpose, we use European call options on November 23, 2006; relevant data are shown in Table 2. We compare prices derived with different methods: the closed formula (3) (column 4), the *ESA* modified with the rejection sampling (column 5), the *ESA* proposed in [5, 6] (column 6), and a Monte Carlo estimator (see (5) in [6]) (column 7). For the *ESA*s, we simulate 100,000 variance paths and 1000 price paths conditional on each variance path and jumps. Prices in column 5 are very similar to those obtained with the closed formula (3) and improve the approximation obtained using the original *ESA*. Our results also confirm that *ESA* is more efficient than a standard Monte Carlo approach, as stated in [5, 6]. The time needed to obtain each price with the *ESA* in column 5 is about 545 seconds with a FORTRAN code running on an AMD Athlon MP 2800+, 2.25 GHz processor. This computational time is shorter than that reported in [5, 6] for a comparable number of simulations.

This is an encouraging result for pricing options that do not have closed-form formulæ such as barrier options and Greeks.

Table 2. Comparison among prices of European options with spot price $S(0) = 4116.40$, time to maturity 1 year, riskless rate $r = 3.78\%$ and dividend yield $d = 3.37\%$ on November 23, 2006

Moneyness %	Strike K	mkt vol σ_{MKT}	Closed formula (3)	ESA (rejection)	ESA (B&K)	MC price
90.0	3704.76	0.1780	530.43	530.65	531.18	529.42
95.0	3910.58	0.1660	383.14	383.65	383.31	382.38
97.5	4013.49	0.1600	316.19	316.61	316.76	315.50
100.0	4116.40	0.1550	255.95	255.98	256.22	255.29
102.5	4219.31	0.1500	201.77	201.45	201.42	201.07
105.0	4322.22	0.1450	154.22	154.62	153.92	153.43
110.0	4528.04	0.1380	83.71	83.45	83.32	82.60
115.0	4733.86	0.1320	40.13	40.03	38.84	38.59

Valuation of barrier options and Greeks

We provide some numerical results on the valuation of discrete "up-and-out" barrier call options whose prices are given by (12), with $M = 2$ monitoring times, and two different barriers $H = \{5000; 5500\}$. To have a sort of benchmark, we use the same data as the European case. Barrier option prices are computed simulating 1200 volatility paths and 40 price paths conditional on each variance path and jumps.

By comparing the prices of European and barrier options ($H = 5000$) with the same moneyness (see Tables 2 and 3), the relative change in price ranges from 24% (moneyness 90%) to 70% (moneyness 115%). The higher the moneyness, the less likely the option will expire with a positive payoff, either because the underlying hits the barrier before the maturity, knocking-down the option, or because the option is, at expiration, out-of-the-money. This feature is also present when the barrier level changes, as in Table 3.

Table 3. Prices of barrier options with two different barriers, spot price $S(0) = 4116.40$, time to maturity 1 year, riskless rate $r = 3.78\%$ and dividend yield $d = 3.37\%$ on November 23, 2006

Moneyness %	Strike K	mkt vol σ_{MKT}	$H = 5000$	$H = 5500$
90.0	3704.76	0.1780	415.0204	519.1443
95.0	3910.58	0.1660	298.2722	374.2468
97.5	4013.49	0.1600	239.5962	308.4141
100.0	4116.40	0.1550	189.5186	249.8215
102.5	4219.31	0.1500	142.6677	196.8798
105.0	4322.22	0.1450	103.9032	149.7763
110.0	4528.04	0.1380	45.5808	79.8648
115.0	4733.86	0.1320	12.1330	35.8308

Finally, Table 4 reports delta and gamma for European and barrier options for different strikes. The overall time required to obtain each barrier option price (Table 3) along with its Greeks (Table 4) is about 1600 seconds.

Table 4. Simulation estimates of Greeks for European and barrier options with the following option parameters: barrier $H = 5000$, spot price $S(0) = 4116.40$, time to maturity $T - t = 1$ (year), riskless rate $r = 3.78\%$ and dividend yield $d = 3.37\%$ on November 23, 2006

Moneyness %	Strike K	Delta (European)	Delta (barrier)	Gamma (European)	Gamma (barrier)
97.50	4013.49	0.61919	0.263209	0.00054727	0.0006277
100.00	4116.40	0.56006	0.241053	0.00060312	0.0005138
102.50	4219.31	0.49546	0.210052	0.00065031	0.0003766
105.00	4322.22	0.42850	0.182986	0.00068271	0.0002149

7 Conclusions

An alternative stochastic volatility jump-diffusion model for option pricing is proposed. To capture all empirical features of spot returns and volatility, we introduce a jump component in both dynamics and we suppose that jumps occur concurrently. This pricing model admits, in the spirit of Heston, a closed-form solution for European-style options. To evaluate path-dependent options, we propose a modified version of the numerical algorithm developed in [5, 6] whose major advantage is the lack of discretisation bias. In particular, we replace the inversion technique proposed by the authors with a rejection sampling procedure to improve the algorithm efficiency. We firstly apply our methodology to price options written on the DJ Euro Stoxx 50 index, and then we compare these prices with values obtained applying the closed-form expression, the Broadie and Kaya algorithm and a standard Monte Carlo simulation (see Table 2). The numerical experiments confirm that prices derived with the *ESA* modified by the rejection sampling provide the most accurate approximation with respect to the closed formula values. On the basis of this result, we perform the valuation of barrier options and Greeks whose values cannot be expressed by explicit expressions.

References

1. Bakshi, G., Cao., C., Chen, Z.: Empirical performance of alternative option pricing models. J. Finan. 52, 2003–2049 (1997)
2. Bates, D.S.: Jumps and stochastic volatility: exchange rate processes Implicit in Deutsche mark options. Rev. Finan. Stud. 9, 69–107 (1996)
3. Bates, D.S.: Post-'87 crash fears in S&P 500 future option market. J. Econometrics 94, 181–238 (2000)

4. Black, F., Scholes, M.: The pricing of options and corporate liabilities. J. Pol. Econ. 81, 637–654 (1973)
5. Broadie, M., Kaya, Ö.: Exact simulation of option Greeks under stochastic volatility and jump diffusion models. In: Ingalls, R.G., Rossetti, M.D., Smith, J.S., Peters, B.A.: Proceedings of the 2004 Winter Simulation Conference. (2004)
6. Broadie, M., Kaya, Ö.: Exact simulation of stochastic volatility and other affine jump diffusion processes. Oper. Res. 2, 217–231 (2006)
7. D'Ippoliti, F., Moretto, E., Pasquali, S., Trivellato, B.: Option valuation in a stochastic bolatility jump-diffusion model. In: CNR-IMATI Technical Report 2007 MI/4 http://www.mi.imati.cnr.it/iami/papers/07-4.pdf (2007)
8. Duffie, D., Pan, J., Singleton, K.: Transform analysis and asset pricing for sffine jump-diffusions. Econometrica 68, 1343–1376 (2000)
9. Eraker, B., Johannes., M., Polson, N.: The impacts of jumps in volatility and returns. J. Finan. 58, 1269–1300 (2003)
10. Hanson, F.B., Yan, G.: Option pricing for a stochastic-volatility jump-diffusion model with log-uniform jump-amplitudes. In: Proceedings of American Control Conference, 2989–2994 (2006).
11. Heston, S.L.: A closed-form solution for options with stochastic volatility with applications to bond and currency options. Rev. Finan. Stud. 6, 327–343 (1993)
12. Lamberton, D., Lapeyre B.: Introduction au Calcul Stochastique Appliqué à la Finance. Second edition. Ellipse, Édition Marketing, Paris (1997)
13. Mania, M., Tevzadze, R., Toronjadze, T.: Mean-variance hedging under partial information. SIAM Journal on Control and Optimization, 47, 2381–2409 (2008)
14. Pan, J.: The jump-risk premia implicit in option: evidence from an integrated zime-series study. J. Finan. Econ. 63, 3–50 (2002)
15. Stein, E.M., Stein, J.C.: Stock price distributions with stochastic volatility: an analytic approach. Rev. Finan. Stud. 4, 727–752 (1991)
16. Willard, G.A.: Calculating prices and sensitivities for path-dependent derivatives securities in multifactor models. J. Deriv. 1, 45–61 (1997)

A skewed GARCH-type model for multivariate financial time series

Cinzia Franceschini and Nicola Loperfido

Abstract. Skewness of a random vector can be evaluated via its third cumulant, i.e., a matrix whose elements are central moments of order three. In the general case, modelling third cumulants might require a large number of parameters, which can be substantially reduced if skew-normality of the underlying distribution is assumed. We propose a multivariate GARCH model with constant conditional correlations and multivariate skew-normal random shocks. The application deals with multivariate financial time series whose skewness is significantly negative, according to the sign test for symmetry.

Key words: financial returns, skew-normal distribution, third cumulant

1 Introduction

Observed financial returns are often negatively skewed, i.e., the third central moment is negative. This empirical finding is discussed in [3]. [7] conjectures that negative skewness originates from asymmetric behaviour of financial markets with respect to relevant news. [6] conclude that \Skewness should be taken into account in the estimation of stock returns".

Skewness of financial returns has been modelled in several ways. [12] reviews previous literature on this topic. [5] models skewness as a direct consequence of the feedback effect. [4] generalises the model to the multivariate case.

All the above authors deal with scalar measures of skewness, even when they model multivariate returns. In this paper, we measure skewness of a random vector using a matrix containing all its central moments of order three. More precisely, we measure and model skewness of a random vector using its third cumulant and the multivariate skew-normal distribution [2], respectively. It is structured as follows. Sections 2 and 3 recall the definition and some basic properties of the multivariate third moment and the multivariate skew-normal distribution. Section 4 introduces a multivariate GARCH-type model with skew-normal errors. Section 5 introduces a negatively skewed financial dataset. Section 6 applies the sign test for symmetry to the same dataset. Section 7 contains some concluding remarks.

2 Third moment

The third moment of a p-dimensional random vector z is defined as $\mu_3(z) = E\left(z \otimes z^T \otimes z^T\right)$, where $\otimes$ denotes the Kronecker (tensor) product and third moment is finite [9, page 177]. The third central moment is defined in a similar way: $\overline{\mu}_3(z) = \mu_3(z - \mu)$, where μ denotes the expectation of z. For a p-dimensional random vector the third moment is a $p \times p^2$ matrix containing $p(p+1)(p+2)/6$ possibly distinct elements. As an example, let $z = (Z_1, Z_2, Z_3)^T$ and $\mu_{ijk} = E\left(Z_i Z_j Z_k\right)$, for $i, j, k = 1, \ldots, 3$. Then the third moment of z is

$$\mu_3(z) = \begin{pmatrix} \mu_{111} & \mu_{112} & \mu_{113} & \mu_{211} & \mu_{212} & \mu_{213} & \mu_{311} & \mu_{312} & \mu_{313} \\ \mu_{121} & \mu_{122} & \mu_{123} & \mu_{221} & \mu_{222} & \mu_{223} & \mu_{321} & \mu_{322} & \mu_{323} \\ \mu_{131} & \mu_{132} & \mu_{133} & \mu_{231} & \mu_{232} & \mu_{233} & \mu_{331} & \mu_{332} & \mu_{333} \end{pmatrix}.$$

In particular, if all components of z are standardised, its third moment is scale-free, exactly like many univariate measures of skewness.

Moments of linear transformations $y = Az$ admit simple representations in terms of matrix operations. For example, the expectation $E(y) = AE(z)$ is evaluated via matrix multiplication only. The variance $V(y) = AV(z)A^T$ is evaluated using both the matrix multiplication and transposition. The third moment $\mu_3(y)$ is evaluated using the matrix multiplication, transposition and the tensor product:

Proposition 1. *Let z be a p-dimensional random vector with finite third moment $\mu_3(z)$ and let A be a $k \times p$ real matrix. Then the third moment of Az is $\mu_3(Az) = A\mu_3(z)\left(A^T \otimes A^T\right)$.*

The third central moment of a random variable is zero, when it is finite and the corresponding distribution is symmetric. There are several definitions of multivariate symmetry. For example, a random vector z is said to be centrally symmetric at μ if $z - \mu$ and $\mu - z$ are identically distributed [15]. The following proposition generalises this result to the multivariate case.

Proposition 2. *If the random vector z is centrally symmetric and the third central moment is finite, it is a null matrix.*

Sometimes it is more convenient to deal with cumulants, rather than with moments. The following proposition generalises to the multivariate case a well known identity holding for random variables.

Proposition 3. *The third central moment of a random vector equals its third cumulant, when they both are finite.*

The following proposition simplifies the task of finding entries of $\mu_3(z)$ corresponding to $E\left(Z_i Z_j Z_k\right)$.

Proposition 4. *Let $z = (Z_1, \ldots, Z_p)^T$ be a random vector whose third moment $\mu_3(z)$ is finite. Then $\mu_3(z) = (M_1, \ldots, M_p)$, where $M_i = E\left(Z_i z z^T\right)$.*

Hence $E\left(Z_i Z_j Z_k\right)$ is in the ith row and in the jth column of the kth matrix M_k. There is a simple relation between the third central moment and the first, second and third moments of a random vector [9, page 187]:

Proposition 5. *Let* $\mu = \mu_1(z)$, $\mu_2(z)$, $\mu_3(z)$ *the first, second and third moments of the random vector z. Then the third central moment of z can be represented by* $\mu_3(z) - \mu_2(z) \otimes \mu^T - \mu^T \otimes \mu_2(z) - \mu\left[\mu_2^V(z)\right]^T + 2\mu \otimes \mu^T \otimes \mu^T$, *where* A^V *denotes the vector obtained by stacking the columns of the matrix A on top of each other.*

3 The multivariate skew-normal distribution

We denote by $z \sim SN_p(\Omega, \alpha)$ a multivariate skew-normal random vector [2] with scale parameter Ω and shape parameter α. Its probability density function is

$$f(z; \Omega, \alpha) = 2\phi_p(z; \Omega) \, \Phi\left(\alpha^T z\right), \quad z, \alpha \in R^p, \Omega \in R^p \times R^p, \qquad (1)$$

where $\Phi(\cdot)$ is the cumulative distribution function of a standard normal variable and $\phi_p(z; \Omega)$ is the probability density function of a p-dimensional normal distribution with mean 0_p and correlation matrix Ω. Expectation, variance and third central moment of $z \sim SN_p(\Omega, \alpha)$ (i.e., its first three cumulants) have a simple analytical form [1, 8]:

$$E(z) = \sqrt{\frac{2}{\pi}}\delta, \quad V(z) = \Omega - \frac{2}{\pi}\delta\delta^T, \quad \bar{\mu}_3(z) = \sqrt{\frac{2}{\pi}}\left(\frac{4}{\pi} - 1\right)\delta \otimes \delta^T \otimes \delta^T, \quad (2)$$

where $\delta = \Omega\alpha/\sqrt{1 + \alpha^T\Omega\alpha}$. As a direct consequence, the third cumulant of $z \sim SN_p(\Omega, \alpha)$ needs only p parameters to be identified, and is a matrix with negative (null) entries if and only if all components of δ are negative (null) too.

The second and third moments of $z \sim SN_p(\Omega, \alpha)$ have a simple analytical form too:

$$\mu_2(z) = E\left(zz^T\right) = \Omega, \qquad (3)$$

$$\mu_3(z) = \sqrt{\frac{2}{\pi}}\left[\delta^T \otimes \Omega + \delta\left(\Omega^V\right)^T + \Omega \otimes \delta^T - \delta\delta^T \otimes \delta^T\right]. \qquad (4)$$

Expectation of zz^T depends on the scale matrix Ω only, due to the invariance property of skew-normal distributions: if $z \sim SN_p(\Omega, \alpha)$ then $zz^T \sim W(\Omega, 1)$, i.e., a Wishart distribution depending on the matrix Ω only [11]. As a direct consequence, the distribution of a function $g(\cdot)$ of z satisfying $g(z) = g(-z)$ does not depend either on α or on δ.

The probability density function of the ith component z_i of $z \sim SN_p(\Omega, \alpha)$ is

$$f(z_i) = 2\phi(z_i) \, \Phi\left(\frac{\delta_i z_i}{\sqrt{1 - \delta_i^2}}\right), \qquad (5)$$

where δ_i is the ith component of δ and ϕ denotes the pdf of a standard normal distribution. The third moment of the corresponding standardised distribution is

$$\sqrt{2}\,(4-\pi)\left(\frac{\delta_i}{\sqrt{\pi-2\delta_i^2}}\right)^3. \tag{6}$$

Hence positive (negative) values of δ_i lead to positive (negative) skewness. Moreover, positive values of δ_i lead to $F_i(0) > 1 - F_i(0)$ when $x > 0$, with F_i denoting the cdf of z_i.

4 A skewed GARCH-type model

In order to describe skewness using a limited number of parameters, we shall introduce the following model for a p-dimensional vector of financial returns x_t:

$$x_t = D_t \varepsilon_t, \quad \varepsilon_t = z_t - E(z_t), \quad z_t \sim SN_p(\Omega, \alpha), \quad D_t = diag\left(\sigma_{1t}, \dots, \sigma_{pt}\right) \tag{7}$$

$$\sigma_{kt}^2 = \omega_{0k} + \sum_{i=1}^{q} \omega_{ik} x_{k,t-i}^2 + \sum_{j=q+1}^{q+p} \omega_{jk} \sigma_{k,t+q-j}^2, \tag{8}$$

where ordinary stationarity assumptions hold and $\{z_t\}$ is a sequence of mutually independent random vectors.

The following proposition gives the analytical moment of the third cumulant $\overline{\mu}_3(x_t)$ of a vector x_t belonging to the above stochastic process. In particular it shows that $\overline{\mu}_3(x_t)$ is negative (null) when all the elements in the vector δ are negative (null) too.

Proposition 6. *Let $\{x_t, t \in Z\}$ be a stochastic process satisfying (10), (11) and $E\left(\sigma_{it}\sigma_{jt}\sigma_{ht}\right) < +\infty$ for $i, j, h = 1, \dots, p$. Then*

$$\mu_3(x_t) = \overline{\mu}_3(x_t) = \sqrt{\frac{2}{\pi}}\left(\frac{4}{\pi}-1\right)\Delta\mu_3(\sigma_t)(\Delta \otimes \Delta), \tag{9}$$

where $\Delta = diag\left(\delta_1, \dots, \delta_p\right)$ and $\sigma_t = \left(\sigma_{1t}, \dots, \sigma_{pt}\right)^T$.

Proof. We shall write $\mu_3(y \mid w)$ and $\overline{\mu}_3(y \mid w)$ to denote the third moment and the third cumulant of the random vector y, conditionally on the random vector w. From the definition of $\{x_t, t \in Z\}$ we have the following identities:

$$\mu_3(x_t \mid \sigma_t) = \mu_3\{D_t [z_t - E(z_t)] \mid \sigma_t\} = \overline{\mu}_3(D_t z_t \mid \sigma_t). \tag{10}$$

Apply now linear properties of the third cumulant:

$$\mu_3(x_t \mid \sigma_t) = D_t \overline{\mu}_3(z_t)(D_t \otimes D_t). \tag{11}$$

By assumption the distribution of z_t is multivariate skew-normal:

$$\mu_3 (x_t \,|\, \sigma_t) = \sqrt{\frac{2}{\pi}} \left(\frac{4}{\pi} - 1 \right) D_t \left(\delta \otimes \delta^T \otimes \delta^T \right) (D_t \otimes D_t). \qquad (12)$$

Consider now the following mixed moments of order three:

$$E \left(X_{it} X_{jt} X_{kt} \,|\, \sigma_t \right) =$$
$$\sqrt{\frac{2}{\pi}} \left(\frac{4}{\pi} - 1 \right) (\delta_i \sigma_{it}) (\delta_j \sigma_{jt}) (\delta_k \sigma_{kt}) \quad i, j, k = 1, \dots, p. \qquad (13)$$

We can use definitions of Δ and σ_t to write the above equations in matrix form:

$$\mu_3 (x_t \,|\, \sigma_t) = \sqrt{\frac{2}{\pi}} \left(\frac{4}{\pi} - 1 \right) (\Delta \sigma_t) \otimes (\Delta \sigma_t)^T \otimes (\Delta \sigma_t)^T. \qquad (14)$$

Ordinary properties of tensor products imply that

$$\mu_3 (x_t \,|\, \sigma_t) = \sqrt{\frac{2}{\pi}} \left(\frac{4}{\pi} - 1 \right) \Delta \left(\sigma_t \otimes \sigma_t^T \otimes \sigma_t^T \right) (\Delta \otimes \Delta). \qquad (15)$$

By assumption $E \left(\sigma_{it} \sigma_{jt} \sigma_{ht} \right) < +\infty$ for $i, j, h = 1, \dots, p$, so that we can take expectations with respect to σ_t:

$$\mu_3 (x_t) = \sqrt{\frac{2}{\pi}} \left(\frac{4}{\pi} - 1 \right) \Delta E \left(\sigma_t \otimes \sigma_t^T \otimes \sigma_t^T \right) (\Delta \otimes \Delta). \qquad (16)$$

The expectation in the right-hand side of the above equation equals $\mu_3 (\sigma_t)$. Moreover, since $P (\sigma_{it} > 0) = 1$, the assumption $E \left(\sigma_{it} \sigma_{jt} \sigma_{ht} \right) < +\infty$ for $i, j, h = 1, \dots, p$ also implies that $E (\sigma_{it}) < +\infty$ for $i = 1, \dots, p$ and that the expectation of x_t equals the null vector. As a direct consequence, the third moment equals the third cumulant of x_t and this completes the proof. $\square$

5 Data analysis

This section deals with daily percent log-returns (i.e., daily log-returns multiplied by 100) corresponding to the indices DAX30 (Germany), IBEX35 (Spain) and S&PMIB (Italy) from 01/01/2001 to 30/11/2007. The mean vector, the covariance matrix and the correlation matrix are

$$\begin{pmatrix} -0.0064 \\ 0.030 \\ 0.011 \end{pmatrix}, \quad \begin{pmatrix} 1.446 & 1.268 & 1.559 \\ 1.268 & 1.549 & 1.531 \\ 1.559 & 1.531 & 2.399 \end{pmatrix} \quad \text{and} \quad \begin{pmatrix} 1.000 & 0.847 & 0.837 \\ 0.847 & 1.000 & 0.794 \\ 0.837 & 0.794 & 1.000 \end{pmatrix}, \qquad (17)$$

respectively. Not surprisingly, means are negligible with respect to standard deviations and variables are positively correlated. Figure 1 shows histograms and scatterplots.

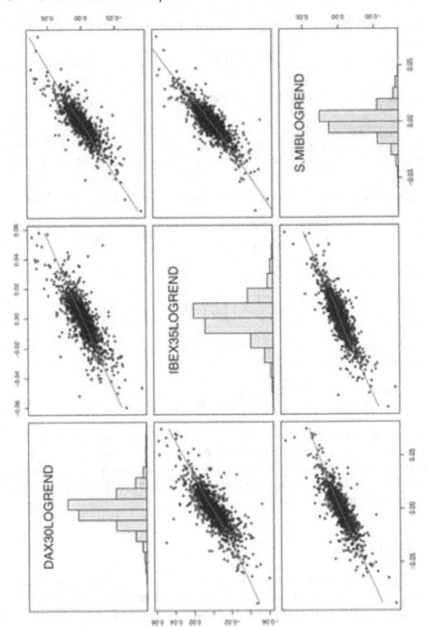

Fig. 1. Scatterplots and histograms for DAX30, IBEX35 and S&PMIB

We shall measure skewness using the following indices, defined as:

$$A_1 = \sum_{i=1}^{n} \left(\frac{x_i - \bar{x}}{s} \right)^3, \quad A_2 = \frac{q_1 - 2q_2 - q_3}{q_3 - q_1} \quad \text{and} \quad A_3 = \frac{3}{s}(x - q_2), \quad (18)$$

where q_i is the ith quartile ($i = 1, 2, 3$). Their values for the three series are reported in Table 1.

Table 1. Skewness coefficients

	A_1	A_2	A_3
S&PMib	-0.227	-1.049	-0.095
Ibex35	-0.046	-1.104	-0.081
Dax30	-0.120	-1.076	-0.089

All indices suggest negative skewness. In order to assess multivariate skewness, we shall consider the third cumulant and the third moment. The third sample cumulant is

$$\bar{m}_3(X) = \frac{1}{n} \sum_{i=1}^{n} (x_i - m) \otimes (x_i - m)^T \otimes (x_i - m)^T, \quad (19)$$

where x_i is the transpose of the ith row of the $n \times p$ data matrix X and m is the mean vector. The third sample cumulants of the above data are

$$- \begin{pmatrix} 0.394 \ 0.201 \ 0.396 \ 0.201 \ 0.092 \ 0.242 \ 0.396 \ 0.242 \ 0.414 \\ 0.201 \ 0.092 \ 0.242 \ 0.092 \ 0.088 \ 0.146 \ 0.242 \ 0.146 \ 0.216 \\ 0.396 \ 0.242 \ 0.414 \ 0.242 \ 0.146 \ 0.216 \ 0.414 \ 0.216 \ 0.446 \end{pmatrix}. \quad (20)$$

The most interesting feature of the above matrix is the negative sign of all its elements. The third moment has a similar structure, since all entries but one are negative:

$$- \begin{pmatrix} 0.422 \ 0.173 \ 0.340 \ 0.173 \ 0.025 \ 0.199 \ 0.340 \ 0.199 \ 0.3940 \\ 0.173 \ 0.025 \ 0.190 \ 0.025 \ \text{-}0.053 \ 0.035 \ 0.199 \ 0.035 \ 0.109 \\ 0.340 \ 0.190 \ 0.390 \ 0.199 \ 0.035 \ 0.109 \ 0.394 \ 0.109 \ 0.365 \end{pmatrix}. \quad (21)$$

We found the same pattern in other multivariate financial time series from small markets.

6 Sign tests for symmetry

This section deals with formal testing procedures for the hypothesis of symmetry. When testing for symmetry, the default choice for a test statistic is the third standardised moment, which might be inappropriate for financial data. Their dependence structure and their heavy tails make it difficult to understand its sampling properties. On the contrary, the sign test for symmetry possesses very appealing sampling properties, when the location parameter is assumed to be known [10, page 247]. When dealing with financial returns, it is realistic to assumed to be known and equal to zero, for theoretical as well as for empirical reasons. From the theoretical point of view, it prevents systematic gains or losses. From the empirical point of view, as can be seen in (17), means of observed returns are very close to zero. The following paragraphs in this section state model's assumptions, describe the sign test for symmetry and apply it to the data described in the previous section.

We shall assume the following model for a p-dimensional vector of financial returns x_t: $x_t = D_t \varepsilon_t$, $E(\varepsilon_t) = 0$, $D_t = diag(\sigma_{1t}, \ldots, \sigma_{pt})$ and

$$\sigma_{kt}^2 = \omega_{0k} + \sum_{i=1}^{q} \omega_{ik} x_{k,t-i}^2 + \sum_{j=q+1}^{q+p} \omega_{jk} \sigma_{k,t+q-j}^2, \qquad (22)$$

where ordinary stationarity assumptions hold and $\{\varepsilon_t\}$ is a sequence of mutually independent random vectors. We shall test the hypotheses

$$H_0^{ijk} : F_{ijk}(0) = 1 - F_{ijk}(0) \quad \text{versus} \quad H_1^{ijk} : F_{ijk}(0) < 1 - F_{ijk}(0) \qquad (23)$$

for $i, j, k = 1, 2, 3$, where F_{ijk} denotes the cdf of $\varepsilon_{t,i} \varepsilon_{t,j} \varepsilon_{t,k}$. Many hypotheses H_a^{ijk} for $i, j, k = 1, 2, 3$ and $a = 0, 1$ are equivalent to each other and can be expressed in a simpler way. For example, H_0^{ijj} and H_1^{ijj} are equivalent to $F_i(0) = 1 - F_i(0)$ and $F_i(0) < 1 - F_i(0)$, respectively, where F_i denotes the cdf of $\varepsilon_{t,i}$. Hence it suffices to test the following systems of hypotheses

$$H_0^i : F_i(0) = 1 - F_i(0) \quad \text{versus} \quad H_1^i : F_i(0) < 1 - F_i(0), i = 1, 2, 3 \qquad (24)$$

and

$$H_0^{123} : F_{123}(0) = 1 - F_{123}(0) \quad \text{versus} \quad H_1^{123} : F_{123}(0) < 1 - F_{123}(0). \qquad (25)$$

Let x^1, x^2 and x^3 denote the column vector of returns in the German, Spanish and Italian markets. The sign test rejects the null hypothesis H_0^{ijk} if the number n_{ijk} of positive elements in the vector $x^i \circ x^j \circ x^k$ is larger than an assigned value, where "$\circ$" denotes the Schur (or Hadamard) product. Equivalently, it rejects H_0^{ijk} if $z_{ijk} = 2\sqrt{n}(f_{ijk} - 0.5)$ is larger than an assigned value, where f_{ijk} is the relative frequency of positive elements in $x_i \circ x_j \circ x_k$ and n is its length. Under H_0^{ijk}, $n_{ijk} \sim Bi(n, 0.5)$ and $z_{ijk} \sim N(0, 1)$, asymptotically.

Table 2 reports the relative frequencies of positive components in x^1, x^2, x^3 and $x^1 \circ x^2 \circ x^3$, together with the corresponding test statistics and p-values.

In all four cases, there is little evidence supporting the null hypothesis of symmetry against the alternative hypothesis of negative asymmetry. Results are consistent with the exploratory analysis in the previous section and motivate models describing multivariate negative skewness.

Table 2. Tests statistics

Indices	Frequency	Statistic	p-value
Dax30	0.542	3.579	<0.001
Ibex35	0.561	5.180	<0.001
S&PMib	0.543	3.673	<0.001
Product	0.544	3.767	<0.001

7 Conclusions

We considered the third cumulant of multivariate financial returns, motivated it through a real data example and modeled it through the multivariate skew-normal distribution. Preliminary studies hint that negative third cumulants might constitute a stylised fact of multivariate financial returns [13], but more studies are needed to confirm or disprove this conjecture. By proposition 2, testing for central symmetry would be a natural way for doing it. [14] gives an excellent overview of the literature on this topic. Multivariate GARCH-type models with skew-normal errors might be helpful in keeping under control the number of parameters, but some caution is needed when using maximum likelihood procedures, since it is well known that sometimes they lead to frontier estimates.

Acknowledgement. Financially supported by the Ministero dell'Università, dell'Istruzione e della Ricerca with grant PRIN No. 2006132978.

References

1. Azzalini, A., Capitanio, A.: Statistical applications of the multivariate skew-normal distributions. J. R. Stat. Soc. B 61, 579–602 (1999)
2. Azzalini, A., Dalla Valle, A.: The multivariate skew-normal distribution. Biometrika 83, 715–726 (1996)
3. Cutler, D.M., Poterba, J.M., Summers, L.H.: What moves stock prices? J. Portfolio Man. 15, 4–12 (1989)
4. De Luca, G., Genton, M.G., Loperfido, N.: A multivariate skew-garch model. In: Terrell, D. (ed.) Advances in Econometrics: Econometric Analysis of Economic and Financial Time Series, Part A (Special volume in honor of Robert Engle and Clive Granger, the 2003 winners of the Nobel Prize in Economics), pp. 33–57. Elsevier (2006)
5. De Luca, G., Loperfido, N.: A skew-in-mean GARCH model for financial returns. In: Genton, M.G. (ed.) Skew-Elliptical Distributions and their Applications: A Journey Beyond Normality, pp. 205–222. CRC/Chapman & Hall (2004)
6. Engle, R.F., Gonzalez Rivera, G.: Semiparametric ARCH models. J. Bus. Econ. Stat. 9, 345–359 (1991)
7. French K.R., Schwert W.G., Stambaugh R.F.: Expected stock returns and volatility. J. Finan. Econ. 19, 3–29 (1987)

8. Genton, M.G., He, L., Liu, X.: Moments of skew-normal random vectors and their quadratic forms. Stat. Prob. Lett. 51, 319–235 (2001)
9. Kollo, T., von Rosen, D.: Advanced Multivariate Statistics with Matrices. Springer, Dordrecht, The Netherlands (2005)
10. Lehmann, E.L., Romano, J.P.: Testing Statistical Hypotheses. Springer, New York (2005)
11. Loperfido, N.: Quadratic forms of skew-normal random vectors. Stat. Prob. Lett. 54, 381–387 (2001)
12. McDonald, J.B.: Probability distributions for financial models. In: Maddala, G.S., Rao, C.R. (eds.) Handbook of Statistics: Statistical Methods in Finance. Amsterdam, Elsevier Science 14, (1996)
13. Rydberg, T.H.: Realistic statistical modelling of financial data. Int. Stat. Rev. 68, 233–258 (2000)
14. Serfling, R.J. Multivariate symmetry and asymmetry. In: Kotz, S., Read, C.B., Balakrishnan, N., Vidakovic, B. (eds.) Encyclopedia of Statistical Sciences, 2nd edition. Wiley, New York (2006)
15. Zuo, Y., Serfling, R.: On the performance of some robust nonparametric location measures relative to a general notion of multivariate simmetry. J. Stat. Planning Inference 84, 55–79 (2000)

Financial time series and neural networks in a minority game context

Luca Grilli, Massimo Alfonso Russo, and Angelo Sfrecola

Abstract. In this paper we consider financial time series from U.S. Fixed Income Market, S&P500, DJ Eurostoxx 50, Dow Jones, Mibtel and Nikkei 225. It is well known that financial time series reveal some anomalies regarding the Efficient Market Hypothesis and some scaling behaviour, such as fat tails and clustered volatility, is evident. This suggests that financial time series can be considered as "pseudo"-random. For this kind of time series the prediction power of neural networks has been shown to be appreciable [10]. At first, we consider the financial time series from the Minority Game point of view and then we apply a neural network with learning algorithm in order to analyse its prediction power. We prove that the Fixed Income Market shows many differences from other markets in terms of predictability as a measure of market efficiency.

Key words: Minority Game, learning algorithms, neural networks, financial time series, Efficient Market Hypothesis

1 Minority games and financial markets

At the very beginning of the last century Bachelier [2] introduced the hypothesis that price fluctuations follow a random walk; this resulted later in the so-called Efficient Market Hypothesis (EMH). In such markets arbitrages are not possible and so speculation does not produce any gain. Later, empirical studies showed that the implications of EMH are too strong and the data revealed some anomalies. Even though these anomalies are frequent, economists base the Portfolio Theory on the assumption that the market is efficient. One of the most important implications of EMH is the rationality of all agents who are gain maximisers and take decisions considering all the information available (which have to be obtained easily and immediately) and in general do not face any transaction costs. Is it realistic? The huge literature on this subject shows that an answer is not easy but in general some anomalies are present in the market. One of the main problems is rationality; as a rule, agents make satisfactory choices instead of optimal ones; they are not deductive in making decisions but inductive in the sense that they learn from experience. As a consequence, rationality hypothesis is often replaced by the so-called "Bounded Rationality"; see [13] for more details. Empirical studies also show cluster formations and other anomalies in financial time series [3,9].

In order to model the inductive behaviour of financial agents, one of the most famous examples is the Minority Game (MG) model. The MG is a simple model of interacting agents initially derived from Arthur's famous El Farol's Bar problem [1]. A popular bar with a limited seating capacity organises a Jazz-music night on Thursdays and a fixed number of potential customers (players) has to decide whether to go or not to go to the bar. If the bar is too crowded (say more than a fixed capacity level) then no customer will have a good time and so they should prefer to stay at home. Therefore every week players have to choose one out of two possible actions: to stay at home or to go to the bar. The players who are in the minority win the game.

Since the introduction of the MG model, there have been, to date, 200 papers on this subject (there is an overview of literature on MG at the Econophysics website).

The MG problem is very simple, nevertheless it shows fascinating properties and several applications. The underlying idea is competition for limited resources and it can be applied to different fields such as stock markets (see [5–7] for a complete list of references). In particular the MG can be used to model a very simple market system where many heterogeneous agents interact through a price system they all contribute to determine. In this market each trader has to take a binary decision every time (say buy/sell) and the profit is made only by the players in the minority group. For instance, if the price increases it means that the minority of traders are selling and they get profit from it. This is a simple market where there is a fixed number of players and only one asset; they have to take a binary decision (buy/sell) in each time step t. When all players have announced their strategies the prices are made according to the basic rule that if the minority decides to sell, then the price grows (the sellers get profit); if the minority decide to buy, then price falls (the buyers get profit).

In this model cooperation is not allowed; players cannot communicate and so they all get information from the global minority. In order to make decisions, players use the global history of the past minorities or, in most cases, a limited number of past minorities that can be considered the time window used by the player. In our case the global history is given by the time series of price fluctuations. Let us consider the set of players $i = \{1, \ldots, N\}$ where $N \in \mathbb{N}$ (odd and fixed). Indicate with t the time step when each player makes a decision. In the market there is one asset and the possible decision in each time step is buy or sell; as a consequence the player i at time t chooses $\sigma_i^t \in \{+1, -1\}$ (buy/sell).

In each time step t, let p_t be the price of the asset at time t; the minority (the winning strategy) is determined by

$$S^t = -\operatorname{sign} \log \left(\frac{p_t}{p_{t-1}} \right).$$

Consequently, the time series of price fluctuations is replaced by a time series consisting of two possible values: $+1$ and -1 (the minority decisions).

In [4] it is shown that often similar results can be obtained by replacing the real history with an artificial one.

The main point is that we suppose that players make their decisions according to a learning rule, as a consequence they follow an inductive behaviour and this affects the time series of the minority decisions that is not simply a random sequence of -1 and $+1$. If this is the case the time series can be defined "pseudo-random" as a consequence of the periodicity derived from the generating rule. This periodicity is not due to the presence of a "trend" which is buried under noise but it is a consequence of the inductive behaviour of players and this is the reason why classical techniques such as simple autocorrelation analysis do not give us information, by definition, on the learning procedure. On the contrary, the neural network with an appropriate learning algorithm can capture such "regularities" very well and consequently can predict the time series as shown in [10]. The main result presented in [10] is that a neural network with an appropriate learning algorithm can predict "pseudo-random" time series very well whatever the learning algorithm. On the other hand the neural network is not able to predict a randomly generated time series. As a consequence, if we apply a similar analysis to financial time series in the MG context presented before, we can test for EMH since bad results in terms of prediction power of the neural network can suggest that EMH is fulfilled and time series are randomly generated. If this is not the case and the prediction power is remarkable then the time series is "pseudo-random" as a consequence of inductive behaviour of the players. The neural network approach also reveals the time window of past decisions that players are considering in order to make their choice. As we will see, it is dependent on the market we consider.

2 Neural network and financial time series

The main issue of this paper is to determine the predictability of financial time series taking into account the imperfection of the market as a consequence of agents' behaviour. In [10] it is shown that, when players make their decisions according to some learning rule, then the time series of the minority decisions is not simply a random sequence of -1 and $+1$. The time series generated with learning algorithms can be defined as "pseudo-random" time series. The reason is that, by construction, it presents a sort of periodicity derived from the generating algorithm. This periodicity is not evident directly from the time series but a neural network with an appropriate learning algorithm can capture such "regularities" and consequently can predict the time series. The authors show that, for three artificial sequences of minority decisions generated according to different algorithms, the prediction power of the neural network is very high.

In this paper we suppose that each player, in order to make her decision, is provided with a neural network. We consider time series from U.S. Fixed Income Market, S&P500, DJ Eurostoxx 50, Dow Jones, Mibtel and Nikkei 225 (all the time series from Jan 2003 to Jan 2008, daily prices, data from the Italian Stock Exchange).

Following the motivations presented in [10], in this paper we consider a neural network that uses the Hebbian Learning algorithm to update the vector of weights. The neural network is able to adjust its parameters, at each round of the game, and so the perceptron is trying to learn the minority decisions. If S indicates the minority decision and superscript $+$ indicates the updated value of a parameter, the vector of minority decision x whose component $x^t \in \{+1, -1\}$ is the entry of the time series at instant t. Let us suppose that each player is provided with such a neural network and makes her decision according to the following rules:

$$\sigma_i = \text{sign}(x \cdot \omega_i)$$

$$\omega_i^+ = \omega_i - \frac{\eta}{M} x \, \text{sign} \left(\sum_{j=1}^{N} \text{sign}(x \cdot \omega_j) \right) = \omega_i + \frac{\eta}{M} x S,$$

where ω is a M-dimensional weight vector from which decisions are made. So each player uses an M-bit window of the past minority decisions as an input in order to adjust the weights and try to capture the "regularities" in the time series. As we can see later, the choice of M is often crucial in order to determine the best prediction power. It is possible to compute the number of predictions as a function of M in order to obtain the value of M for which it is maximum. In this case the parameter M indicates how many past price fluctuations are considered by the agent in order to make a decision. The parameter η is the learning rate.

In [10] it is shown that it is crucial to select the window of past entries to consider as an input for current decision correctly, that is the choice of parameter M. The authors show that the number of corrected predictions is maximum if the neural network uses the same M as the sequence generator. This suggests that, if an M value exists for which the neural network predictions are maximum then it is possible to infer that the sequence of minority decisions is generated by a learning algorithm with exactly the same value M. If we apply the same arguments to financial markets time series, the presence of a value M for which the number of corrected predictions is maximum indicates that the time series is generated by a learning algorithm with that parameter M, that is the length of the time window used by the investor, and this is key information derived exclusively with this approach.

Moreover, to determine this value we analyse the number of predictions of the neural network as a function of M. Figures 1, 2, 4, 5, 6 and 7 show the results of these simulations. The result is different according to the market considered; in particular the case of U.S. Treasury Bond seems to be the most interesting. In this market the maximum is reached for $M = 32$, that is the dimension of the temporal window of the past minority decisions to consider as an input of the neural network. The case of S&P500, DJ Eurostoxx 50, Dow Jones, Mibtel and Nikkei 225 is completely different; the maximum value for M is, in general, very low ($M = 3 - 5$). This can suggest that in these markets investors look at the very recent past to make decisions and do not consider what has happened in the remote past. On the other hand, Fixed Income Market presents a different situation and it seems to be the most predictable since the number of predicted entries is the highest one (about 60% of corrected predictions). This can be explained according to some features that make this market different since it must follow common laws dictated by macroeconomic variables. As a consequence, the data present a strong positive correlation between bonds [9]. Another reason is that usually only large investors (like insurance companies or mutual funds) are interested

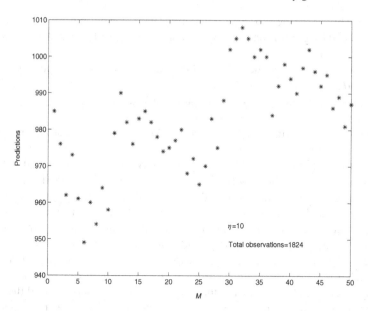

Fig. 1. Number of corrected predictions as a function of M in the case of U.S. Treasury Bond. The maximum is reached for $M = 32$

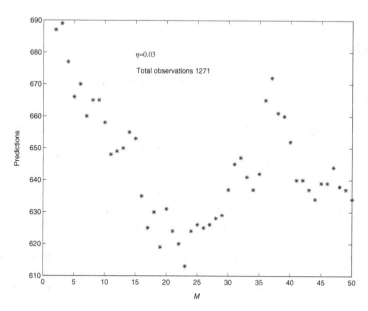

Fig. 2. Number of corrected predictions as a function of M in the case of S&P500. The maximum is reached for $M = 3$

in long-term bonds, so the expectations about the market evolution are so similar that the behaviour of long-term bond prices does not reflect any difference in the *perceived value* of such assets [3].

The analysis shows that the number of corrected predictions is dependent on the parameter M; it is not the same for the parameter η since the number of corrected predictions remains quite constant (we report it in Fig. 3).

The neural network approach has shown the presence of a value M for which the prediction power is maximum and this is a signal in the direction that the time series is "pseudo-random" and agents use M as the time window. This can be interpreted in terms of lack of EMH, but this is only partially true since the neural network cannot predict in a significant way (in terms of number of corrected predictions) the financial time series considered and this can indicate that these time series are randomly generated and so these markets are efficient. This result is not surprising since all the markets considered in this paper present a huge number of transactions and huge volumes, and information provided to agents is immediately and easily available. The neural network can predict, for these time series, slightly more than 50%, which is the expected value of corrected predictions in cases where choices are randomly made, which is a signal in the direction that these markets fulfil the EMH and results in other directions can be considered "anomalies". A comparative analysis reveals that the Fixed Income Markets seems to the least efficient since the number of predictions is maximal.

3 Conclusions

In [10] the authors show that in an MG framework, a neural network that use a Hebbian algorithm can predict almost every minority decision in the case in which the sequence of minority decisions follows a "pseudo"-random distribution. The neural network can capture the "periodicity" of the time series and then predict it. On the other hand they show that the prediction power is not so good when the time series is randomly generated. In this paper we consider financial time series from U.S. Fixed Income Market, S&P500, DJ Eurostoxx 50, Dow Jones, Mibtel and Nikkei 225. If agents make satisfactory choices instead of optimal ones, they are inductive in the sense that they learn from experience and MG is a very good model for inductive behaviour of financial agents. If financial time series are generated by some learning procedure, then we can consider financial time series as "pseudo"-random time series and in this case the prediction of neural networks is appreciable. So we consider the financial time series from the Minority Game point of view and then we apply a neural network with learning algorithm in order to analyse its prediction power as a measure of market efficiency.

We show that the case of U.S. Treasury Bond seems to be the most interesting since the time window of the past minorities considered by the investor is $M = 32$, which is very high with respect to other markets, and for this time series the neural network can predict about 60% of entries. This is a signal in the direction that the Fixed Income Market is more predictable as a consequence of features that make

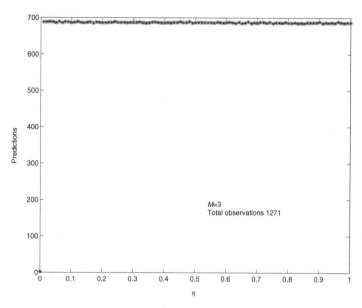

Fig. 3. Number of corrected predictions as a function of η in the case of S&P500. The number of predictions is quite constant

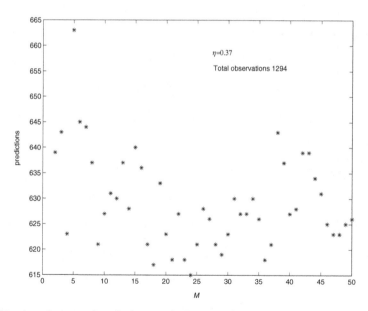

Fig. 4. Number of corrected predictions as a function of M in the case of Mibtel. The maximum is reached for $M = 5$

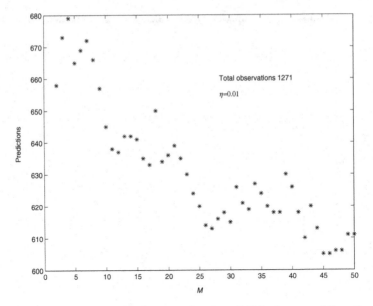

Fig. 5. Number of corrected predictions as a function of M in the case of Dow Jones. The maximum is reached for $M = 4$

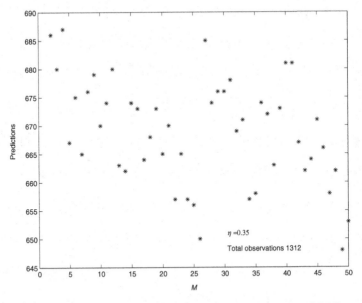

Fig. 6. Number of corrected predictions as a function of M in the case of DJ Eurostoxx 50. The maximum is reached for $M = 4$

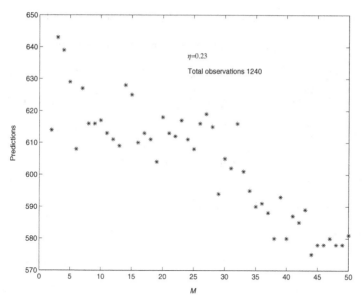

Fig. 7. Number of corrected predictions as a function of M in the case of Nikkei 225. The maximum is reached for $M = 3$

this market different. On the other hand the case of S&P500, DJ Eurostoxx 50, Dow Jones, Mibtel and Nikkei 225 is completely different, as these markets' investors consider only the very recent past since $M = 2 - 4$ and the neural network can predict slightly more than 50% of entries. This can lead us to consider these time series as randomly generated and so consider these markets more efficient. In both cases the neural network shows the presence of a value M for which the number of predictions is maximum and this is the number of past entries that agents consider in order to make decisions. This information is derived directly from the data.

References

1. Arthur, W.B.: Inductive reasoning and bounded rationality. Am. Econ. Rev. 84, 406–411 (1994)
2. Bachelier, L.: Theorie de la speculation. Paris (1900). Reprinted by MIT Press, Cambridge (1964)
3. Bernaschi, M., Grilli, L., Vergni, D.: Statistical analysis of fixed income market. Phys. A Stat. Mech. Appl. 308, 381–390 (2002)
4. Cavagna A.: Irrelevance of memory in the minority game. Phys. Rev. E, 59:R3783 (1999)
5. Challet, D.: Inter-pattern speculation: beyond minority, majority and $-games. cond-mat/05021405 (2005)
6. Challet, D., Chessa, A., Marsili, M., Zhang, Y.-C.: From Minority Games to real markets. cond-mat/0011042 (2000)
7. Coolen, A.C.C.: Generating funcional analysis of Minority Games with real market histories. cond-mat/0410335 (2004)

8. Hart, M., Jefferies, P., Johnson, N.F., Hui, P.M.: Crowd-anticrowd model of the Minority Game. cond-mat/00034860 (2000)
9. Grilli, L.: Long-term fixed income market structure. Phys. A Stat. Mech. Appl. 332, 441–447 (2004)
10. Grilli, L., Sfrecola, A.: A Neural Networks approach to Minority Game. Neural Comput. Appl. 18, 109–113 (2009)
11. Kinzel, W., Kanter, I.: Dynamics of interacting neural networks. J. Phys. A. 33, L141–L147 (2000)
12. Metzler, R., Kinzel, W., Kanter, I. Interacting neural networks. Phys. Rev. E. 62, 2555 (2000)
13. Simon, H.: Models of Bounded Rationality. MIT Press, Cambridge (1997)

Robust estimation of style analysis coefficients

Michele La Rocca and Domenico Vistocco

Abstract. Style analysis, as originally proposed by Sharpe, is an asset class factor model aimed at obtaining information on the internal allocation of a financial portfolio and at comparing portfolios with similar investment strategies. The classical approach is based on a constrained linear regression model and the coefficients are usually estimated exploiting a least squares procedure. This solution clearly suffers from the presence of outlying observations. The aim of the paper is to investigate the use of a robust estimator for style coefficients based on constrained quantile regression. The performance of the novel procedure is evaluated by means of a Monte Carlo study where different sets of outliers (both in the constituent returns and in the portfolio returns) have been considered.

Key words: style analysis, quantile regression, subsampling

1 Introduction

Style analysis, as widely described by Horst et al. [12], is a popular and important tool in portfolio management. Firstly, it can be used to estimate the relevant factor exposure of a financial portfolio. Secondly, it can be a valuable tool in performance measurement since the style portfolio can be used as a benchmark in evaluating the portfolio performance. Finally, it can be used to gain highly accurate future portfolio return predictions since it is well known from empirical studies [12] that factor exposures seem to be more relevant than actual portfolio holdings.

The method, originally proposed by Sharpe [25], is a return-based analysis aimed at decomposing portfolio performance with respect to the contribution of different constituents composing the portfolio. Each sector is represented by an index whose returns are available. The model regresses portfolio returns on constituent returns in order to decompose the portfolio performance with respect to each constituent. Indeed, in the framework of classical regression, the estimated coefficients mean the sensitivity of portfolio expected returns to constituent returns. The classical approach is based on a linear regression model, estimated by using least squares, but different constraints can be imposed on the coefficients.

Following Horst et al. [12], we distinguish three types of style models:

i.   *weak style analysis*: the coefficients are estimated using an unconstrained regression model;
ii. *semi-strong style analysis*: the coefficients are imposed to be positive;
iii. *strong style analysis*: the coefficients are imposed to be positive and to sum up to one.

The three types of style model are typically estimated as regression through the origin.

The use of the double constraint (strong style analysis) and the absence of the intercept allow the interpretation of the regression coefficients in terms of composition quotas and the estimation of the internal composition of the portfolio [8, 9]. Notwithstanding, classical inferential procedures should be interpreted with caution, due to the imposition of inequality linear constraints [13]. Some general results are available for the normal linear regression model [11]; a different approach based on Bayesian inference is formulated in [10].

In the framework of style analysis, a commonly applied solution is the approximation proposed by Lobosco and Di Bartolomeo [21]. These authors obtain an approximate solution for the confidence intervals of style weights using a second-order Taylor approximation. The proposed solution works well except when the parameters are on the boundaries, i.e., when one or more parameters are near 0 and/or when a parameter falls near 1. Kim et al. proposes two approximate solutions for this special case [14] based on the method of Andrews [1] and on the Bayesian method proposed by Geweke [11]. A different Bayesian approach is instead discussed by Christodoulakis [6, 7].

As they are essentially based on a least-squares estimation procedure, common solutions for the estimation of the style analysis coefficients suffer from the presence of outliers. In this paper we investigate the use of quantile regression [18] to estimate style coefficients. In particular we compare the classical solution for the strong style model with robust estimators based on constrained median regression. Different sets of outliers have been simulated both in constituent returns and in portfolio returns. The estimators are then compared with respect to efficiency and some considerations on the consistency of the median regression estimator is provided too. The use of the quantile regression approach allows a further gain in efficiency as an L-estimator [15, 19] can be easily obtained using linear combinations of quantile estimators, i.e., for different conditional quantiles.

The paper is organised as follows: in the next section the classical Sharpe-style model is briefly introduced along with the basic notation. In Section 3 the quantile regression approach to style analysis is described. The simulation schema and the main results are discussed in Section 4. Finally, some concluding remarks and possible further developments are provided in Section 5.

2 Sharpe-style regression model

The Sharpe-style analysis model regresses portfolio returns on the returns of a variety of investment class returns. The method thus identifies the portfolio style in the time series of its returns and of constituent returns [12]. The use of past returns is a Hobson's choice as typically there is no other information available to external investors.

Let us denote by $\mathbf{r}^{port}$ the random vector of portfolio returns along time and by $\mathbf{R}^{const}$ the matrix containing the returns along time of the ith portfolio constituent on the ith column ($i = 1, \ldots, n$). Data refer to T subsequent time periods. The style analysis model regresses portfolio returns on the returns of the n constituents:

$$\mathbf{r}^{port} = \mathbf{R}^{const}\mathbf{w}^{const} + \mathbf{e} \qquad \text{s.t.: } \mathbf{w}^{const} \geq 0, \mathbf{1}^T\mathbf{w}^{const} = 1.$$

The random vector $\mathbf{e}$ can be interpreted as the tracking error of the portfolio, where $\mathbb{E}(\mathbf{R}^{const}\mathbf{e} = \mathbf{0})$.

Style analysis models can vary with respect to the choice of style indexes as well as with respect to the specific location of the response conditional distribution they are estimating. The classical style analysis model is based on a constrained linear regression model estimated by least squares [25, 26]. This model focuses on the conditional expectation of portfolio returns distribution $\mathbb{E}(\mathbf{r}^{port} \mid \mathbf{R}^{const})$: estimated compositions are interpretable in terms of sensitivity of portfolio expected returns to constituent returns.

The presence of the two constraints imposes the coefficients to be exhaustive and non-negative, thus allowing their interpretation in terms of compositional data: the estimated coefficients mean constituent quotas in composing the portfolio. The $\mathbf{R}^{const}\mathbf{w}^{const}$ term of the equation can be interpreted as the return of a weighted portfolio: the portfolio with optimised weights is then a portfolio with the same style as the observed portfolio. It differs from the former as estimates of the internal composition are available [8,9]. We refer the interested reader to the paper of Kim et al. [14] for the assumptions on portfolio returns and on constituent returns commonly adopted in style models.

In the following we restrict our attention to the strong style analysis model, i.e., the model where both the above constraints are considered for estimating style coefficients. Even if such constraints cause some problems for inference, the strong style model is nevertheless widespread for the above-mentioned interpretation issues.

3 A robust approach to style analysis

Quantile regression (QR), as introduced by Koenker and Basset [18], can be viewed as an extension of classical least-squares estimation of conditional mean models to the estimation of a set of conditional quantile functions. For a comprehensive review of general quantile modelling and estimation, see [16].

The use of QR in the style analysis context was originally proposed in [5] and revisited in [2] and [3]. It offers a useful complement to the standard model as it allows discrimination of portfolios that would be otherwise judged equivalent [4].

In the classical approach, portfolio style is determined by estimating the influence of style exposure on expected returns. Extracting information at places other than the expected value should provide useful insights as the style exposure could affect returns in different ways at different locations of the portfolio returns distribution. By exploiting QR, a more detailed comparison of financial portfolios can then be achieved as QR coefficients are interpretable in terms of sensitivity of portfolio conditional quantile returns to constituent returns [5]. The QR model for a given conditional quantile θ can be written as:

$$Q_\theta(\mathbf{r}^{port} \mid \mathbf{R}^{const}) = \mathbf{R}^{const}\mathbf{w}^{const}(\theta) \quad \text{s.t.: } \mathbf{w}^{const}(\theta) \geq 0, \mathbf{1}^T\mathbf{w}^{const}(\theta) = 1, \forall\theta,$$

where θ ($0 < \theta < 1$) denotes the particular quantile of interest.

As for the classical model, the $w^{const_i}(\theta)$ coefficient of the QR model can be interpreted as the rate of change of the θth conditional quantile of the portfolio returns distribution for a unit change in the ith constituent returns holding the values of $R^{const}_{.j, j\neq i}$ constant.

The conditional quantiles are estimated through an optimisation function minimising a sum of weighted absolute deviation where the choice of the weight determines the particular conditional quantile to estimate. We refer to Koenker and Ng [20] for computing inequality constrained quantile regression.

The use of absolute deviations ensures that conditional quantile estimates are robust. The method is nonparametric in the sense that it does not assume any specific probability distribution of the observations. In the following we use a semiparametric approach as we assume a linear model in order to compare QR estimates with the classical style model. Moreover we restrict our attention to the median regression by setting $\theta = 0.5$. As previously stated, it is worthwhile to mention that the use of different values of θ allows a set of conditional quantile estimators to be obtained that can be easily linearly combined in order to construct an L-estimator, in order to gain efficiency [15, 19].

4 Simulation results

In this section the finite sample properties of the proposed procedure are investigated via a Monte Carlo study. Artificial fund returns are simulated using the following data-generation process:

$$\mathbf{r}_t^{port} = \mathbf{r}_t^{const\prime}\mathbf{w}^{const} + \sigma\mathbf{e}_t, \quad t = 1, 2, \ldots, T.$$

In particular, we considered a portfolio with 5 constituents generated by using GARCH(1,1) processes to simulate the behaviour of true time series returns. The true style weights have been set to $w_i^{const} = 0.2, i = 1, 2, \ldots, 5$, thus mimicking a typical "buy and hold" strategy. This allows a better interpretation of simulation results whereas the extension to different management strategies does not entail particular difficulties. The scaling factor σ has been fixed in order to have R^2 close to 0.90 while $e_t \sim N(0, 1)$. We considered additive outliers at randomly chosen positions both in the constituent series (in X) and in the portfolio returns (in Y). The positions of outlier contamination has been set both to 1% and to 5%. We considered median regression ($\theta = 0.5$) and we used T=250, 500, 1000 as sample sizes. We carried out 1000 Monte Carlo runs for each simulation of the experimental set up.

Figure 1 depicts the impact of outlying observations on LS and QR estimators. Each row of the panel graph refers to a portfolio constituent ($i = 1, \ldots, 5$) while the columns show the different cases of presence of outliers: no outliers, outliers in portfolio returns (in Y), outliers in constituent returns (in X), and outliers both in portfolio returns and in constituent returns (in X and Y). In each panel the left boxplot refers to the LS estimator while the right one depicts the QR estimator behaviour. As expected, the impact of outlying observation can be very serious on LS estimates of style coefficients, especially when considering outliers in the constituent series. It is worth noticing that the variability of LS estimates increases very much and this can have serious practical drawbacks since the style coefficients vary in the unit interval: a large variability of the estimates induces results with limited practical utility. Clearly, when no outlying observations are present in the data, the LS estimates are more efficient than quantile estimates. However, the differences between the two distributions are not so evident. Although it is well known that quantile regression estimators are robust only in Y [16], the simulation study shows more evidence of robustness in the case of outliers in constituent returns (third column of panels in Fig. 1). A possible explanation can be given by considering the presence of the double constraint, which forces each estimated coefficient to be inside the unit interval. However, a formal study based on the influence function of the constrained estimators is not available at the moment. This issue is still under investigation.

In order to obtain information on consistency of the constrained median estimators, we use different values for the length of time series. Figure 2 depicts the behaviour of the QR estimator for $T = 250$ (left boxplot in each panel), $T = 500$ (middle boxplot) and $T = 1000$ (right boxplot). As in the previous figure, the rows of the plot refer to the different constituents while the columns report the different cases treated in our simulation with respect to the presence of outlying observations. It is evident that in any case efficiency increases as sample size increases.

Figures 1 and 2 are built using a percentage of outlier contamination set to 1%. Similar patterns have been noticed for the case of 5% contamination and so the related plots are not reported here for the sake of brevity. Using a percentage of outlier contamination set to 5%, as expected, an increase in the variability of the QR estimator is observed, although there is only a very limited difference between the two cases. For the sake of space we do not include any results for the comparison between the different cases of outlier contamination considered. It is straightforward to note, anyway, that increases in the variability of the QR estimator due to an increase in the percentage of outlier contamination are counterbalanced moving the number of observations from $T = 250$ to $T = 500$ and then to $T = 1000$.

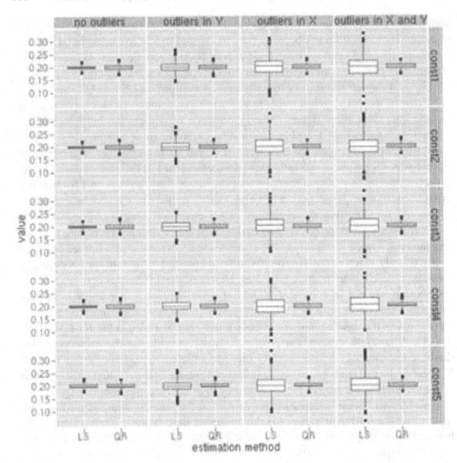

Fig. 1. Comparison of the least-squares (LS) estimators and of the median estimators through quantile regression (QR) for $T = 250$. The different subpanels of the two plots refer to the portfolio constituents (rows) and to the different cases of presence of outliers (columns). In particular the first column depicts the situation with no outlying observation, the second and third columns refer, respectively, to the presence of outliers in portfolio returns and outliers in constituent returns, while the last column depicts the behaviour of LS and QR estimates when outliers are considered both in portfolio returns and in constituent returns. In each panel the left boxplot depicts the sampling distribution of the LS estimator while the right one refers to the sampling distribution of the QR estimator

5 Conclusions and further issues

Style analysis is widely used in financial practice in order to decompose portfolio performance with respect to a set of indexes representing the market in which the portfolio invests. The classical Sharpe method is commonly used for estimating purposes but requires corrections in case of the presence of outliers. In this paper we compare this classical procedure with a robust procedure based on a constrained me-

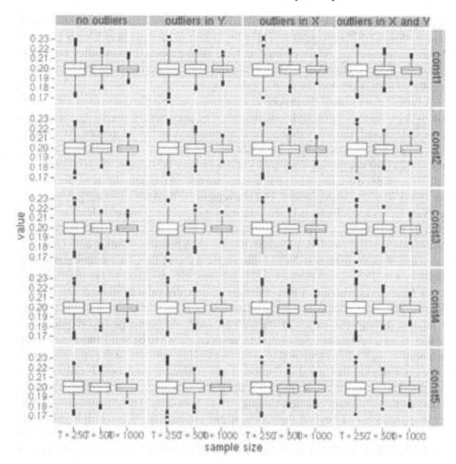

Fig. 2. Comparison of the median estimators through quantile regression (QR) for $T = 250$, $T = 500$ and $T = 1000$. The different subpanels of the two plots refer to the portfolio constituents (rows) and to the different cases of the presence of outliers (columns). In particular the first column depicts the situation with no outlying observation, the second and third columns refer, respectively, to the presence of outliers in portfolio returns and outliers in constituent returns, while the last column depicts the case when outliers are considered both in portfolio returns and in constituent returns. The boxplots depict the sampling distributions of the QR estimators for $T = 250$ (left boxplot), $T = 500$ (middle boxplot) and $T = 1000$ (right boxplot)

dian regression, showing some empirical results for efficiency and consistency of the robust estimators. The results of the simulation study encourage us to further investigate this approach. A topic deserving further attention is a formal study of the robustness of the constrained median regression estimator in the presence of outliers in the X series based on the influence function of the constrained robust estimator.

It is worthwhile to point out that further gain in estimator efficiency can be obtained as the median regression has been estimated through quantile regression. Such a technique allows a simple extension toward L-estimators (defined as weighted linear combinations of different quantiles) in order to gain an increase in efficiency [15,19]. Moreover, many other robust estimators have been proposed and studied for linear regression models. However a comparison of their relative merits in the framework considered here is beyond the scope of this paper.

A further extension of the proposed approach concerns the use of quantile regression to draw inferences on style coefficients. The presence of inequality constraints in the style model, indeed, requires some caution in drawing inferences. Among the different proposals appearing in the literature, the Lobosco–Di Bartolomeo approximation [21] for computing corrected standard errors is widespread and it performs well for regular cases, i.e., when parameters are not on the boundaries of the parameter space. This proposal, indeed, is a convenient method for estimating confidence intervals for style coefficients based on a Taylor expansion. Nevertheless, as it is essentially based on a least-squares estimation procedure, the Lobosco–Di Bartolomeo solution also suffers from the presence of outliers. A possible solution could relate to a joint use of quantile regression and subsampling theory [23]. Subsampling was first introduced by Politis and Romano [22] and can be considered as the most general theory for the construction of first-order asymptotically valid confidence intervals or regions. The basic idea is to approximate the sampling distribution of the statistic of interest through the values of the statistic (suitably normalised) computed over smaller subsets of the data. Subsampling has been shown to be valid under very weak assumptions and, when compared to other resampling schemes such as the bootstrap, it does not require that the distribution of the statistic is somehow locally smooth as a function of the unknown model. Indeed, the subsampling is applicable even in situations that represent counterexamples to the bootstrap. These issues are still under investigation and beyond the scope of this paper. Here it is worth highlighting that preliminary results appear promising and encourage us to further investigate this approach: confidence intervals based on the joint use of QR and subsampling show better performance with respect to both coverage error and length of the intervals. The next step should concern an empirical analysis with real financial series.

Acknowledgement. The authors wish to thank the two anonymous referees for their helpful comments and suggestions on a previous draft of the paper; they helped to improve the final version of the work.

All computations and graphics were done in the R language [24], using the basic packages and the additional packages: fGarch [29], ggplot2 [27], mgcv [28] and quantreg [17].

The research work of Michele La Rocca benefits from the research structures of the STAT-LAB at the Department of Economics and Statistics, University of Salerno and of CFEPSR, Portici (Na). The research work of Domenico Vistocco is supported by Laboratorio di Calcolo ed Analisi Quantitativa, Department of Economics, University of Cassino.

References

1. Andrews, D.W.K.: Estimation when the parameter are on the boundary. Econometrica 67, 1341–1383 (1999)

2. Attardi, L., Vistocco, D.: Comparing financial portfolio style through quantile regression. Stat. Appl. 18, 6–16 (2006)
3. Attardi, L., Vistocco, D.: On estimating portfolio conditional returns distribution through style analysis models. In: Perna, C., Sibillo, M. (eds) Quantitative Methods for Finance and Insurance, pp. 11–17. Springer-Verlag, Milan (2007)
4. Attardi, L., Vistocco, D.: An index for ranking financial portfolios according to internal turnover. In: Greenacre, M., Lauro, N.C., Palumbo, F. (eds.) Studies in Classification, Data Analysis, and Knowledge Organization. Springer, Berlin-Heidelberg (in press)
5. Basset, G.W., Chen, H.L.: Portfolio style: return-based attribution using quantile regression. In: Fitzenberger, B., Koenker, R., Machado, J.A.F. (eds.) Economic Applications of Quantile Regression (Studies in Empirical Economics), pp. 293–305. Physica-Verlag, Heidelberg (2001)
6. Christodoulakis, G.A.: Sharpe style analysis in the msci sector portfolios: a Monte Carlo integration approach. Oper. Res. Int. J. 2, 123–137 (2003)
7. Christodoulakis, G.A.: Sharpe style analysis in the msci sector portfolios: a Monte Carlo integration approach. In: Knight, J., Satchell, S. (eds.) Linear Factor Models in Finance, pp. 83–94. Elsevier (2005)
8. Conversano, C., Vistocco, D.: Model based visualization of portfolio style analysis. In: Antock, J. (eds.) Proceedings of the International Conference "COMPSTAT 2004", pp. 815–822. Physica-Verlag (2004)
9. Conversano, C., Vistocco, D.: Analysis of mutual fund management styles: a modeling, ranking and visualizing approach. J. Appl. Stat. (accepted)
10. Davis, W.W.: Bayesian analysis of the linear model subject to linear inequality constraints. J. Am. Stat. Assoc. 78, 573–579 (1978)
11. Geweke, J.: Exact inference in the inequality constrained normal linear regression model. J. Appl. Econometrics 1, 127–141 (1986)
12. Horst, J.R., Nijman, T.E., de Roon, F.A.: Evaluating style analysis. J. Empirical Finan. 11, 29–51 (2004)
13. Judge, G.G., Takayama, T.: Inequality restrictions in regression analysis. J. Am. Stat. Assoc. 61, 166–181 (1966)
14. Kim, T.H., White, H., Stone, D.: Asymptotic and bayesian confidence intervals for sharpe-style weights. J. Finan. Econometrics 3, 315–343 (2005)
15. Koenker R.: A note on L-estimator for linear models. Stat. Prob. Lett. 2, 323–325 (1984)
16. Koenker, R.: Quantile regression. Econometric Soc. Monographs (2005)
17. Koenker, R.: Quantreg: quantile regression. R package version 4.10 (2007) http://www.r-project.org.
18. Koenker R., Basset, G.W.: Regression quantiles. Econometrica 46, 33–50 (1978)
19. Koenker R., Basset, G.W.: L-estimation for linear models. J. Am. Stat. Assoc. 82, 851–857 (1987) 20
20. Koenker R., Ng, P.: Inequality constrained quantile regression. Sankhya Ind. J. Stat. 67, 418–440 (2005) 21
21. Lobosco, A., Di Bartolomeo, D.: Approximating the confidence intervals for sharpe style weights. Finan. Anal. J. 80–85 (1997)
22. Politis, D.N., Romano, J.P.: Large sample confidence regions based on subsamples under minimal assumptions. Ann. Stat. 22, 2031–2050 (1994)
23. Politis, D.N., Romano, J.P., Wolf, M.: Subsampling. Springer-Verlag, New York (1999)
24. R Development Core Team.: R: A Language and Environment for Statistical Computing. R Foundation for Statistical Computing, Vienna. http://www.R-project.org.
25. Sharpe, W.: Asset allocation: management styles and performance measurement. J. Portfolio Man. 7–19 (1992)

26. Sharpe, W.: Determining a fund's effective asset mix. Invest. Man. Rev. 2, 59–69 (1998)
27. Wickham, H.: ggplot2: An implementation of the Grammar of Graphics. R package version 0.5.7, (2008) http://had.co.nz/ggplot2/.
28. Wood, S.N.: Generalized Additive Models: An Introduction with R. Chapman and Hall, CRC (2006)
29. Wuertz, D., Chalabi, Y., Miklovic, M. et al.: fGarch: Rmetrics – Autoregressive Conditional Heteroschedastic Modelling. R package version 260.72 (2007) http://www.rmetrics.org

Managing demographic risk in enhanced pensions

Susanna Levantesi and Massimiliano Menzietti

Abstract. This paper deals with demographic risk analysis in Enhanced Pensions, i.e., long-term care (LTC) insurance cover for the retired. Both disability and longevity risks affect such cover. Specifically, we concentrate on the risk of systematic deviations between projected and realised mortality and disability, adopting a multiple scenario approach. To this purpose we study the behaviour of the random risk reserve. Moreover, we analyse the effect of demographic risk on risk-based capital requirements, explaining how they can be reduced through either safety loading or capital allocation strategies. A profit analysis is also considered.

Key words: long term care covers, enhanced pension, demographic risks, risk reserve, solvency requirements

1 Introduction

The "Enhanced Pension" (EP) is a long-term care (LTC) insurance cover for the retired. It offers an immediate life annuity that is increased once the insured becomes LTC disabled and requires a single premium. EP is affected by demographic risks (longevity and disability risks) arising from the uncertainty in future mortality and disability trends that cause the risk of systematic deviations from the expected values. Some analyses of these arguments have been performed by Ferri and Olivieri [1] and Olivieri and Pitacco [6]. To evaluate such a risk we carry out an analysis taking into account a multiple scenario approach. To define a set of projected scenarios we consider general population statistics of mortality and disability.

We firstly analyse the behaviour of the risk reserve, then we define the capital requirements necessary to guarantee the solvency of the insurer. Finally we study the portfolio profitability. Such an analysis cannot be carried out by analytical tools, but requires a Monte Carlo simulation model.

The paper is organised as follows. In Section 2 we define the actuarial framework for EPs. In Section 3 we develop nine demographic scenarios and describe through a suitable model how they can change over time. In Section 4 we present a risk theory model based on the portfolio risk reserve and the Risk Based Capital requirements necessary to preserve the insurance company from failures with a fixed confidence

level. Section 5 deals with the profit analysis of the portfolio according to the profit
profile. Simulation results are analysed in Section 6, while some concluding remarks
are presented in Section 7.

2 Actuarial model for Enhanced Pensions

The probabilistic framework of an EP is defined consistently with a continuous and
inhomogeneous multiple state model (see Haberman and Pitacco [2]). Let $S(t)$ represent the random state occupied by the insured at time t, for any $t \geq 0$, where t is
the policy duration and 0 the time of entry. The possible realisations of $S(t)$ are: 1 =
"active" (or healthy), 2 = "LTC disabled" or 3 = "dead". We disregard the possibility
of recovery from the LTC state due to the usually chronic character of disability and
we assume $S(0) = 1$. Let us define transition probabilities and intensities:

$$P_{ij}(t,u) = \Pr\{S(u) = j \mid S(t) = i\} \quad 0 \leq t \leq u, \quad i,j \in \{1,2,3\}, \tag{1}$$

$$\mu_{ij}(t) = \lim_{u \to t} \frac{P_{ij}(t,u)}{u-t} \quad t \geq 0, \quad i,j \in \{1,2,3\}, \quad i \neq j. \tag{2}$$

EPs are single premium covers providing an annuity paid at an annual rate $b_1(t)$ when
the insured is healthy and an enhanced annuity paid at an annual rate $b_2(t) > b_1(t)$
when the insured is LTC disabled. Let us suppose all benefits to be constant with
time. Let ω be the maximum policy duration related to residual life expectancy at age
x and let $v(s,t) = \prod_{h=s+1}^{t} v(h-1,h)$ be the value at time s of a monetary unit at
time t; the actuarial value at time 0 of these benefits, $\Pi(0,\omega)$, is given by:

$$\Pi(0,\omega) = b_1 a_{11}(0,\omega) + b_2 a_{12}(0,\omega), \tag{3}$$

where: $a_{ij}(t,u) = \sum_{s=t}^{u-t-1} P_{ij}(t,s)v(s,t)$ for all $i, j \in 1, 2$. Assuming the equivalence
principle, the gross single premium paid in $t = 0$, Π^T is defined as:

$$\Pi^T = \frac{\Pi(0,\omega)}{1 - \alpha - \beta - \gamma \left[a_{11}(0,\omega) + a_{21}(0,\omega)\right]}, \tag{4}$$

where α, β and γ represent the premium loadings for acquisition, premium earned
and general expenses, respectively.

3 Demographic scenarios

Long-term covers, such as the EPs, are affected by demographic trends (mortality and
disability). A risk source in actuarial evaluations is the uncertainty in future mortality
and disability; to represent such an uncertainty we adopt different projected scenarios.
    We start from a basic scenario, H_B, defined according to the most recent statistical
data about people reporting disability (see ISTAT [3]) and, consistent with this data,

the Italian Life-Table SIM-1999. Actives' mortality $\mu_{13}(t)$ is approximated by the Weibull law, while transition intensities $\mu_{12}(t)$ are approximated by the Gompertz law (for details about transition intensities' estimation see Levantesi and Menzietti [5]). Disabled mortality intensity $\mu_{23}(t)$ is expressed in terms of $\mu_{13}(t)$ according to the time-dependent coefficient $K(t)$, $\mu_{23}(t) = K(t)\mu_{13}(t)$. Values of $K(t)$, coming from the experience data of an important reinsurance company, are well approximated by the function $\exp(c_0 + c_1 t + c_2 t^2)$.

Mortality of projected scenarios has been modelled evaluating a different set of Weibull parameters (α, β) for each ISTAT projection (low, main and high hypothesis, see ISTAT [4]). Furthermore, the coefficient $K(t)$ is supposed to be the same for all scenarios. Regarding transition intensity, $\mu_{12}(t)$, three different sets of Gompertz parameters have been defined starting from a basic scenario to represent a 40% decrease (Hp. a), a 10% decrease (Hp. b) and a 20% increase (Hp. c) in disability trend, respectively. By combining mortality and disability projections we obtain nine scenarios.

We assume that possible changes in demographic scenarios occur every k years, e.g., in numerical implementation 5 years is considered a reasonable time to capture demographic changes. Let $H(t)$ be the scenario occurring at time t ($t = 0, k, 2k, \ldots$). It is modelled as a time-discrete stochastic process. Let $\bar{P}(t)$ be the vector of scenario probabilities at time t and $\mathbf{M}(t)$ the matrix of scenario transition probabilities between t and $t + k$. The following equation holds: $\bar{P}(t + k) = \bar{P}(t) \cdot \mathbf{M}(t)$.

We suppose that at initial time the occurring scenario is the central one. We assume that the stochastic process $H(t)$ is time homogeneous ($\mathbf{M}(t) = \mathbf{M}, \forall t$) and the scenario probability distribution, $\bar{P}(t)$, is stationary after the first period, so that $\bar{P}(t) = \bar{P}$, $\forall t \geq k$. Note that $\bar{P}$ is the left eigenvector of the transition matrix $\mathbf{M}$ corresponding to the eigenvalue 1. Values of $\bar{P}$ are assigned assuming the greatest probability of occurrence for the central scenario and a correlation coefficient between mortality and disability equal to 75%:

$$\bar{P} = (\,0.01\ \ 0.03\ \ 0.16\ \ 0.03\ \ 0.54\ \ 0.03\ \ 0.16\ \ 0.03\ \,0.01\,).$$

Further, we assume that transitions between strongly different scenarios are not possible in a single period and consistently with supposed correlation between mortality and disability, some transitions are more likely than others.

Resulting scenarios' transition probabilities are reported in the matrix below.

$$\mathbf{M} = \begin{pmatrix}
0.1650 & 0.1775 & 0.0000 & 0.1775 & 0.4800 & 0.0000 & 0.0000 & 0.0000 & 0.0000 \\
0.0492 & 0.1850 & 0.0933 & 0.0592 & 0.5100 & 0.1033 & 0.0000 & 0.0000 & 0.0000 \\
0.0000 & 0.0100 & 0.4250 & 0.0000 & 0.5550 & 0.0100 & 0.0000 & 0.0000 & 0.0000 \\
0.0492 & 0.0592 & 0.0000 & 0.1850 & 0.5100 & 0.0000 & 0.0933 & 0.1033 & 0.0000 \\
0.0100 & 0.0300 & 0.1600 & 0.0300 & 0.5400 & 0.0300 & 0.1600 & 0.0300 & 0.0100 \\
0.0000 & 0.1033 & 0.0933 & 0.0000 & 0.5100 & 0.1850 & 0.0000 & 0.0592 & 0.0492 \\
0.0000 & 0.0000 & 0.0000 & 0.0100 & 0.5550 & 0.0000 & 0.4250 & 0.0100 & 0.0000 \\
0.0000 & 0.0000 & 0.0000 & 0.1033 & 0.5100 & 0.0592 & 0.0933 & 0.1850 & 0.0492 \\
0.0000 & 0.0000 & 0.0000 & 0.0000 & 0.4800 & 0.1775 & 0.0000 & 0.1775 & 0.1650
\end{pmatrix}$$

4 A risk theory model

Demographic risk analysis is carried out on a portfolio of EPs with $N_i(t)$ contracts in state i at time t, closed to new entries. The random risk reserve is adopted as risk measure. It represents the insurer's ability to meet liabilities, therefore it can be considered a valid tool to evaluate the insurance company solvency and, more generally, in the risk management assessment. Let $U(0)$ be the value of the risk reserve at time 0; the risk reserve at the end of year t is defined as:

$$U(t) = U(t-1) + P^T(t) + J(t) - E(t) - B(t) - \Delta V(t) - K(t), \qquad (5)$$

where:

- $P^T(t)$ is the gross single premiums income;
- $J(t)$ are the investment returns on assets, $A(t)$, where the assets are defined as $A(t) = A(t-1) + P^T(t) - E(t) - B(t) + J(t) - K(t)$;
- $E(t)$ are the expenses: $E(t) = \sum_{i=1,2} N_i(t-1)\epsilon_i(t)$;
- $B(t)$ is the outcome for benefits: $B(t) = \sum_{i=1,2} N_i(t-1)b_i$;
- $\Delta V(t)$ is the annual increment in technical provision, $V(t) = \sum_{i=1,2} N_i(t)V_i(t)$, and $V_i(t)$ is the technical provision for an insured in state i;
- $K(t)$ are the capital flows; if $K(t) > 0$ the insurance company distributes dividends and if $K(t) < 0$ stockholders invest capital.

We assume that premiums, benefits, expenses and capital flows are paid at the beginning of each year. To compare outputs of different scenarios and portfolios we use the ratio between risk reserve and total single premium income

$$u(t) = \frac{U(t)}{\Pi(0, \omega)N_1(0)}.$$

The risk analysis is performed according to a multiple scenarios approach that considers each scenario as a possible state of the stochastic process $H(t)$, according to the probability vector $\bar{P}$, allowing evaluation of the risk of systematic deviations in biometric functions (see Olivieri and Pitacco [6] and Levantesi and Menzietti [5]).

The demographic pricing basis is defined according to the central scenario with a safety loading given by a reduction of death probabilities. We disregard financial risk, adopting a deterministic and constant interest rate. We assume a financial pricing basis equal to the real-world one. Technical provision is reviewed every 5 years consistently with the scenario change period. Further, the insurance company perceives scenario changes with a delay of one period.

4.1 Risk-based capital requirements

Risk-based capital (RBC) is a method for assessing the solvency of an insurance company; it consists in computing capital requirements that reflect the size of overall risk exposures of an insurer. Let us consider RBC requirements based on risk reserve distribution. We calculate RBC requirements with different time horizons and confidence levels. Let us define the finite time ruin probability as the probability of being in a ruin state in at least one of the time points $1, 2. . ., T$, for a given $U(0) = u$:

$$\Psi_u(0, T) = 1 - Pr\left\{\bigcap_{t=1}^{T} U(t) \geq 0 \middle| U(0) = u\right\}. \tag{6}$$

RBC requirements for the time horizon $(0, T)$ with a $(1 - \epsilon)$ confidence level are defined as follows:

$$RBC_{1-\epsilon}(0, T) = \inf\left\{U(0) \geq 0 \middle| \Psi_0(0, T) < \epsilon\right\}. \tag{7}$$

Note that the risk reserve must be not negative for all $t \in (0, T)$.

To make data comparable, results are expressed as a ratio between RBC requirements and total single premium income

$$rbc_{1-\epsilon}(0, T) = \frac{RBC_{1-\epsilon}(0, T)}{\Pi(0, \omega)N_1(0)}.$$

An alternative method to calculate RBC requirements is based on the Value-at-Risk (VaR) of the U-distribution in the time horizon $(0, T)$ with a $(1 - \epsilon)$ confidence level: $VaR_{1-\epsilon}(0, T) = -U_\epsilon(T)$, where $U_\epsilon(t)$ is the ϵ-th quantile of the U-distribution at time t. Hence RBC requirements are given by:

$$RBC_{1-\epsilon}^{VaR}(0, T) = VaR_{1-\epsilon}(0, T)v(0, T). \tag{8}$$

If an initial capital $U(0)$ is given, the $RBC_{1-\epsilon}^{VaR}(0, T)$ requirements increase by the amount $U(0)$. Values are reported in relative terms as

$$rbc_{1-\epsilon}^{VaR}(0, T) = \frac{RBC_{1-\epsilon}^{VaR}(0, T)}{\Pi(0, \omega)N_1(0)}.$$

5 Profit analysis

In this section we analyse the annual profit, $Y(t)$, emerging from the management of the portfolio. In order to capture the profit sources, $Y(t)$ can be broken down into insurance profit, $Y^I(t)$, and profit coming from investment income on shareholders' funds (which we call "patrimonial profit"), $Y^P(t)$.

$$Y^I(t) = (1 + i(t - 1, t))[V(t - 1) + P^T(t) - E(t) - B(t)] - V(t) \tag{9}$$

$$Y^P(t) = U(t - 1)i(t - 1, t) \tag{10}$$

The following relation holds: $Y(t) = Y^I(t) + Y^P(t)$. The sequence $\{Y(t)\}_{t \geq 1}$ is called profit profile. Let ρ be the rate of return on capital required by the shareholders; the present value of future profits discounted at rate ρ (with $\rho > i$), $Y(0, T)$ is given by:

$$Y(0, T) = \sum_{t=1}^{T} Y(t)v_\rho(0, t), \tag{11}$$

while $Y^I(0, T) = \sum_{t=1}^{T} Y^I(t)v_\rho(0, t)$ and $Y^P(0, T) = \sum_{t=1}^{T} Y^P(t)v_\rho(0, t)$ are the present value of the future insurance and patrimonial profits, respectively.

6 Portfolio simulation results

Let us consider a cohort of 1000 policyholders, males, with the same age at policy issue, $x = 65$, same year of entry (time 0), a maximum policy duration $\omega = 49$, expense loadings $\alpha = 5\%$, $\beta = 2\%$, $\gamma = 0.7\%$ and a constant interest rate $i(0, t) = i = 3\%$ $\forall t$. The annual benefit amounts are distributed as in Table 1.

Table 1. Annual benefit amounts distribution (euros)

b_1	b_2	$fr(\%)$
6,000	12,000	40
9,000	18,000	30
12,000	24,000	15
15,000	27,000	10
18,000	30,000	5

Results of 100,000 simulations are reported in the following tables, assuming a safety loading on demographic pricing bases given by a 10% reduction of healthy and disabled death probabilities and an initial capital $K(0) = RBC_{99.5\%}(0, 1)$. Simulated values of $u(t)$ are shown in Figure 1. The figure highlights the strong variability of the risk reserve distribution, especially when $t > 5$, as a consequence of demographic scenario changes. Even though the risk reserve has a positive trend due to safety loading, lower percentiles are negative. Economic consequences of such an aspect are

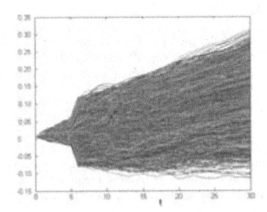

Fig. 1. $u(t)$ with safety loading = 10% reduction of death probabilities, initial capital $K(0) = RBC_{99.5\%}(0, 1)$

Table 2. Moments of $u(t)$ and the finite time ruin probability with initial capital $K(0) = RBC_{99.5\%}(0, 1)$, safety loading = 10% reduction of death probabilities

$u(T)$	$T = 1$	$T = 5$	$T = 10$	$T = 20$	$T = 30$
Mean (%)	0.79	1.51	2.75	6.61	10.94
Std Dev (%)	0.37	1.26	4.58	5.27	6.36
Coeff Var	0.4673	0.8372	1.6614	0.7974	0.5817
Skew	0.4495	0.2867	0.0010	−0.0194	−0.0002
$\Psi_u(0, T)(\%)$	0.50	15.06	28.98	40.67	40.84

relevant for the insurer solvency and will be quantified through solvency requirements. In Table 2 we report the values of the $u(t)$ moments and the coefficient of variation as well as the finite time ruin probability. It can be noticed that expected values of $u(t)$ are always positive and increase with time as well as the standard deviation. Looking at the coefficient of variation we observe an increase of relative variability up to $t = 10$; thereafter it decreases. Such a behaviour demonstrates that demographic risk is mainly caused by the scenario changes (perceived with a delay of 5 years) affecting the evaluation of technical provisions. When technical provisions decrease, the coefficient of variation of $u(t)$ becomes steady. The risk tendency to become stable is confirmed by the finite time ruin probability values that increase with time. As expected, $\Psi_u(0, 1)$ is consistent with the initial capital provision, $K(0) = RBC_{99.5\%}(0, 1)$.

Table 3 shows the values of RBC requirements for three different confidence levels: 98%, 99% and 99.5%. RBC values rise with time and become steady in $T = 20$ only if RBC is computed on a time horizon $(0, T)$, rather than at time T. On the other hand, if we look at RBC computed according to VaR, we obtain lower values with respect to the previous ones, especially for $T > 10$. Results show that the initial capital should be increased by about 6% of the single premium income to guarantee the insurance solvency on the portfolio time horizon.

Table 3. Risk-based capital with safety loading = 10% reduction of death probabilities, initial capital $K(0) = RBC_{99.5\%}(0, 1)$

$rbc_{1-\epsilon}(0, T)$	$T = 1$	$T = 5$	$T = 10$	$T = 20$	$T = 30$
$\epsilon = 0.5\%$	0.67%	1.78%	6.52%	6.57%	6.57%
$\epsilon = 1.0\%$	0.62%	1.62%	6.20%	6.24%	6.24%
$\epsilon = 2.0\%$	0.55%	1.43%	5.91%	5.94%	5.94%
$rbc_{1-\epsilon}^{VaR}(0, T)$	$T = 1$	$T = 5$	$T = 10$	$T = 20$	$T = 30$
$\epsilon = 0.5\%$	0.67%	1.77%	6.03%	4.07%	2.63%
$\epsilon = 1.0\%$	0.62%	1.60%	5.75%	3.61%	2.12%
$\epsilon = 2.0\%$	0.55%	1.39%	5.30%	3.01%	1.48%

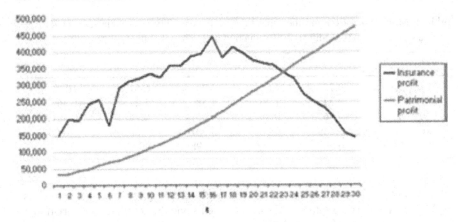

Fig. 2. Expected value of annual insurance and patrimonial profit with safety loading = 10% reduction of death probabilities, initial capital $K(0) = RBC_{99.5\%}(0, 1)$

Table 4. Moments of $u(t)$ and the finite time ruin probability with initial capital $K(0) = 0$, safety loading = 10% reduction of death probabilities

$u(T)$	$T = 1$	$T = 5$	$T = 10$	$T = 20$	$T = 30$
Mean (%)	0.10	0.73	1.85	5.40	9.31
Std Dev (%)	0.37	1.26	4.58	5.27	6.36
Skew	0.4495	0.2867	0.0010	−0.0194	−0.0002
$\Psi_u(0, T)$ (%)	42.06	58.42	62.37	67.72	67.79

Figure 2 shows the expected values of annual profit components as stated in Section 5. The insurance profit line shows greater variability, being affected by demographic risks. Meanwhile, the patrimonial profit line is more regular due to the absence of financial risk, and increases with time, depending on investments of risk reserve (return produced by the investment of risk reserve).

In order to evaluate the effect of different initial capital provisions, we fix $K(0) = 0$. Further, according to ISVAP (the Italian insurance supervisory authority), which shares the minimum solvency margin in life insurance to face demographic and financial risk in 1% and 3% of technical provisions, respectively, we fix $K(0) = 1\% V(0^+)$. The moments of $u(t)$ distribution and the ruin probability are reported in Tables 4 and 5. They can be compared with the results of Table 2.

Note that $K(0)$ values do not affect the standard deviation and skewness of $u(t)$ distribution, while they do influence the $u(t)$ expected value, which increases when $K(0)$ rises. Now, let us consider the highest safety loading given by a 20% reduction of death probabilities for both healthy and disabled people. Values of the moments of $u(t)$ and finite time ruin probabilities are reported in Table 6. If we compare these values with the ones in Table 2 (where the safety loading is equal to a 10% reduction of death probabilities), we find that safety loading strongly affects the expected values of $u(t)$, but does not significantly affect the standard deviation and skewness. In other

Table 5. Moments of $u(t)$ and the finite time ruin probability with initial capital $K(0) = 1\%V(0^+)$, safety loading = 10% reduction of death probabilities

$u(T)$	$T = 1$	$T = 5$	$T = 10$	$T = 20$	$T = 30$
Mean (%)	1.13	1.89	3.20	7.21	11.74
Std Dev (%)	0.37	1.26	4.58	5.27	6.36
Skew	0.4495	0.2867	0.0010	−0.0194	−0.0002
$\Psi_u(0, T)$ (%)	0.00	7.02	24.18	36.69	36.82

Table 6. Moments of capital ratio and the finite time ruin probability with initial capital $K(0) = RBC_{99.5\%}(0, 1)$, safety loading = 20% reduction of death probabilities

$u(T)$	$T = 1$	$T = 5$	$T = 10$	$T = 20$	$T = 30$
Mean (%)	0.78	2.07	4.49	12.01	20.39
Std Dev (%)	0.37	1.26	4.50	5.25	6.20
Skew	0.4544	0.2892	0.0120	−0.0197	−0.0038
$\Psi_u(0, T)$ (%)	0.50	7.07	22.21	30.56	30.56

Table 7. Risk-based capital with safety loading = 20% reduction of death probabilities, initial capital $K(0) = RBC_{99.5\%}(0, 1)$

$rbc_{1-\epsilon}(0, T)$	$T = 1$	$T = 5$	$T = 10$	$T = 20$	$T = 30$
$\epsilon = 0.5\%$	0.57%	1.23%	5.39%	5.39%	5.39%
$\epsilon = 1.0\%$	0.52%	1.09%	5.14%	5.14%	5.14%
$\epsilon = 2.0\%$	0.46%	0.92%	4.88%	4.88%	4.88%
$rbc_{1-\epsilon}^{VaR}(0, T)$	$T = 1$	$T = 5$	$T = 10$	$T = 20$	$T = 30$
$\epsilon = 0.5\%$	0.57%	1.18%	4.52%	0.97%	−1.52%
$\epsilon = 1.0\%$	0.52%	1.02%	4.24%	0.49%	−2.02%
$\epsilon = 2.0\%$	0.46%	0.81%	3.78%	−0.12%	−2.65%

words, safety loading reduces the probability of risk reserve to become negative (as proved by the $\Psi_u(0, T)$ values), but does not lower its variability. Moreover, required capital decreases with safety loading increase (see Table 7 compared with Table 3). Note that in the long term the rbc^{VaR} are negative, therefore an initial capital is not necessary to guarantee the insurance solvency. Nonetheless, the requirement reduction is financed by the policyholders through a premium increase – due to a higher safety loading – making the insurance company less competitive on the market. Therefore, it is important to combine a solvency target with commercial policies.

Now, let us consider the safety loading impact on $Y(0, T)$ discounted at a rate $\rho = 5\% > i$ (see (11)). As expected, $Y(0, T)$ rises with an increase of safety loading: the values – expressed as a ratio to total single premium income – move from 6.21% to 11.85% when the safety loading rises. It is worth noting that looking at the profit

Table 8. Present value of future profits as ratio to total single premium income, $Y(0, T)/(\Pi(0, \omega)N_1(0))$ with initial capital $K(0) = RBC_{99.5\%}(0, 1)$

	SL = 10%	SL = 20%	Δ%
Total	6.21%	11.85%	91%
Insurance	3.03%	6.17%	104%
Patrimonial	3.18%	5.69%	79%

sources separately, we have a higher increase in $Y^I(0, T)$ than in $Y^P(0, T)$: 104% compared to 79%.

7 Conclusions

This paper focuses on disability and longevity risks arising from issues in the estimate of residual life expectancy of healthy and disabled people. Our analysis highlights that the EPs are affected by a significant demographic risk caused by systematic deviations between expected and realised demographic scenarios. The results confirm that such a risk is difficult to control, depending on uncertainty in the future evolution of biometric functions. The risk reserve distribution shows a strong variability due to the demographic scenario changes affecting the technical provision valuation. Since such a risk is systematic, the $u(t)$ variability does not lessen when either safety loading or initial capital increase. Nonetheless, they are useful tools in managing demographic risk because they significantly reduce the ruin probability of the insurance company as far as the RBC requirements necessary to ensure the insurer solvency at a fixed confidence level. In this paper we take into account an initial capital only, reducing the probability of incurring losses. However, the risk of systematic deviations persists, requiring an appropriate capital allocation strategy. This topic will be the subject of future research together with a suitable reinsurance strategy.

References

1. Ferri, S., Olivieri, A.: Technical bases for LTC covers including mortality and disability projections. In: Proceedings of the 31st ASTIN Colloquium, Porto Cervo, Italy (2000)
2. Haberman, S., Pitacco, E.: Actuarial Models for Disability Insurance. Chapman and Hall, London (1999)
3. ISTAT: Le condizioni di salute e ricorso ai servizi sanitari. ISTAT, Rome (2000)
4. ISTAT: Previsioni della popolazione residente per sesso eta'e regione dal 1.1.2001 al 1.1.2051. ISTAT, Rome (2002)
5. Levantesi, S., Menzietti, M.: Longevity and disability risk analysis in enhanced life annuities. In: Proceedings of the 1st LIFE Colloquium, Stockholm (2007)
6. Olivieri, A., Pitacco, E.: Facing LTC risks. In: Proceedings of the 32nd ASTIN Colloquium, Washington (2001)

Clustering mutual funds by return and risk levels

Francesco Lisi and Edoardo Otranto

Abstract. Mutual funds classifications, often made by rating agencies, are very common and sometimes criticised. In this work, a three-step statistical procedure for mutual funds classification is proposed. In the first step fund time series are characterised in terms of returns. In the second step, a clustering analysis is performed in order to obtain classes of homogeneous funds with respect to the risk levels. In particular, the risk is defined starting from an Asymmetric Threshold-GARCH model aimed to describe minimum, normal and turmoil risk. The third step merges the previous two. An application to 75 European funds belonging to 5 different categories is presented.

Key words: clustering, GARCH models, financial risk

1 Introduction

The number of mutual funds has grown dramatically over recent years. This has led to a number of classification schemes that should give reliable information to investors on features and performance of funds. Most of these classifications are produced by national or international rating agencies. For example, Morningstar groups funds into categories according to their actual investment style, portfolio composition, capitalisation, growth prospects, etc. This information is then used, together with that related to returns, risks and costs, to set up a more concise classification commonly referred to as Star Rating (see [11] for details). Actually, each rating agency has a specific owner evaluation method and also national associations of mutual funds managers keep and publish their own classifications.

Problems arise as, in general, classes of different classifications do not coincide. Also, all classification procedures have some drawback; for example, they are often based on subjective information and require long elaboration time (see, for example, [15]).

In the statistical literature, classification of financial time series has received relatively little attention. In addition, to the best of our knowledge, there are no comparisons between different proposed classifications and those of the rating agencies. Some authors use only returns for grouping financial time series. For example, [15] propose a classification scheme that combines different statistical methodologies (principal component analysis, clustering analysis, Sharpe's constrained regression) applied on past returns of the time series. Also, the clustering algorithm proposed by [9], referring to different kinds of functions, is based only on return levels. Other authors based their classifications only on risk and grouped the assets according to the distance between volatility models for financial time series [2, 8, 12–14]. Risk-adjusted returns, i.e., returns standardised through standard deviation, are used for clustering time series by [4]. This approach is interesting, but using the unconditional variance as a measure of risk and ignoring the dynamics of volatility seems too simplistic.

In this paper, a classification based only on the information contained in the net asset value (NAV) time series is considered. It rests on the simple and largely agreed idea that two very important points in evaluation of funds are return and risk levels. In order to measure the return level, the mean annual net period return is considered. As regards the riskiness, in the time series literature, it is commonly measured in terms of conditional variance (volatility) of a time series. As is well known, volatility is characterised by a time-varying behaviour and clustering effects, which imply that quiet (low volatility) and turmoil (high volatility) periods alternate. In order to account both for the time-varying nature of volatility and for its different behaviour in quiet and turmoil periods, an asymmetric version of the standard Threshold GARCH model [5, 17], is considered in this work.

The whole classification scheme consists of three steps: the first groups funds with respect to returns whereas the second groups them with respect to riskiness. In particular, the whole risk is broken down into constant minimum risk, time-varying standard risk and time-varying turmoil risk. Following [12,13] and [14], the clustering related to volatility is based on a distance between GARCH models, which is an extension of the AR metric introduced by [16]. Lastly, the third step merges the results of the first two steps to obtain a concise classification.

The method is applied to 75 funds belonging to five categories: aggressive balanced funds, prudential balanced funds, corporate bond investments, large capitalisation stock funds and monetary funds. In order to make a comparison with the classification implied by the Morningstar Star Rating, which ranges from 1 to 5 stars, our clustering is based on 5 "stars" as well. As expected, our classification does not coincide with the Morningstar Rating because it is only partially based on the same criteria. Nevertheless, in more than 82% of the considered funds the two ratings do not differ for more than one star.

The paper is organised as follows. Section 2 describes how the risk is defined. Section 3 contains an application and the comparison of our clustering with the Morningstar Rating classification. Section 4 concludes.

2 Risk modelling

In this section the reference framework for fund riskiness modelling is described. Let y_t be the time series of the NAV of a fund and r_t the corresponding log-return time series. We suppose that the return dynamics can be described by the following model:

$$r_t = \mu_t + \varepsilon_t = \mu_t + h_t^{1/2} u_t, \qquad t = 1, \ldots, T$$

$$\varepsilon_t \mid I_{t-1} \sim N(0, h_t),$$

$$(1)$$

where $\mu_t = E_{t-1}(r_t)$ is the conditonal expectation and u_t is an i.i.d. zero-mean and unit variance innovation. The conditional variance h_t follows an asymmetric version of the Threshold GARCH(1,1) process [5, 17], which stresses the possibility of a different volatility behaviour in correspondence with high negative shocks. We refer to it as the Asymmetric Threshold GARCH (AT-GARCH) model. Formally, the conditional variance can be described as:

$$h_t = \gamma + \alpha \varepsilon_{t-1}^2 + \beta h_{t-1} + \delta S_{t-1} \varepsilon_{t-1}^2$$

$$S_t = \begin{cases} 1 & \text{if } \varepsilon_t < \varepsilon_t^* \\ 0 & \text{otherwise} \end{cases},$$

$$(2)$$

where $\gamma, \alpha, \beta, \delta$ are unknown parameters, whereas ε_t^* is a threshold identifying the turmoil state. The value of ε_t^* could represent a parameter to be estimated, but in this work we set it equal to the first decile of the empirical distribution of ε. On the whole, this choice maximises the likelihood and the number of significant estimates of δ. Also, the first decile seems suitable because it provides, through the parameter δ, the change in the volatility dynamics when high – but not extreme – negative returns occur.

The purpose of this work is to classify funds in terms of gain and risk. While the net period return is the most common measure of gain, several possible risk measures are used in the literature. However, most of them look at specific aspects of riskiness: standard deviation gives a medium constant measure; Value-at-Risk tries to estimate an extreme risk; the time-varying conditional variance in a standard GARCH model focuses on the time-varying risk, and so on.

In this paper we make an effort to jointly look at risk from different points of view. To do this, following [13], we consider the squared disturbances ε_t^2 as a proxy of the instantaneous volatility of r_t. It is well known that ε_t^2 is a conditionally unbiased, but very noisy, estimator of the conditional variance and that realised volatility and intra-daily range are, in general, better estimators [1, 3, 10]. However, the adoption of ε_t^2 in our framework is justified by practical motivations because intra-daily data are not available for mutual funds time series and, thus, realised volatility or range are not feasible. Starting from (2), after simple algebra, it can be shown that, for an AT-GARCH(1,1), ε_t^2 follows the ARMA(1, 1) model:

$$\varepsilon_t^2 = \gamma + \left(\alpha + \delta S_{t-j} + \beta \right) \varepsilon_{t-1}^2 - \beta \left(\varepsilon_{t-1}^2 - h_{t-1} \right) + \left(\varepsilon_t^2 - h_t \right), \qquad (3)$$

where $(\varepsilon_t^2 - h_t)$ are uncorrelated, zero-mean errors.

The AR(∞) representation of (3) is:

$$\varepsilon_t^2 = \frac{\gamma}{1 - \beta} + \sum_{j=1}^{\infty} (\alpha + \delta S_{t-j}) \beta^{j-1} \varepsilon_{t-j}^2 + \left(\varepsilon_t^2 - h_t \right), \qquad (4)$$

from which it is easy to derive the expected value at time t given past information

$$E_{t-1}(\varepsilon_t^2) = \frac{\gamma}{1-\beta} + \sum_{j=1}^{\infty} (\alpha + \delta S_{t-1}) \beta^{j-1} \varepsilon_{t-j}^2. \qquad (5)$$

This representation splits the expected volatility, $E_{t-1}(\varepsilon_t^2)$, considered as a whole measure of risk, into three positive parts: a constant part, $\gamma/(1-\beta)$, representing the minimum risk level which can be reached given the model; the time-varying standard risk $(\sum_{j=1}^{\infty} \alpha \beta^{j-1} \varepsilon_{t-j}^2)$ and the time-varying turmoil risk $(\sum_{j=1}^{\infty} \delta S_{t-j} \beta^{j-1} \varepsilon_{t-j}^2)$, the last two being dependent on past information. Of course, the estimation of expression (5) requires a finite truncation.

In order to classify funds with respect to all three risk components, we propose considering the distance between an homoskedastic model and a GARCH(1,1) model. Using the metric introduced by [12] and re-considered by [14], in the case of specification (2) this distance is given by:

$$\frac{\alpha + \delta S_{t-1}}{\sqrt{(1-\beta^2)}}. \qquad (6)$$

The previous analytical formulation allows us to provide a vectorial description of the risk of each fund. In particular, we characterise the *minimum constant risk* through the distance between the zero-risk case ($\gamma = \alpha = \beta = \delta = 0$) and the $\alpha = \delta = 0$ case

$$v_m = \frac{\gamma}{1-\beta}. \qquad (7)$$

The *time-varying standard risk* is represented, instead, by the distance between a GARCH(1,1) model ($\delta = 0$) and the corresponding homoskedastic model ($\alpha = \beta = \delta = 0$)

$$v_s = \frac{\alpha}{\sqrt{(1-\beta^2)}}. \qquad (8)$$

Lastly, the *turmoil risk* is described by the difference of the distance between an AT-GARCH model, and the homoskedastic model and the distance measured by (8):

$$v_t = \frac{\delta}{\sqrt{(1-\beta^2)}}. \qquad (9)$$

The whole risk is then characterised by the vector $[v_m, v_s, v_t]'$. If an extra element, accounting for the return level, $\bar{r}$, is considered, each fund may be featured by the vector:

$$\mathbf{f} = [\bar{r}, v_m, v_s, v_t]'.$$

In order to obtain groups of funds with similar return and risk levels, some clustering algorithm can be easily applied directly to $\mathbf{f}$ or to some function of the elements of $\mathbf{f}$. For example, in the next section risk will be defined as the average of v_m, v_s and v_t.

3 An application

As an application of the previously described procedure, the daily time series of NAV of 75 funds of the Euro area and belonging to five different categories were considered. The five typologies are the aggressive balanced, prudential balanced, corporate bond investments, large capitalisation stock and monetary funds. Data, provided by Bloomberg, range from 1/1/2002 to 18/2/2008, for a total of 1601 observations for each series.

Our experiment consists in providing a classification of these funds, characterising each group in terms of return and riskiness (following the definitions of constant minimum, time-varying standard and time-varying turmoil risk) and comparing our classification with that produced by the Morningstar star rating.

For each fund the return time series was considered and for each calendar year the net percentage return was computed; finally the average of the one-year returns, $\bar{r}$, was used to represent the gain.

To describe riskiness, first model (1)–(2) was estimated for each fund. When parameters were not significant at the 5% level, they were set equal to zero and the corresponding constrained model was estimated. Of course, before accepting the model the absence of residual ARCH effects in the standardised residuals was checked. Parameter estimation allowed us to calculate the risks defined as in (7), (8) and (9) and to characterise the funds by the elements $\bar{r}, v_m, v_s, v_t$ or by some functions of them.

With these vectors a clustering analysis was performed. In the clustering, a classical hierarchical algorithm with the Euclidean distance was used, whereas distances between clusters are calculated following the average-linkage criterion (see, for example, [7]).[1] In particular, the classification procedure followed three steps:

1. The series were classified into three groups, referring only to the degree of gain, i.e., $\bar{r}$ low, medium and high.

2. The series were classified into three groups only with respect to the degree of risk (low, medium and high). To summarise the different kinds of risk, the average of the three standardised risks was computed for each series. Standardisation is important because of the different magnitudes of risks; for example, minimum risk generally has an order of magnitude lower than that of the other two risks.

3. The previous two classifications were merged, combining the degree or gain and risk so as to obtain a rating from 1 to 5 "stars"; in particular, denoting with h, m and l the high, medium and low levels respectively and with the couple (a, b) the levels of gain and risk (with $a, b = h, m, l$), stars were assigned in the following way:

   1 star for (l, h) (low gain and high risk);
   2 stars for (l, m), (l, l) (low gain and medium risk, low gain and low risk);
   3 stars for (m, h), (m, m), (h, h) (medium gain and high risk, medium gain and medium risk, high gain and high risk);

---

[1] The clustering was performed also using the Manhattan distance and the single-linkage and complete-linkage criteria. Results are very similar.

4 stars for (m, l), (h, m) (medium gain and low risk, high gain and medium risk);
5 stars for (h, l) (high gain and low risk).
Of course, this is a subjective definition of stars, nevertheless it seemed reasonable to us.

As in [13], the quality of the clustering was measured using the C-index [6]. This index assumes values in the interval [0, 1], assuming small values when the quality of the clustering is good. In our experiments, we always obtained $C \leq 0.1$.

Table 1 lists the step-by-step results of the classification procedure for the group of monetary funds. The left part of the table shows the classification based on the risk evaluation and the global rating provided by Morningstar. The central part lists the elements characterising the funds (one-year average return, constant minimum, time-varying standard and time-varying turmoil risks). Note that v_m assumes very small values (due to the small values of $\hat{\gamma}$) and that only the last fund presents a turmoil risk.[2] The right part of the table shows the results of the three-step classification procedure. The *Gain* column contains the classification in high, medium and low gain obtained by the clustering of step 1; the *Risk* column contains the classification in high, medium and low risk obtained by the grouping of step 2; lastly, the *Stars* column shows the five-group classification described in step 3.

The differences with respect to the Morningstar rating are not large: the classification is the same in 8 cases over 15, in 6 cases it does not differ for more than one star and only in one case (the 14th fund) the two classifications differ for 2 stars.

Table 1. Monetary funds: Morningstar classification and details of the clustering procedure

Morningstar						Clustering		
Risk	Stars	Return	v_m	v_s	v_t	Gain	Risk	Stars
Low	3	2.25	4.77E-09	0	0	Medium	Low	4
Below average	5	2.66	0	0.087	0	High	Low	5
Below average	3	2.08	7.01E-08	0.171	0	Low	Medium	2
Below average	3	2.26	5.71E-08	0	0	Medium	Low	4
Below average	3	2.34	0	0.180	0	Medium	Medium	3
Below average	3	2.26	0	0.231	0	Medium	Medium	3
Average	2	1.70	1.93E-07	0	0	Low	Low	2
Average	2	1.87	0	0.144	0	Low	Medium	2
Average	4	2.41	0	0.208	0	Medium	Medium	3
Above average	2	2.05	0	0.155	0	Low	Medium	2
Above average	4	2.71	1.91E-07	0.385	0	High	High	3
Above average	2	2.10	0	0.234	0	Low	Medium	2
Above average	3	1.96	0	0.145	0	Low	Medium	2
High	5	2.28	1.29E-06	0	0	Medium	Medium	3
High	1	1.80	0	0.151	0.333	Low	High	1

---

[2] On the whole, instead, parameter δ was significant in 11 cases (about 14% of funds).

Table 2. Comparison of Morningstar and Clustering Classification

Stars		1	2	3	4	5	0	1	2	3
							Differences in stars			
Aggr. Bal.	Clustering	2	4	7	1	1	7	6	2	
	Morningstar	0	3	8	4	0				
Prud. Bal.	Clustering	0	2	6	7	0	4	9	2	
	Morningstar	0	2	9	4	0				
Corp. Bond	Clustering	1	0	3	2	9	3	5	5	2
	Morningstar	0	3	10	2	0				
Stock	Clustering	0	1	2	11	1	4	10	1	
	Morningstar	1	3	6	5	0				
Monetary	Clustering	1	6	5	2	1	8	6	1	
	Morningstar	1	4	6	2	2				

Table 3. Empirical probability and cumulative distribution functions of differences in stars (percentages)

Empirical probability function					
0	1	2	3	4	5
34.7	48.0	14.7	2.6	0.0	0.0
Empirical cumulative distribution function					
0	1	2	3	4	5
34.7	82.7	97.4	100	100	100

The same procedure was applied to the other four categories and results are summarised and compared with the Morningstar classification in Table 2. Clearly, the classifications are different because they are based on different criteria and definitions of gain and risk. However, in 82.7% of cases the two classifications do not differ for more than one star. This is evident looking at Table 3, in which the empirical probability function of the differences in stars and the corresponding cumulative distribution function are shown. Moreover, excluding the Corporate Bond Investments, which present the largest differences between the two classifications, the percentage of differences equal to or less than 1 increases up to 90% while the remaining 10% differs by two stars. In particular, the classifications relative to the Aggressive Balanced and the Monetary funds seem very similar between the two methodologies.

4 Some concluding remarks

In this paper a clustering procedure to classify mutual funds in terms of gain and risk has been proposed. It refers to a purely statistical approach, based on few tools to characterise return and risk. The method is model-based, in the sense that the definition of risk is linked to the estimation of a particular Threshold GARCH model, which characterises quiet and turmoil states of financial markets.

The risk is evaluated simply considering an equally weighted average of three different kinds of risk (constant minimum risk, time-varying standard risk and time-varying turmoil risk). Different weights could also be considered but at the cost of introducing a subjectivity element.

Surprisingly, in our application, this simple method provided a classification which does not show large differences with respect to the Morningstar classification. Of course, this exercise could be extended to compare our clustering method with other alternative classifications and to consider different weighting systems. For example, it would be interesting to link weights to some financial variable. As regards applications, instead, the main interest focuses on using this approach in asset allocation or portfolio selection problems.

Acknowledgement. We are grateful to the participants of the MAF 2008 conference, in particular to Alessandra Amendola, Giuseppe Storti and Umberto Triacca. This work was supported by Italian MIUR under Grant 2006137221_001 and by the University of Padua by Grant CPDA073598.

References

1. Andersen, T.G., Bollerslev, T.: Answering the skeptics: yes, standard volatility models do provide accurate forecasts. Int. Econ. Rev. 39, 885–905 (1998)
2. Caiado, J., Crato, N.: A GARCH-based method for clustering of financial time series: International stock markets evidence. MPRA Paper 2074, (2007)
3. Christensen K., Podolskji M.: Realized range-based estimation of integrated variance. J. Econometrics 141, 323–349 (2007)
4. Da Costa, Jr. N., Cunha, S., Da Silva, S.: Stock selection based on cluster analysis. Econ. Bull. 13, 1–9 (2005)
5. Glosten, L.R., Jagannathan, R., Runkle, D.E.: On the relation between expected value and the nominal excess return on stocks. J. Finan. 48, 1779–1801 (1993)
6. Hubert, L. and Schultz, J.: Quadratic assignment as a general data-analysis strategy. Br. J. Math. Stat. Psychol. 29, 190–241 (1976)
7. Johnson, S.C.: Hierarchical clustering schemes. Psychometrika 2, 241–254 (1967)
8. Kalantzis, T., Papanastassiou, D.: Classification of GARCH time series: an empirical investigation. Appl. Finan. Econ. 18, 1–6 (2008)
9. Lytkin, N.I., Kulikowski, C.A., Muchnik, I.B.: Variance-based criteria for clustering and their application to the analysis of management styles of mutual funds based on time series of daily returns. DIMACS Technical Report 2008-01 (2008)
10. Martens M., van Dijk D.: Measuring volatility with realized range. J. Econ. 138, 181–207 (2007)
11. Morningstar: The Morningstar rating methodology. Morningstar Methodology Paper (2007).
12. Otranto, E.: Classifying the markets volatility with ARMA distance measures. Quad. Stat. 6, 1–19 (2004)

13. Otranto, E.: Clustering heteroskedastic time series by model-based procedures. Comput. Stat. Data Anal. 52, 4685–4698 (2008)
14. Otranto, E., Trudda, A.: Evaluating the risk of pension funds by statistical procedures. In Lakatos, G.M. (ed.) Transition economies: 21st century issues and challenges, pp. 189–204. Nova Science Publishers, Hauppauge (2008)
15. Pattarin, F., Paterlini, S., Minerva, T.: Clustering financial time series: an application to mutual funds style analysis. Comput. Stat. Data Anal. 47, 353–372 (2004)
16. Piccolo, D.: A distance measure for classifying ARIMA models. J. Time Series Anal. 11, 153–164 (1990)
17. Zakoian, J.M.: Threshold heteroskedastic models. J. Econ. Dyn. Control 18, 931–955 (1994)

Multivariate Variance Gamma and Gaussian dependence: a study with copulas[*]

Elisa Luciano and Patrizia Semeraro

Abstract. This paper explores the dynamic dependence properties of a Lévy process, the Variance Gamma, which has non-Gaussian marginal features and non-Gaussian dependence. By computing the distance between the Gaussian copula and the actual one, we show that even a non-Gaussian process, such as the Variance Gamma, can "converge" to linear dependence over time. Empirical versions of different dependence measures confirm the result over major stock indices data.

Key words: multivariate variance Gamma, Lévy process, copulas, non-linear dependence

1 Introduction

Risk measures and the current evolution of financial markets have spurred the interest of the financial community towards models of asset prices which present both non-Gaussian marginal behaviour and non-Gaussian, or non-linear, dependence. When choosing from the available menu of these processes, one looks for parsimoniousness of parameters, good fit of market data and, possibly, ability to capture their dependence and the evolution of the latter over time. It is difficult to encapsulate all of these features – dynamic dependence, in particular – in a single model. The present paper studies an extension of the popular Variance Gamma (VG) model, named α-VG, which has non-Gaussian features both at the marginal and joint level, while succeeding in being both parsimonious and accurate in data fitting. We show that dependence "converges" towards linear dependence over time. This represents good news for empirical applications, since over long horizons one can rely on standard dependence measures, such as the linear correlation coefficient, as well as on a standard analytical copula or dependence function, namely the Gaussian one, even starting from data which do not present the standard Gaussian features of the Black Scholes or log-normal model. Let us put the model in the appropriate context first and then outline the difficulties in copula towards dynamic dependence description then.

In the financial literature, different univariate Lévy processes – able to capture non-normality – have been applied in order to model stock returns (as a reference for Lévy processes, see for example Sato [10]). Their multivariate extensions are still under investigation and represent an open field of research. One of the most popular Lévy processes in Finance is the Variance Gamma introduced by Madan and Seneta [8]. A multivariate extension has been introduced by Madan and Seneta themselves. A generalisation of this multivariate process, named α-VG, has been introduced by Semeraro [11]. The generalisation is able to capture independence and to span a wide range of dependence. For fixed margins it also allows various levels of dependence to be modelled. This was impossible under the previous VG model. A thorough application to credit analysis is in Fiorani et al. [5]. The α-VG process depends on three parameters for each margin ($\mu_j, \sigma_j, \alpha_j$) and an additional common parameter a. The linear correlation coefficient is known in closed formula and its expression is independent of time. It can be proved [7] that the process also has non-linear dependence.

How can we study dynamic dependence of the α-VG process? Powerful tools to study non-linear dependence between random variables are copulas. In a seminal paper, Embrechts et al. [4] invoked their use to represent both linear and non-linear dependence. Copulas, which had been introduced in the late 1950s in statistics and had been used mostly by actuaries, do answer *static* dependence representation needs. However, they hardly cover all the *dynamic* representation issues in finance. For Lévy processes or the distributions they are generated from, the reason is that, for given infinitely divisible margins, the conditions that a copula has to satisfy in order to provide an infinitely divisible joint distribution are not known [3].

In contrast, if one starts from a multivariate stochastic process as a primitive entity, the corresponding copula seldom exists in *closed form* at every point in time. Indeed, copula knowledge at a single point in time does not help in representing dependence at later maturities. Apart from specific cases, such as the traditional Black Scholes process, the copula of the process is time dependent. And reconstructing it from the evolution equation of the underlying process is not an easy task. In order to describe the evolution of dependence over time we need a family of copulas $\{C_t, t \geq 0\}$. Most of the time, as in the VG case, it is neither possible to derive C_t from the expression of C_1 nor to get C_1 in closed form. However, via Sklar's Theorem [12], a numerical version of the copula at any time t can be obtained. The latter argument, together with the fact that for the α-VG process the linear correlation is constant in time, leads us to compare the α-VG empirical copula for different tenures t with the Gaussian closed form one. We study the evolution over time of the distance between the empirical and the Gaussian copula as a measure of the corresponding evolution of non-linear dependence.

The paper is organised as follows: Section 2 reviews the VG model and its dependence; it illustrates how we reconstruct the empirical copula. Section 3 compares the approximating (analytical) and actual (numerical) copula, while Section 4 concludes.

2 VG models

The VG *univariate* model for financial returns $X(t)$ has been introduced by Madan and Seneta [8]. It is a natural candidate for exploring multivariate extensions of Lévy processes and copula identification problems outside the Black Scholes case for a number of reasons:

- it can be written as a time-changed Wiener process: its distribution at time t can be obtained by conditioning;
- it is one of the simplest Lévy processes that present non-Gaussian features at the marginal level, such as asymmetry and kurtosis;
- there is a well developed tradition of risk measurement implementations for it.

Formally, let us recall that the VG is a three-parameter Lévy process (μ, σ, α) with characteristic function

$$\psi_{X_{VG}(t)}(u) = [\psi_{X_{VG}(1)}(u)]^t = \left(1 - iu\mu\alpha + \frac{1}{2}\sigma^2\alpha u^2\right)^{-\frac{t}{\alpha}}. \tag{1}$$

The VG process has been generalised to the *multivariate* setting by Madan and Seneta themselves [8] and calibrated on data by Luciano and Schoutens [7]. This multivariate generalisation has some drawbacks: it cannot generate independence and it has a dependence structure determined by the marginal parameters, one of which (α) must be common to each marginal process.

To overcome the problem, the multivariate VG process has been generalised to the α-VG process [11]. The latter can be obtained by time changing a multivariate Brownian motion with independent components by a multivariate subordinator with gamma margins.

Let Y_i, $i = 1, \ldots, n$ and Z be independent real gamma processes with parameters respectively

$$(\frac{1}{\alpha_i} - a, \frac{1}{\alpha_i}), i = 1, \ldots, n$$

and $(a, 1)$, where $\alpha_j > 0$ $j = 1, \ldots, n$ are real parameters and $a \leq \frac{1}{\alpha_i}$ $\forall i$. The multivariate subordinator $\{\mathbf{G}(t), t \geq 0\}$ is defined by the following

$$\mathbf{G}(t) = (G_1(t), \ldots, G_n(t))^T = (Y_1(t) + \alpha_1 Z(t), \ldots, Y_n(t) + \alpha_n Z(t))^T. \tag{2}$$

Let W_i be independent Brownian motions with drift μ_i and variance σ_i. The $\mathbf{R}^n$ valued process $\mathbf{X} = \{\mathbf{X}(t), t > 0\}$ defined as:

$$\mathbf{X}(t) = (W_1(G_1(t)), \ldots, W_n(G_n(t)))^T \tag{3}$$

where $\mathbf{G}$ is independent from $\mathbf{W}$, is an α-VG process.

It depends on three marginal parameters $(\mu_j, \sigma_j, \alpha_j)$ and an additional common parameter a. Its characteristic function is the following

$$\psi_{\mathbf{X}(t)}(\mathbf{u}) = \prod_{j=1}^{n}\left(1 - \alpha_j\left(i\mu_j u_j - \frac{1}{2}\sigma_j^2 u_j^2\right)\right)^{-t\left(\frac{1}{\alpha_j}-a\right)}$$

$$\left(1 - \sum_{j=1}^{n}\alpha_j\left(i\mu_j u_j - \frac{1}{2}\sigma_j^2 u_j^2\right)\right)^{-ta}. \tag{4}$$

The usual multivariate VG obtains for $\alpha_j = \alpha$, $j = 1, \ldots, n$ and $a = \frac{1}{\alpha}$.

For the sake of simplicity, from now on we consider the bivariate case.

Since the marginal processes are VG, the corresponding distributions at time t, F_t^1 and F_t^2 can be obtained in a standard way, i.e., conditioning with respect to the marginal time change:

$$F_t^i(x_i) = \int_0^{+\infty} \Phi\left(\frac{x_i - \mu_i(w_i + \alpha_i z)}{\sigma_i\sqrt{w_i + \alpha_i z}}\right) f_{G(t)}^i(z)dz, \tag{5}$$

where Φ is a standard normal distribution function and $f_{G(t)}^i$ is the density of a gamma distribution with parameters $\left(\frac{t}{\alpha_i}, \frac{t}{\alpha_i}\right)$. The expression for the joint distribution at time t, $F_t = F_{\mathbf{X}(t)}$, is:

$$F_t(x_1, x_2) = \int_0^{\infty}\int_0^{\infty}\int_0^{\infty} \Phi\left(\frac{x_1 - \mu_1(w_1 + \alpha z)}{\sigma_1\sqrt{w_1 + \alpha_1 z}}\right) \Phi\left(\frac{x_2 - \mu_2(w_2 + \beta z)}{\sigma_2\sqrt{w_2 + \alpha_2 z}}\right) \tag{6}$$

$$\cdot f_{Y_1(t)}(w_1) f_{Y_2(t)}(w_2) f_{Z(t)}(z) dw_1 dw_2 dz, \tag{7}$$

where $f_{Y_1(t)}$, $f_{Y_2(t)}$, $f_{Z(t)}$ are densities of gamma distributions with parameters respectively: $\left(t\left(\frac{1}{\alpha_1} - a\right), \frac{1}{\alpha_1}\right)$, $\left(t\left(\frac{1}{\alpha_2} - a\right), \frac{1}{\alpha_2}\right)$ and $(ta, 1)$ [11].

2.1 Dependence structure

In this section, we investigate the dependence or association structure of the α-VG process.

We know from Sklar's Theorem that there exists a copula such that any joint distribution can be written in terms of the marginal ones:

$$F_t(x_1, x_2) = C_t(F_t^1(x_1), F_t^2(x_2)). \tag{8}$$

The copula C_t satisfies:

$$C_t(u_1, u_2) = F_t((F_t^1)^{-1}(u_1), (F_t^2)^{-1}(u_2)), \tag{9}$$

where $(F_t^i)^{-1}$ is the generalised inverse of F_t^i, $i = 1, 2$.

Since the marginal and joint distributions in (5) and (6) cannot be written in closed form, the copula of the α-VG process and the ensuing non-linear dependence measures, such as Spearman's rho and Kendall's tau, cannot be obtained analytically.

The only measure of dependence one can find in closed form is the linear correlation coefficient:

$$\rho^{X(t)} = \frac{\mu_1\mu_2 a_1 a_2 a}{\sqrt{(\sigma_1^2 + \mu_1^2 a_1)(\sigma_2^2 + \mu_2^2 a_j)}}. \tag{10}$$

This coefficient is independent of time, but depends on both the marginal and the common parameter a. For given marginal parameters the correlation is increasing in the parameter a if $\mu_1\mu_2 > 0$, as is the case in most financial applications. Since a has to satisfy the following bounds: $0 \le a \le min\left(\frac{1}{a_1}, \frac{1}{a_2}\right)$ the maximal correlation allowed by the model corresponds to $a = min\left(\frac{1}{a_1}, \frac{1}{a_2}\right)$.

However, it can be proved that linear dependence is not exhaustive, since even when $\rho = 0$ the components of the process can be dependent [7]. In order to study the whole dependence we should evaluate empirical versions of the copula obtained from (9) using the integral expression of the marginal and joint distributions in (5) and (6). A possibility which is open to the researcher, in order to find numerically the copula of the process at time t, is then the following:

- fix a grid (u_i, v_i), $i = 1, \dots, N$ on the square $[0, 1]^2$;
- for each $i = 1, \dots, N$ compute $(F_t^1)^{-1}(u_i)$ and $(F_t^2)^{-1}(v_i)$ by numerical approximation of the integral expression: let $(\tilde{F}_t^1)^{-1}(u_i)$ and $(\tilde{F}_t^2)^{-1}(v_i)$ be the numerical results;
- find a numerical approximation for the integral expression (6), let it be $\hat{F}_t(x_i, y_i)$;
- find the approximated value of $C_t(u_i, v_i)$:

$$\hat{C}_t(u_i, v_i) = \hat{F}((\tilde{F}_t^1)^{-1}(u_i), (\tilde{F}_t^1)^{-1}(u_i)), \quad i = 1, \dots, N.$$

We name the copula $\hat{C}_t$ numerical, empirical or actual copula of the α-VG distribution at time t.

In order to discuss the behaviour of non-linear dependence we compare the empirical copula and the Gaussian one with the same linear correlation coefficient, for different tenors t. We use the classical L^1 distance:

$$d_t(C_t, C_t') = \int_0^1 |C_t(u, v) - C_t'(u, v)| du dv. \tag{11}$$

It is easy to demonstrate that the distance d is consistent with concordance order, i.e., $C_t \prec C_t' \prec C_t''$ implies $d(C_t, C_t') \le d(C_t, C_t'')$ [9]. It follows that the nearer the copulas are in terms of concordance, the nearer they are in terms of d_t. Observe that the maximal distance between two copulas is $\frac{1}{6}$, i.e., the distance between the upper and lower Fréchet bounds.

Therefore for each t we:

- fix the marginal parameters and a linear correlation coefficient;

- find the numerical copula $\hat{C}_t$ of the process over the prespecified grid;
- compute the distance between the numerical and Gaussian copula.[1] Please note that, since the linear correlation $\rho^{\mathbf{X}(t)}$ in (10) is independent of time, the Gaussian copula remains the same too: ($C_t = C'_t$ in (11)).

3 Empirical investigation

3.1 Data

The procedure outlined above has been applied to a sample of seven major stock indices: S&P 500, Nasdaq, CAC 40, FTSE 100, Nikkei 225, Dax and Hang Seng. For each index we estimated the marginal VG parameters under the risk neutral measure, using our knowledge of the (marginal) characteristic function, namely (4). From the characteristic function, call option theoretical prices were obtained using the Fractional Fast Fourier Transform (FRFT) in Chourdakis [2], which is more efficient than the standard Fast Fourier Transform (FFT). The data for the corresponding observed prices are Bloomberg quotes of the corresponding options with three months to expiry. For each index, six strikes (the closest to the initial price) were selected, and the corresponding option prices were monitored over a one-hundred-day window, from 7/14/06 to 11/30/06.

3.2 Selection of the α-VG parameters

We estimated the marginal parameters as follows: using the six quotes of the first day only, we obtained the parameter values which minimised the mean square error between theoretical and observed prices, the theoretical ones being obtained by FRFT. We used the results as guess values for the second day, the second day results as guess values for the third day, and so on. The marginal parameters used here are the average of the estimates over the entire period. The previous procedure is intended to provide marginal parameters which are actually not dependent on an initial arbitrary guess and are representative of the corresponding stock index price, under the assumption that the latter is stationary over the whole time window. The marginal values for the VG processes are reported in Table 1.

We performed our analysis using the marginal parameters reported above and the maximal correlation allowed by the model. The idea is indeed that positive and large dependence must be well described. For each pair of assets, Table 2 gives the maximal possible value of a, namely $a = min\{\frac{1}{a_1}, \frac{1}{a_2}\}$ (lower entry) and the corresponding correlation coefficient ρ (upper entry), obtained using (10) in correspondence to the maximal a.

---

[1] Since we have the empirical copula only on a grid we use the discrete version of the previous distance.

Table 1. Calibrated parameters for the α-VG price processes for the stock indices in the sample

Asset i	μ_i	σ_i	α_i
S&P	−0.65	0.22	0.10
Nasdaq	−0.67	0.11	0.13
CAC 40	−0.46	0.10	0.11
FTSE	−0.59	0.045	0.031
Nikkei	−0.34	0.16	0.10
DAX	−0.27	0.13	0.14
Hang Seng	−1.68	0.8	0.03

Table 2. Maximal correlation and a-parameter (in parentheses) for the calibrated α-VG models, all stock indices

	S & P	Nasdaq	CAC 40	FTSE	Nikkei	Dax
Nasdaq	0.803 (7.590)					
CAC 40	0.795 (9.020)	0.701 (7.590)				
FTSE	0.505 (9.791)	0.410 (7.590)	0.406 (9.020)			
Nikkei	0.556 (9.593)	0.461 (7.590)	0.457 (9.020)	0.284 (9.593)		
Dax	0.512 (7.092)	0.536 (7.092)	0.447 (7.092)	0.261 (7.092)	0.294 (7.092)	
Hang Seng	0.500 (9.791)	0.406 (7.590)	0.403 (9.020)	0.834 (31.976)	0.282 (9.593)	0.259 (7.092)

3.3 Copula results

We computed the empirical copula $\hat{C}_t$ for the following tenors: $t = 0.1, 1, 10, 100$. We report in Table 3 the distances d_t corresponding to each pair of stocks and each time t.

In order to give a qualitative idea of the distances obtained we also provide a graphical representation of the copula level curves for the pair Nasdaq and S&P at time $t = 1$.

We observe that the distance in Table 3 is very low and decreasing in time. The plot (and similar, unreported ones, for other couples and tenors) reinforces the conclusion. Therefore the Gaussian copula seems to be a good approximation of the true copula, at least for long horizons.

Table 3. Distances between the Gaussian and empirical copula for calibrated α-VG price processes, over different stock indices and over time t (expressed in years)

Pair	t 0.1	1	10	100
S&P/Nasdaq	0.015	0.0098	0.0098	0.0097
S&P/CAC 40	0.022	0.0098	0.0097	0.0097
S&P/FTSE	0.0101	0.0085	0.0085	0.0085
S&P/Nikkei	0.037	0.0094	0.0091	0.0089
S&P/DAX	0.034	0.0092	0.0088	0.0087
S&P/Hang Seng	0.011	0.0083	0.0084	0.0084
Nasdaq/CAC 40	0.020	0.0095	0.0095	0.0094
Nasdaq/FTSE	0.010	0.0079	0.0079	0.0079
Nasdaq/Nikkei	0.0263	0.0088	0.0085	0.0083
Nasdaq/DAX	0.035	0.0092	0.0088	0.0087
Nasdaq/Hang Seng	0.010	0.0078	0.0079	0.0079
CAC 40/FTSE	0.010	0.0079	0.0079	0.0079
CAC 40/Nikkei	0.0261	0.0088	0.0085	0.0085
CAC 40/DAX	0.0273	0.0088	0.0085	0.0083
CAC 40/Hang Seng	0.010	0.0078	0.0079	0.0079
FTSE/Nikkei	0.0170	0.0078	0.0074	0.0072
FTSE/DAX	0.0165	0.0077	0.0072	0.0071
FTSE/Hang Seng	0.0097	0.0098	0.0098	0.0098
Nikkei/DAX	0.0201	0.0078	0.0074	0.0073
Nikkei/Hang Seng	0.012	0.0071	0.0071	0.0071
DAX/Hang Seng	0.0115	0.0069	0.0069	0.0069

3.4 Measures of dependence

In order to confirm our results we also compare two non-linear dependence measures obtained simulating the copula with the corresponding ones of the Gaussian copula.

For $t = 0.1, 1, 10, 100$ we computed the simulated values of Spearman's rho, $\tilde{\rho}_S(t)$, and Kendall's tau, $\tilde{\tau}(t)$, obtained from the empirical copulas. The methodology is described in Appendix A.

We found the analytical values of the Gaussian copula corresponding to each pair, by means of the relationships:

$$\rho_S = \frac{6}{\pi} \arcsin \frac{\rho}{2}; \quad \tau = \frac{2}{\pi} \arcsin \rho. \tag{12}$$

The results obtained are consistent with respect to the copula distances, as expected. They confirm the "tendency" towards Gaussian dependence as t increases. We report below the results for the first index pair, namely S & P-Nasdaq. The others behave in a similar way.

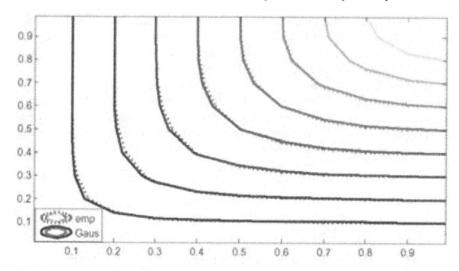

Fig. 1. Level curves of the Gaussian (Gaus) and empirical (emp) copula of the α-VG calibrated price processes, S & P - Nasdaq, after one year

Table 4. Simulated values of $\tilde{\rho}_s(t)$ and $\tilde{\tau}(t)$ for the numerical $(\hat{C})$ and Gaussian copula (Gauss) over different horizons. S & P/Nasdaq pair

Pair		$\hat{C}_{0.1}$	$\hat{C}_1$	$\hat{C}_{10}$	$\hat{C}_{100}$	Gauss
S&P/Nasdaq	$\tilde{\rho}_s$	0.74	0.78	0.79	0.79	0.79
	$\tilde{\tau}$	0.54	0.59	0.58	0.59	0.59

4 Conclusions and further research

This paper measures the non-linear dependence of the α-VG process, calibrated to a set of stock market data, by means of a distance between its empirical copula at time t and the corresponding Gaussian one, which is characterised by the (constant) correlation coefficient of the process.

Our empirical analysis suggests that non-linear dependence is "decreasing" in time, since the approximation given by the Gaussian copula improves in time. As expected, non-linear dependence coefficients confirm the result. The tentative conclusion is that, similarly to marginal non-Gaussianity, which is usually stronger on short-horizon than on long-horizon returns, joint non-linear dependence and non-Gaussianity fade over time.

This represents an important point of departure for practical, large-scale implementations of the α-VG model and of its subcase, the traditional VG model. Any multivariate derivative price or portfolio risk measure indeed is based on the joint distribution of returns. If we use a time-varying empirical copula in order to re-assess prices and risk measures over time, and we want the results to be reliable, extensive and time-consuming simulations are needed. If, on the contrary, we can safely approximate the actual copula with the Gaussian one, at least over long horizons, life becomes much easier. Closed or semi-closed formulas exist for pricing and risk measurement in the presence of the Gaussian copula (see for instance [1]). Standard linear correlation can be used for the model joint calibration.

In a nutshell, one can adopt an accurate, non-Gaussian model and safely ignore non-linear (and non-analytical) dependence, in favour of the linear dependence represented by the familiar Gaussian copula, provided the horizon is long enough. In the stock market case analysed here, one year was quite sufficient for non-Gaussian dependence to be ignored.

Appendix

Simulated measure of dependence

The simulated version of Spearman's rho at time t, $\tilde{\rho}_S(t)$, can be obtained from a sample of N realisations of the processes at time t $(x_1^i(t), x_2^i(t))$, $i = 1, \ldots, N$:

$$\tilde{\rho}_S(t) = 1 - 6\frac{\sum_{i=1}^{N}(R_i - S_i)^2}{N(N^2 - 1)}, \tag{13}$$

where $R_i = Rank(x_1^i(t))$ and $S_i = Rank(x_2^i(t))$. Similarly for Kendall's tau, $\tilde{\tau}_C(t)$:

$$\tilde{\tau}_C(t) = \frac{c - d}{\binom{N}{2}}, \tag{14}$$

where c is the number of concordance pairs of the sample and d the number of discordant ones. A pair $(x_1^i(t), x_2^i(t))$ is said to be discordant [concordant] if $x_1^i(t)x_2^i(t) \leq 0$ $[x_1^i(t)x_2^i(t) \geq 0]$. The N realisations of the process are obtained as follows:

- Simulate N realisations from the independent laws $\mathcal{L}(Y_1)$, $\mathcal{L}(Y_2)$, $\mathcal{L}(Z)$; let them be respectively y_1^n, y_2^n, z^n for $n = 1, \ldots, N$.
- Obtain N realisations (g_1^n, g_2^n) of $\mathbf{G}$ through the relations $G_1 = Y_1 + Z$ and $G_2 = Y_2 + Z$.
- Generate N independent random draws from each of the independent random variables M_1 and M_2 with laws $N(0, G_1)$ and $N(0, G_2)$. The draws for M_1 in turn are obtained from N normal distributions with zero mean and variance g_1^n, namely

$$M_1(n) = N(0, g_1^n).$$

The draws for M_2 are from normal distributions with zero mean and variance g_2^n, namely

$$M_2(n) = N(0, g_2^n).$$

- Obtain N realisations (x_1^n, x_2^n) of $\mathbf{X}(1)$ by means of the relations

$$x_1^n = \mu_1 g_1^n + \sigma_1 M_1(n)$$
$$x_2^n = \mu_2 g_2^n + \sigma_2 M_2(n).$$

Acknowledgement. We thank Claudio Patrucco for having provided the marginal calibrations and Ksenia Rulik for computational assistance. Financial support from MIUR (grant 20061327131) is gratefully acknowledged.

References

1. Cherubini, U., Luciano, E., Vecchiato W.: Copula Methods in Finance, John Wiley, Chichester (2004)
2. Chourdakis, K.: Option pricing using the fractional FFT. J. Comput. Finan. 8, 1–18 (2005)
3. Cont, R., Tankov, P.: Financial modelling with jump processes. Chapman and Hall-CRC financial mathematics series (2004)
4. Embrechts, P., McNeil, A.J., Straumann, D.: Correlation and dependency in risk management: properties and pitfalls. In: Dempster, M.A.H. (ed.) Value at Risk and Beyond. Cambridge University Press (2002)
5. Fiorani, F., Luciano, E., Semeraro, P.: Single and joint default in a structural model with purely discontinuous assets prices. Quant. Finan. (2009)
6. Luciano, E., Schoutens, W.: A multivariate Jump-Driven Financial Asset Model. Quant. Finan. 6, 385–402 (2005)
7. Luciano, E., Semeraro, P.: Extending Time-Changed Lévy Asset Models Through Multivariate Subordinators, working paper, Collegio Carlo Alberto (2007)
8. Madan, D.B., Seneta, E.: The VG model for share market returns. J. Busin. 63, 511–524 (1990)
9. Nelsen, R.B.: An introduction to copulas. Lecture Notes in Statistics 139, Springer (1999)
10. Sato, K.I.: Lévy processes and infinitely divisible distributions. Studies in Advanced Mathematics. Cambridge University Press (2003)
11. Semeraro, P.: A multivariate variance gamma model for financial application. J. Theoret. Appl. Finan. 11, 1–18 (2006)
12. Sklar A.: Fonctions de repartition à n dimensions et leurs marges. Publication Inst. Statist. Univ. Paris 8, 229–231 (1959)

A simple dimension reduction procedure for corporate finance composite indicators[*]

Marco Marozzi and Luigi Santamaria

Abstract. Financial ratios provide useful quantitative financial information to both investors and analysts so that they can rate a company. Many financial indicators from accounting books are taken into account. Instead of sequentially examining each ratio, one can analyse together different combinations of ratios in order to simultaneously take into account different aspects. This may be done by computing a composite indicator. The focus of the paper is on reducing the dimension of a composite indicator. A quick and compact solution is proposed, and a practical application to corporate finance is presented. In particular, the liquidity issue is addressed. The results suggest that analysts should take our method into consideration as it is much simpler than other dimension reduction methods such as principal component or factor analysis and is therefore much easier to be used in practice by non-statisticians (as financial analysts generally are). Moreover, the proposed method is always readily comprehended and requires milder assumptions.

Key words: dimension reduction, composite indicator, financial ratios, liquidity

1 Introduction

Financial ratios provide useful quantitative financial information to both investors and analysts so that they can rate a company. Many financial indicators from accounting books are taken into account. In general, ratios measuring profitability, liquidity, solvency and efficiency are considered.

Instead of sequentially examining each ratio, one can analyse different combinations of ratios together in order to simultaneously take into account different aspects. This can be done by computing a composite indicator.

Complex variables can be measured by means of composite indicators. The basic idea is to break down a complex variable into components which are measurable by means of simple (partial) indicators. The partial indicators are then combined to obtain the composite indicator. To this end one should

---

[*] The paper has been written by and the proposed methods are due to M. Marozzi. L. Santamaria gave helpful comments to present the application results.

- possibly transform the original data into comparable data through a proper function $T(\cdot)$ and obtain the partial indicators;
- combine the partial indicators to obtain the composite indicator through a proper link (combining) function $f(\cdot)$.

If $X_1, \ldots, X_K$ are the measurable components of the complex variable, then the composite indicator is defined as

$$M = f(T_1(X_1), \ldots, T_K(X_K)). \tag{1}$$

Fayers and Hand [3] report extensive literature on the practical application of composite indicators (the authors call them multi-item measurement scales). In practice, the simple weighted or unweighted summations are generally used as combining functions. See Aiello and Attanasio [1] for a review on the most commonly used data transformations to construct simple indicators.

The purpose of this paper is to reduce the dimensions of a composite indicator for the easier practice of financial analysts. In the second section, we discuss how to construct a composite indicator. A simple method to simplify a composite indicator is presented in Section 3. A practical application to the listed company liquidity issue is discussed in Section 4. Section 5 concludes with some remarks.

2 Composite indicator computation

Let X_{ik} denote the kth financial ratio (partial component), $k = 1, \ldots, K$, for the ith company, $i = 1, \ldots, N$. Let us suppose, without loss of generality, that the partial components are positively correlated to the complex variable. To compute a composite indicator, first of all one should transform the original data into comparable data in order to obtain the partial indicators. Let us consider linear transformations. A linear transformation LT changes the origin and scale of the data, but does not change the shape

$$LT(X_{ik}) = a + bX_{ik}, \ a \in]-\infty, +\infty[, \ b > 0. \tag{2}$$

Linear transformations allow us to maintain the same ratio between observations (they are proportional transformations).

The four linear transformations most used in practice are briefly presented [4]. The first two linear transformations are defined as

$$LT_1(X_{ik}) = \frac{X_{ik}}{max_i(X_{ik})} \tag{3}$$

and

$$LT_2(X_{ik}) = \frac{X_{ik} - min_i(X_{ik})}{max_i(X_{ik}) - min_i(X_{ik})}, \tag{4}$$

which correspond to LT where $a = 0$ and $b = \frac{1}{max_i(X_{ik})}$, and where $a = \frac{-min_i(X_{ik})}{max_i(X_{ik}) - min_i(X_{ik})}$ and $b = \frac{1}{max_i(X_{ik}) - min_i(X_{ik})}$ respectively. LT_1 and LT_2 cancel the measurement units and force the results into a short and well defined range: $\frac{min_i(X_{ik})}{max_i(X_{ik})} \leq LT_1(X_{ik}) \leq 1$ and $0 \leq LT_2(X_{ik}) \leq 1$ respectively. LT_1 and LT_2 are readily comprehended.

The third and fourth linear transformations are defined as

$$LT_3(X_{ik}) = \frac{X_{ik} - E(X_k)}{SD(X_k)} \tag{5}$$

and

$$LT_4(X_{ik}) = \frac{X_{ik} - MED(X_k)}{MAD(X_k)}, \tag{6}$$

which correspond to LT where $a = \frac{-E(X_k)}{SD(X_k)}$ and $b = \frac{1}{SD(X_k)}$, and where $a = \frac{-MED(X_k)}{MAD(X_k)}$ and $b = \frac{1}{MAD(X_k)}$, respectively. $LT_3(X_{ik})$ indicates how far X_{ik} lies from the mean $E(X_k)$ in terms of the standard deviation $SD(X_k)$. LT_4 is similar to LT_3 and uses the median $MED(X_k)$ instead of the mean as location measure and the median absolute deviation $MAD(X_k)$ instead of the standard deviation as scale measure.

By means of LT_h (where the subscript $h=1, 2, 3$ or 4 denotes the various methods) the original data are transformed into comparable data. The composite indicator is then defined using the sum as the combining function, in accordance with general practice (see [1])

$$M_{h,i} = \sum_{k=1}^{K} LT_h(X_{ik}), \ h = 1, 2, 3, 4. \tag{7}$$

$M_{h,i}$s are used to rank the units. Note that the first and second method may be applied also to ordered categorical variables, or to mixed variables, partly quantitative and partly ordered categorical, with the unique concern of how to score the ordered categories.

In Section 4 we analyse a data set about listed companies, in particular we consider four different liquidity ratios. For the ith company we denote these ratios by $X_{i1}, X_{i2}, X_{i3}, X_{i4}$. Note that $T(X_{i1}), T(X_{i2}), T(X_{i3}), T(X_{i4})$ are partial financial indicators since they correspond to a unique financial ratio: $T(X_{ik}) > T(X_{jk})$ lets the analyst conclude that company i is better than company j for what concerns financial ratio X_k (since of course $T(X_{ik}) > T(X_{jk}) \Leftrightarrow X_{ik} > X_{jk}$), whereas M_i is a composite financial indicator since it simultaneously considers every financial ratio. $M_1, \ldots, M_N$ allow the analyst to rank the companies since $M_i > M_j$ means that company i is better than company j regarding all the financial ratios together. There is reason to believe that financial ratios are correlated. This central question is addressed in the next section: a simple method for reducing the number of partial indicators underlying a composite indicator is proposed.

It is important to emphasise that in this paper we do not consider composite indicators based on non-linear transformations since Arboretti and Marozzi (2005) showed that such composite indicators perform better than those based on linear transformations only when distributions of $X_1, \ldots, X_K$ parent populations are very heavy-tailed. Preliminary analyses on our data show that parent distributions are not heavy-tailed. Composite indicators based on non-linear transformations may be based for example on the rank transformation of X_{ik}s or on Lago and Pesarin's [6] Nonparametric Combination of Dependent Rankings. For details on this matter see [7] and [8].

3 To reduce the dimension of composite indicators

Let $\underline{R}_K(X_k, k \in \{1, \ldots, K\}) = \underline{R}_K$ denote the vector of ranks obtained following the composite financial indicator

$$M_{4,i} = \sum_{k=1}^{K} \frac{X_{ik} - MED(X_k)}{MAD(X_k)}, \tag{8}$$

computed for $i = 1, \ldots, N$, which combines all the partial financial indicators $X_1, \ldots, X_K$. We consider the fourth method because the median absolute deviation is the most useful ancillary estimate of scale [5, p. 107]. Suppose now that X_h is excluded from the analysis. Let $_h\underline{R}_{K-1}(X_k, k \in \{1, \ldots, K\} - h) = {}_h\underline{R}_{K-1}$ denote the corresponding rank vector. If $\underline{R}_K$ and $_h\underline{R}_{K-1}$ are very similar, it follows that the exclusion of X_h does not affect the ranking of the companies much. On the contrary, if the two rank vectors are very different, by leaving out X_h the ranking process is greatly influenced. To estimate the importance of X_h we compute the Spearman correlation coefficient between $\underline{R}_K$ and $_h\underline{R}_{K-1}$

$$s(\underline{R}_K, {}_h\underline{R}_{K-1}) = 1 - \frac{6\sum_{i=1}^{N}(\underline{R}_K[i] - {}_h\underline{R}_{K-1}[i])^2}{N(N^2 - 1)}, \tag{9}$$

where $\underline{R}_K[i]$ and $_h\underline{R}_{K-1}[i]$ are the ith element of the corresponding vector. The closer s is to 1, the less important X_h is. The idea is to leave out the partial indicator X_h that brings the greatest $s(\underline{R}_K, {}_h\underline{R}_{K-1})$. The procedure may be repeated for the $K - 2$ rankings obtained by leaving out one more partial indicator. Let X_l be the next indicator that is excluded from the ranking process. We compute $_{l,h}\underline{R}_{K-2}(X_k, k \in \{1, \ldots, K\} - \{l, h\}) = {}_{l,h}\underline{R}_{K-2}$ and $s(_h\underline{R}_{K-1}, {}_{l,h}\underline{R}_{K-2})$ for $l = 1, \ldots, K, l \neq h$. The partial indicator that brings the greatest s should be excluded, and so on.

Even if the whole procedure naturally lasts until only one partial indicator is left to be used by financial analysts, a natural question arises: when should the partial indicator exclusion procedure be stopped? That is, how many partial financial indicators should be excluded? Within this framework, it is assumed that the best ranking is the one based on all the partial indicators. Of course, there is a trade-off between information and variable number reduction. A natural stopping rule is: stop the procedure as soon as the correlation coefficient is less than a fixed value.

4 A practical application

We present an application of the procedure for reducing the dimension of corporate finance composite indicators. More precisely, the liquidity issue is considered. The aim is to rate a set of companies on the basis of the following liquidity ratios.

The *current ratio*

$$X_1 = \frac{total\ current\ assets}{total\ current\ liabilities}, \tag{10}$$

indicates the company's ability to meet short-term debt obligations; the higher the ratio, the more liquid the company is. If the current assets of a company are more than twice the current liabilities, then that company is generally considered to have good short-term financial strength. If current liabilities exceed current assets, then the company may have problems meeting its short-term obligations.

The *quick ratio*

$$X_2 = \frac{total\ current\ assets - inventory}{total\ current\ liabilities}, \tag{11}$$

is a measure of a company's liquidity and ability to meet its obligations. It expresses the true working capital relationship of its cash, accounts receivables, prepaids and notes receivables available to meet the company's current obligations. The higher the ratio, the more financially strong the company is: a quick ratio of 2 means that for every euro of current liabilities there are two euros of easily convertible assets.

The *interest coverage ratio*

$$X_3 = \frac{earnings\ before\ interest\ and\ taxes}{interest\ expenses}. \tag{12}$$

The lower the interest coverage ratio, the larger the debt burden is on the company. It is a measure of a company ability to meet its interest payments on outstanding debt. A company that sustains earnings well above its interest requirements is in a good position to weather possible financial storms.

The *cash flow to interest expense ratio*

$$X_4 = \frac{cash\ flow}{interest\ expenses}. \tag{13}$$

The meaning is clear: a cash flow to interest expense ratio of 2 means that the company had enough cash flow to cover its interest expenses two times over in a year.

These ratios are important in measuring the ability of a company to meet both its short-term and long-term obligations. To address company liquidity, one may sequentially examine each ratio that addresses the problem from a particular (partial) point of view. For example, the current ratio as well as the quick ratio are regarded as a test of liquidity for a company, but while the first one expresses the working capital relationship of current assets available to meet the company's current obligations, the second one expresses the true working capital relationship of current assets available to meet current obligations since it eliminates inventory from current assets. This is particularly important when a company is carrying heavy inventory as part of its current assets, which might be obsolete. However, it should be noted that in the literature the order of their importance is not clear. For more details see for example [9].

A dataset about 338 companies listed on the main European equity markets has been analysed. We consider listed companies because they have to periodically send out financial information following standard rules. First

$$M_i = \sum_{k=1}^{4} \frac{X_{ik} - MED(X_k)}{MAD(X_k)}, \quad i = 1, \ldots, 338 \tag{14}$$

is computed. This is a composite indicator of liquidity for company i which takes into account simultaneously the partial liquidity indicators X_1, X_2, X_3, X_4. Then, the corresponding rank vector $\underline{R}_4(X_1, X_2, X_3, X_4) = \underline{R}_4$ is computed. This vector has been compared to the vectors corresponding to the consideration of three partial indicators: $_4\underline{R}_3(X_1, X_2, X_3) = {}_4\underline{R}_3$, $_3\underline{R}_3(X_1, X_2, X_4) = {}_3\underline{R}_3$, $_2\underline{R}_3(X_1, X_3, X_4) = {}_2\underline{R}_3$ and $_1\underline{R}_3(X_2, X_3, X_4) = {}_1\underline{R}_3$, through the Spearman correlation coefficient. In the first step of the procedure the quick ratio X_2 left the analysis since we have

$$s(\underline{R}_4, {}_4\underline{R}_3) = 0.9664, \quad s(\underline{R}_4, {}_3\underline{R}_3) = 0.9107,$$

$$s(\underline{R}_4, {}_2\underline{R}_3) = 0.9667, \quad s(\underline{R}_4, {}_1\underline{R}_3) = 0.9600.$$

In the second step, we compare $_2\underline{R}_3(X_1, X_3, X_4) = {}_2\underline{R}_3$ with $_{4,2}\underline{R}_2(X_1, X_3) = {}_{4,2}\underline{R}_2$, $_{3,2}\underline{R}_2(X_1, X_4) = {}_{3,2}\underline{R}_2$ and $_{1,2}\underline{R}_2(X_3, X_4) = {}_{1,2}\underline{R}_2$. The cash flow to interest expense ratio X_4 left the analysis since we have

$$s(_2\underline{R}_3, {}_{4,2}\underline{R}_2) = 0.956, \quad s(_2\underline{R}_3, {}_{3,2}\underline{R}_2) = 0.909, \quad s(_2\underline{R}_3, {}_{1,2}\underline{R}_2) = 0.905.$$

In the last step, the current ratio X_1 left the analysis since it is $s(_{4,2}\underline{R}_2, {}_{3,4,2}\underline{R}_1) = 0.672$ and $s(_{4,2}\underline{R}_2, {}_{1,4,2}\underline{R}_1) = 0.822$.

We conclude that the ranking obtained by considering together X_1, X_2, X_3, X_4 is similar to that based on the interest coverage ratio X_3, and then the analyst is suggested to focus on X_3 in addressing the liquidity issue of the companies. Our method reduces the information included in the original data by dropping the relatively unimportant financial data. These dropped financial data, however, might have important information in comparing a certain set of companies. For example, the quick ratio X_2 has been excluded in the first step of the procedure, and then the inventory has not become an aspect for the analyst to decide whether to invest in a company or not. But depending on Rees [9, p. 195], the market reaction to earnings disclosure of small firms is great. If the inventory becomes large, the smaller firms might go bankrupt because they cannot stand its cost, whereas the larger firms endure it. Moreover there may be a lot of seasonality effect on sales because monthly sales may differ greatly. This affects small firms deeply; in fact many studies have suggested that the bulk of the small firm effect is concentrated in certain months of the year [9, p. 180]. Therefore it might not be possible to apply the results of this paper to smaller firms without taking into account the inventory issue. Moreover, the importance of the financial data available differs among industries. For example, the importance of inventory might be different between the manufacturing industry and the financial industry. In general, the importance of financial data may vary between the comparison among the whole set of companies and the selected set of certain companies. For financial analysts the comparison should be done to selected sets of companies. The analysis of variance can help in evaluating the bias generated from this method.

To evaluate if our result depends on the company capitalisation, we divide the companies into two groups: the 194 companies with a capitalisation less than EUR five billion and the remaining 146 with a capitalisation greater than EUR five billion. We adopted the same criterion used by the anonymous financial firm that gave us the data. For the "small cap" companies we obtained the following results

$$s'(\underline{R}_{4,4}\,\underline{R}_3) = 0.954, \quad s'(\underline{R}_{4,3}\,\underline{R}_3) = 0.903, \quad s'(\underline{R}_{4,2}\,\underline{R}_3) = 0.969, \quad s'(\underline{R}_{4,1}\,\underline{R}_3) = 0.953;$$

$$s'(_2\underline{R}_3,_{4,2}\,\underline{R}_2) = 0.944, \quad s'(_2\underline{R}_3,_{3,2}\,\underline{R}_2) = 0.909, \quad s'(_2\underline{R}_3,_{1,2}\,\underline{R}_2) = 0.907;$$

$$s'(_{2,4}\underline{R}_2,_{3,2,4}\,\underline{R}_1) = 0.689, \quad s'(_{2,4}\underline{R}_2,_{1,2,4}\,\underline{R}_1) = 0.817;$$

therefore the first liquidity ratio that is excluded is the quick ratio X_2, the second is the cash flow to interest expense ratio X_4 and finally the current ratio X_1. The procedure suggests focusing on the interest coverage ratio X_3 when ranking the small cap companies.

For the "large cap" companies we obtained the following results

$$s''(\underline{R}_{4,4}\,\underline{R}_3) = 0.977, \quad s''(\underline{R}_{4,3}\,\underline{R}_3) = 0.931, \quad s''(\underline{R}_{4,2}\,\underline{R}_3) = 0.974, \quad s''(\underline{R}_{4,1}\,\underline{R}_3) = 0.967;$$

$$s''(_4\underline{R}_3,_{3,4}\,\underline{R}_2) = 0.781, \quad s''(_4\underline{R}_3,_{2,4}\,\underline{R}_2) = 0.967, \quad s''(_4\underline{R}_3,_{1,4}\,\underline{R}_2) = 0.959;$$

$$s''(_{2,4}\underline{R}_2,_{3,2,4}\,\underline{R}_1) = 0.698, \quad s''(_{2,4}\underline{R}_2,_{1,2,4}\,\underline{R}_1) = 0.808.$$

These results are again similar to those obtained before, both for all the companies and for the small cap ones. The conclusion is that the dimension reduction procedure is not much affected by the fact that a company is a large cap one or a small cap one. It should be cautioned that this result (as well as the other ones) applies only to the data set that has been considered in the paper, but the analysis may be easily applied to other data sets or to other financial ratios (efficiency, profitability, ...). Moreover, attention should be paid to the industry sector the companies belong to. For example, as we have already noted, the role of the inventory might be different between the manufacturing industry and the financial industry. Therefore we suggest financial analysts to group the companies on the basis of the industry sector before applying the reduction procedure. This question is not addressed here and requires further research.

The data have been reanalysed through principal component analysis, which is the most used dimension reduction method. Principal component analysis suggests that there are two principal components, the first explains 62.9% and the second 30.2% of the variance. The first component is a weighted mean of the liquidity ratios with similar weights so that it may be seen as a sort of generic indicator for company liquidity. The loadings on component one are 0.476 for X_1, 0.488 for X_2, 0.480 for X_3 and 0.552 for X_4. The loadings on component two are respectively 0.519, 0.502, -0.565 and -0.399. Note that the loadings are positive for X_1 and X_2, which compare assets with liabilities, while they are negative for X_3 and X_4, which are measures of company ability to meet its interest payments on outstanding debt. The correlation between the ranking based on X_3 and that based on the first principal component is 0.936. Therefore the rankings are very similar, but the method proposed in this paper is simpler to understand and be employed by financial analysts, who do not do not usually have a strong background in statistics. From the practical point of view, our method is more natural since it imitates what many analysts implicitly do in practice by focusing on the most important aspects, discarding the remaining ones. It is always readily comprehended, while principal components are often quite difficult to be actually interpreted. From the theoretical point of view, a unique and very mild assumption should be fulfilled for using our method: that financial ratios follow the larger the better rule. We do not have to assume other hypotheses, that on the contrary should be generally assumed by other dimension reduction methods such as principal component (think for example about the hypothesis of linearity) or factor analysis. Moreover, it is important to emphasise that, if one considers the first or second linear transformation method, the composite indicator simplifying procedure may be applied also to ordered categorical variables, or to mixed ones, partly quantitative and partly ordered categorical, with the unique concern of how to score the ordered categories.

5 Conclusions

When a financial analyst rates a company, many financial ratios from its accounting books are considered. By computing a composite indicator the analyst can analyse different combinations of ratios together instead of sequentially considering each ratio independently from the other ones. This is very important since ratios are generally correlated. A quick and compact procedure for reducing the number of ratios at the basis of a composite financial indicator has been proposed. A practical application to the liquidity issue has been discussed. We ranked a set of listed companies by means of composite indicators that considered the following liquidity ratios: the current ratio, the quick ratio, the interest coverage ratio and the cash flow to interest expense ratio. The results suggest that analysts should focus on the interest coverage ratio in addressing the liquidity issue of the companies. By applying also principal component analysis to the data at hand we showed that our dimension reduction method should be preferred because it is always readily comprehended and much simpler. Moreover it requires a unique and very mild assumption: that financial ratios follow the larger the better rule. However, financial analysts should pay attention to the industry sector the companies belong to. We suggest that financial analysts should group the companies on the basis of the industry sector before applying our reduction procedure.

References

1. Aiello, F., Attanasio, M.: How to transform a batch of simple indicators to make up a unique one? In: Atti della XLII Riunione Scientifica della Società Italiana di Statistica, Sessioni Plenarie e Specializzate, CLEUP, Padova, pp. 327–338 (2004)
2. Arboretti Giancristofaro, R., Marozzi, M.: A comparison of different methods for the construction of composite indicators. In: Atti del IV Convegno Modelli Complessi e Metodi Computazionali Intensivi per la Stima e la Previsione, CLEUP, Padova, pp. 109–114 (2005)
3. Fayers, P.M., Hand, D.J.: Casual variables, indicator variables and measurement scales: an example from quality of life. J. R. Stat. Soc. A 165, 233–261 (2002)

4. Hoaglin, D.C., Mosteller, F., Tukey, J.W.: Understanding Robust and Exploratory Data Analysis. Wiley, New York (1983)
5. Huber, P.J.: Robust Statistics. Wiley, New York (1981)
6. Lago, A., Pesarin, F.: Non parametric combination of dependent rankings with application to the quality assessment of industrial products. Metron, LVIII 39–52 (2000)
7. Marozzi, M., Bolzan, M.: A non-parametric index of accessibility of services for households. In: Towards Quality of Life Improvement. The Publishing House of the Wroclaw University of Economics, Wroclaw, pp. 152–167 (2006)
8. Marozzi, M., Santamaria, L.: Composite indicators for finance. Ital. J. Appl. Stat. 19, 271–278 (2007)
9. Rees, B.: Financial Analysis. Prentice Hall, Harlow (1995)

The relation between implied and realised volatility in the DAX index options market

Silvia Muzzioli

Abstract. The aim of this paper is to investigate the relation between implied volatility, historical volatility and realised volatility in the DAX index options market. Since implied volatility varies across option type (call versus put) we run a horse race of different implied volatility estimates: implied call and implied put. Two hypotheses are tested in the DAX index options market: unbiasedness and efficiency of the different volatility forecasts. Our results suggest that both implied volatility forecasts are unbiased (after a constant adjustment) and efficient forecasts of future realised volatility in that they subsume all the information contained in historical volatility.

Key words: volatility forecasting, Black-Scholes Implied volatility, put-call parity

1 Introduction

Volatility is a key variable in option pricing models and risk management techniques and has drawn the attention of many theoretical and empirical studies aimed at assessing the best way to forecast it. Among the various models proposed in the literature in order to forecast volatility, we distinguish between option-based volatility forecasts and time series volatility models. The former models use prices of traded options in order to unlock volatility expectations while the latter models use historical information in order to predict future volatility (following [17], in this set we group predictions based on past standard deviation, ARCH conditional volatility models and stochastic volatility models). Many empirical studies have tested the forecasting power of implied volatility versus a time series volatility model.

Some early contributions find evidence that implied volatility (IV) is a biased and inefficient forecast of future realised volatility (see e.g., [2, 6, 14]). Although the results of some of these studies (e.g., [6, 14]) are affected by overlapping samples, as recalled by [4], or mismatching maturities between the option and the volatility forecast horizon, they constitute early evidence against the unbiasedness and information efficiency of IV. More recently, several papers analyse the empirical performance of IV in various option markets, ranging from indexes, futures or individual stocks and find that IV is unbiased and an efficient forecast of future realised volatility. In the index options market, Christensen and Prabhala [5] examine the relation between IV and realised volatility using S&P100 options, over the time period 1983–1995. They find that IV is a good predictor of future realised volatility. Christensen et al. [4] use options on the S&P100 and non-overlapping samples and find evidence for the efficiency of IV as a predictor of future realised volatility. In the futures options market Ederington and Guan [8] analyse the S&P500 futures options market and find that IV is an efficient forecast of future realised volatility. Szakmary et al. [19] consider options on 35 different futures contracts on a variety of asset classes. They find that IV, while not a completely unbiased estimate of future realised volatility, has more informative power than past realised volatility. In the stock options market, Godbey and Mahar [10] analyse the information content of call and put IV extracted from options on 460 stocks that compose the S&P500 index. They find that IV contains some information on future realised volatility that is superior both to past realised volatility and to a GARCH(1,1) estimate.

Option IV differs depending on strike price of the option (the so called smile effect), time to maturity of the option (term structure of volatility) and option type (call versus put). As a consequence, in the literature there is an open debate about which option class is most representative of market volatility expectations. As for the moneyness dimension, most of the studies use at the money options (or close to the money options) since they are the most heavily traded and thus the most liquid. As for the time to maturity dimension, the majority of the studies use options with time to maturity of one month in order to make it equal to the sampling frequency and the estimation horizon of realised volatility. As for the option type, call options are more used than put options. As far as we know, there is little evidence about the different information content of call or put prices. Even if, theoretically, call and put are linked through the put-call parity relation, empirically, given that option prices are observed with measurement errors (stemming from finite quote precision, bid-ask spreads, non-synchronous observations and other measurement errors), small errors in any of the input may produce large errors in the output (see e.g., [12]) and thus call IV and put IV may be different. Moreover, given that put options are frequently bought for portfolio insurance, there is a substantial demand for puts that is not available for the same call options. Also, in [15] we have proved that the use of both call and put options improves the pricing performance of option implied trees, suggesting that call and put may provide different information. Fleming [9] investigates the implied-realised volatility relation in the S&P100 options market and finds that call IV has slightly more predictive power than put IV. In the same market, Christensen and Hansen [3] find that both call and put IV are informative of future realized volatility, even if call IV performs slightly better than put IV. Both studies use American options and need the estimation of the dividend yield. These two aspects influence call and put options in a different manner and may alter the comparison if not properly addressed.

The aim of the paper is to explore the relation between call IV, put IV, historical volatility and realised volatility in the DAX index option market. The market is chosen for two main reasons: (i) the options are European, therefore the estimation of the early exercise premium is not needed and cannot influence the results; (ii) the DAX index is a capital weighted performance index composed of 30 major German stocks and is adjusted for dividends, stocks splits and changes in capital. Since dividends are assumed to be reinvested into the shares, they do not affect the index value. The plan of the paper is the following. In Section 2 we illustrate the data set used, the sampling procedure and the definition of the variables. In Section 3 we describe the methodology used in order to address the unbiasedeness and efficiency of the different volatility forecasts. In Section 4 we report the results of the univariate and encompassing regressions and we test our methodology for robustness. Section 5 concludes.

2 The data set, the sampling procedure and the definition of the variables

Our data set consists of daily closing prices of at the money call and put options on the DAX index, with one-month maturity recorded from 19 July 1999 to 6 December 2006. The data source is DATASTREAM. Each record reports the strike price, expiration month, transaction price and total trading volume of the day separately for call and put prices. We have a total of 1928 observations. As for the underlying we use the DAX index closing prices recorded in the same time period. As a proxy for the risk-free rate we use the one-month Euribor rate. DAX options are European options on the DAX index, which is a capital weighted performance index composed of 30 major German stocks and is adjusted for dividends, stock splits and changes in capital. Since dividends are assumed to be reinvested into the shares, they do not affect the index value, therefore we do not have to estimate the dividend payments. Moreover, as we deal with European options, we do not need the estimation of the early exercise premium. This latter feature is very important since our data set is by construction less prone to estimation errors if compared to the majority of previous studies that use American-style options. The difference between European and American options lies in the early exercise feature. The Black-Scholes formula, which is usually used in order to compute IV, prices only European-style options. For American options adjustments have to be made: for example, Barone-Adesi and Whaley [1] suggest a valuation formula based on the decomposition of the American option into the sum of a European option and a quasi-analytically estimated early exercise premium. However, given the difficulty in implementing the Barone-Adesi and Whaley model, many papers (see e.g., [5]) use the Black and Scholes formula also for American options. Given that American option prices are generally higher than European ones, the use of the Black-Scholes formula will generate an IV that overstates the true IV.

In order to avoid measurement errors, the data set has been filtered according to the following filtering constraints. First, in order not to use stale quotes, we eliminate dates with trading volume less than ten contracts. Second, we eliminate dates with option prices violating the standard no arbitrage bounds. After the application of the filters, we are left with 1860 observations out of 1928. As for the sampling procedure, in order to avoid the telescoping problem described in [4], we use monthly non-overlapping samples. In particular, we collect the prices recorded on the Wednesday that immediately follows the expiry of the option (third Saturday of the expiry month) since the week immediately following the expiration date is one of the most active. These options have a fixed maturity of almost one month (from 17 to 22 days to expiration). If the Wednesday is not a trading day we move to the trading day immediately following. The IV, provided by DATASTREAM, is obtained by inverting the Black and Scholes formula as a weighted average of the two options closest to being at the money and is computed for call options (σ_c) and for put options (σ_p). IV is an ex-ante forecast of future realised volatility in the time period until the option expiration. Therefore we compute the realised volatility (σ_r) in month t as the sample standard deviation of the daily index returns over the option's remaining life:

$$\sigma_r = \sqrt{\frac{1}{n-1}\sum_{i=1}^{n}(R_i - \overline{R})^2},$$

where R_i is the return of the DAX index on day i and $\overline{R}$ is the mean return of the DAX index in month t. We annualise the standard deviation by multiplying it by $\sqrt{252}$.

In order to examine the predictive power of IV versus a time series volatility model, following prior research (see e.g., [5, 13]), we choose to use the lagged (one month before) realised volatility as a proxy for historical volatility (σ_h). Descriptive statistics for volatility and log volatility series are reported in Table 1. We can see that on average realised volatility is lower than both IV estimates, with call IV being slightly higher than put IV. As for the standard deviation, realised volatility is slightly higher than both IV estimates. The volatility series are highly skewed (long right tail) and leptokurtic. In line with the literature (see e.g., [13]) we decided to use the natural logarithm of the volatility series instead of the volatility itself in the empirical analysis for the following reasons: (i) log-volatility series conform more closely to normality than pure volatility series: this is documented in various papers and it is the case in our sample (see Table 1); (ii) natural logarithms are less likely to be affected by outliers in the regression analysis.

Table 1. Descriptive statistics

Statistic	σ_c	σ_p	σ_r	$\ln \sigma_c$	$\ln \sigma_p$	$\ln \sigma_r$
Mean	0.2404	0.2395	0.2279	−1.51	−1.52	−1.6
Std dev	0.11	0.11	0.12	0.41	0.41	0.49
Skewness	1.43	1.31	1.36	0.49	0.4	0.41
Kurtosis	4.77	4.21	4.37	2.73	2.71	2.46
Jarque Bera	41.11	30.28	33.68	3.69	2.68	3.54
p-value	0.00	0.00	0.00	0.16	0.26	0.17

3 The methodology

The information content of IV is examined both in univariate and in encompassing regressions. In univariate regressions, realised volatility is regressed against one of the three volatility forecasts (call IV (σ_c), put IV (σ_p), historical volatility (σ_h)) in order to examine the predictive power of each volatility estimator. The univariate regressions are the following:

$$\ln(\sigma_r) = \alpha + \beta \ln(\sigma_i), \tag{1}$$

where σ_r is realised volatility and σ_i is volatility forecast, $i = h, c, p$. In encompassing regressions, realised volatility is regressed against two or more volatility forecasts in order to distinguish which one has the highest explanatory power. We choose to compare pairwise one IV forecast (call, put) with historical volatility in order to see if IV subsumes all the information contained in historical volatility. The encompassing regressions used are the following:

$$\ln(\sigma_r) = \alpha + \beta \ln(\sigma_i) + \gamma \ln(\sigma_h), \tag{2}$$

where σ_r is realised volatility, σ_i is implied volatility, $i = c, p$ and σ_h is historical volatility. Moreover, we compare call IV and put IV in order to understand if the information carried by call (put) prices is more valuable than the information carried by put (call) prices:

$$\ln(\sigma_r) = \alpha + \beta \ln(\sigma_p) + \gamma \ln(\sigma_c), \tag{3}$$

where σ_r is realised volatility, σ_c is call IV and σ_p is put IV.

Following [4], we tested three hypotheses in the univariate regressions (2). The first hypothesis concerns the amount of information about future realised volatility contained in the volatility forecast. If the volatility forecast contains some information, then the slope coefficient should be different from zero. Therefore we test if $\beta = 0$ and we see whether it can be rejected. The second hypothesis is about the unbiasedness of the volatility forecast. If the volatility forecast is an unbiased estimator of future realised volatility, then the intercept should be zero and the slope coefficient should be one (H_0: $\alpha = 0$ and $\beta = 1$). In case this latter hypothesis is rejected, we see if at least the slope coefficient is equal to one (H_0: $\beta = 1$) and, if not rejected, we interpret the volatility forecast as unbiased after a constant adjustment. Finally if IV is efficient then the error term should be white noise and uncorrelated with the information set. In encompassing regressions there are three hypotheses to be tested. The first is about the efficiency of the volatility forecast: we test whether the volatility forecast (call IV, put IV) subsumes all the information contained in historical volatility. In affirmative case the slope coefficient of historical volatility should be equal to zero, (H_0: $\gamma = 0$). Moreover, as a joint test of information content and efficiency we test if the slope coefficients of historical volatility and IV (call, put) are equal to zero and one respectively (H_0: $\beta = 1$ and $\gamma = 0$). Following [13], we ignore the intercept in the latter null hypothesis, and if our null hypothesis is not rejected, we interpret the volatility forecast as unbiased after a constant adjustment. Finally we investigate the different information content of call IV and put IV. To this end we test, in augmented regression (4), if $\gamma = 0$ and $\beta = 1$, in order to see if put IV subsumes all the information contained in call IV.

In contrast to other papers (see e.g., [3,5]) that use American options on dividend paying indexes, our data set of European-style options on a non-dividend paying index is free of measurement errors that may arise in the estimation of the dividend yield and the early exercise premium. Nonetheless, as we are using closing prices for the index and the option that are non-synchronous (15 minutes' difference) and we are ignoring bid ask spreads, some measurement errors may still affect our estimates. Therefore we adopt an instrumental variable procedure, we regress call (put) IV on an instrument (in univariate regressions) and on an instrument and any other exogenous variable (in encompassing and augmented regressions) and replace fitted values in the original univariate and encompassing regressions. As the instrument for call (put) IV we use both historical volatility and past call (put) IV as they are possibly correlated to the true call (put) IV, but unrelated to the measurement error associated with call (put) IV one month later. As an indicator of the presence of errors in variables we use the Hausman [11] specification test statistic. The Hausman specification test is defined as: $\frac{\beta_{TSLS} - \beta_{OLS}}{VAR(\beta_{TSLS}) - VAR(\beta_{OLS})}$ where: β_{TSLS} is the beta obtained through the Two Stages Least Squares procedure, β_{OLS} is the beta obtained through the Ordinary Least Squares (OLS) procedure and $Var(x)$ is the variance of the coefficient x. The Hausman specification test is distributed as a $\chi^2(1)$.

4 The results

The results of the OLS univariate (equation (2)), encompassing (equation (3)), and augmented (equation (4)) regressions are reported in Table 2. In all the regressions the residuals are normal, homoscedastic and not autocorrelated (the Durbin Watson statistic is not significantly different from two and the Breusch-Godfrey LM test confirms no autocorrelation up to lag 12). First of all, in the three univariate regressions all the beta coefficients are significantly different from zero: this means that all three volatility forecasts (call IV, put IV and historical) contain some information about future realised volatility. However, the null hypothesis that any of the three volatility forecasts is unbiased is strongly rejected in all cases. In particular, in our sample, realised volatility is on average a little lower than the two IV forecasts, suggesting that IV overpredicts realised volatility. The adjusted R^2 is the highest for put IV, closely followed by call IV. Historical volatility has the lowest adjusted R^2. Therefore put IV is ranked first in explaining future realised volatility, closely followed by call IV, while historical volatility is the last. The null hypothesis that β is not significantly different from one cannot be rejected at the 10% critical level for the two IV estimates, while it is strongly rejected for historical volatility. Therefore we can consider both IV estimates as unbiased after a constant adjustment given by the intercept of the regression.

In encompassing regressions (3) we compare pairwise call/put IV forecast with historical volatility in order to understand if IV subsumes all the information contained in historical volatility. The results are striking and provide strong evidence for both the unbiasedness and efficiency of both IV forecasts. First of all, from the comparison of univariate and encompassing regressions, the inclusion of historical volatility does not improve the goodness of fit according to the adjusted R^2. In fact, the slope coefficient of historical volatility is not significantly different from zero at the 10% level in the encompassing regressions (3), indicating that both call and put IV subsume all the information contained in historical volatility. The slope coefficients of both call and put IV are not significantly different from one at the 10% level and the joint test of information content and efficiency ($\gamma = 0$ and $\beta = 1$) does not reject the null hypothesis, indicating that both IV estimates are efficient and unbiased after a constant adjustment.

In order to see if put IV has more predictive power than call IV, we test in augmented regression (3) if $\gamma = 0$ and $\beta = 1$. The joint test $\gamma = 0$ and $\beta = 1$ does not reject the null hypothesis. We see that the slope coefficient of put IV is significantly different from zero only at the 5% level, while the slope coefficient of call IV is not significantly different from zero. As an additional test we regress $\ln(\sigma_c)$ on $\ln(\sigma_p)$ ($\ln(\sigma_p)$ on $\ln(\sigma_c)$) and retrieve the residuals. Then we run univariate regression (2) for $\ln(\sigma_c)$ ($\ln(\sigma_p)$) using as an additional explanatory variable the residuals retrieved from the regression of $\ln(\sigma_c)$ on $\ln(\sigma_p)$ ($\ln(\sigma_p)$ on $\ln(\sigma_c)$). The residuals are significant only in the regression of $\ln(\sigma_r)$ on $\ln(\sigma_c)$, pointing to the fact that put IV contains slightly more information on future realised volatility than call IV.

A possible concern is the problem of data snooping, which occurs when the properties of a data set influence the choice of the estimator or test statistic (see e.g., [7]) and may arise in a multiple regression model, when a large number of explanatory variables are compared and the selection of the candidate variables is not based on a financial theory (e.g., in [20] 3654 models are compared to a given benchmark, in [18] 291 explanatory variables are used in a multiple regression). This is not the case in our regressions, since (i) we do not have any parameter to estimate, (ii) we use only three explanatory variables: historical volatility, call IV and put IV, that are compared pairwise in the regressions and (iii) the choice has been made on the theory that IV, being derived from option prices, is a forward-looking measure of ex post realised volatility and is deemed as the market's expectation of future volatility. Finally, in order to test the robustness of our results and see if IV has been measured with errors, we adopt an instrumental variable procedure and run a two-stage least squares. The Hausman [11] specification test, reported in the last column of Table 2, indicates that the errors in variables problem is not significant in univariate regressions (2), in encompassing regressions (3) or in augmented regression (4).[1] Therefore we can trust the OLS regression results.

In our sample both IV forecasts obtain almost the same performance, with put IV marginally better than call IV. These results are very different from the ones obtained both in [3] and in [9]. The difference can possibly be attributed to the option exercise feature, which in our case is European and not American, and to the underlying index

---

[1] In augmented regression (4) the instrumental variables procedure is used for the variable $\ln(\sigma_p)$.

Table 2. OLS regressions

Intercept	$\ln(\sigma_c)$	$\ln(\sigma_p)$	$\ln(\sigma_h)$	$Adj R^2$	DW	χ^{2a}	χ^{2b}	Hausman Test
−0.01		1.05***		0.77	1.73	13.139		0.10021
(0.915)		(0.000)				(0.00)		
−0.018	1.047***			0.76	1.77	13.139		0.25128
(0.853)	(0.000)					(0.00)		
−0.29			0.82	0.65	2.12	7.517		
(0.008)			(0.000)			(0.02)		
−0.02	0.938***		0.10^{+++}	0.76	1.87		1.288	0.47115
(0.850)	(0.000)		(0.400)				(0.53)	
−0.01		0.9631***	0.082^{+++}	0.77	1.80		1.158	0.95521
(0.915)		(0.000)	(0.489)				(0.56)	
0.0006	0.372	0.6861***		0.77	1.74		2.04	0.14977
(0.994)	(0.244)	(0.033)					(0.35)	

^{a} Note: The numbers in brackets are the p-values. The χ^{2a} column reports the statistic of a χ^2 test for the joint null hypothesis $\alpha = 0$ and $\beta = 1$ in the following univariate regressions $\ln(\sigma_r) = \alpha + \beta \ln(\sigma_i)$ where σ_r =realized volatility and σ_i= volatility forecast, $i = h, c, p$. The χ^{2b} column reports the statistic of a χ^2 test for the joint null hypothesis $\gamma = 0$ and $\beta = 1$ in the following regressions: $\ln(\sigma_r) = \alpha + \beta \ln(\sigma_i) + \gamma \ln(\sigma_h)$, $\ln(\sigma_r) = \alpha + \beta \ln(\sigma_p) + \gamma \ln(\sigma_c)$, where σ_r = realized volatility, σ_i= volatility forecast, $i = c, p$ and σ_h = historical volatility. The superscripts ***, **, * indicate that the slope coefficient is not significantly different from one at the 10%, 5% and 1% critical level respectively. The superscripts $^{+++}$, $^{++}$, $^+$ indicate that the slope coefficient is not significantly different from zero at the 10%, 5% and 1% critical level respectively. The last column reports the Hausman [11] specification test statistic (one degree of freedom), where the 5% critical level is equal to 3.841.

features, which in our case do not require the dividend payment estimation. Another possible explanation stems from the characteristics of the data set used. In particular in our case put IV was on average lower than call IV, while in [3] the opposite is true. As IV usually overpredicts realised volatility, if a choice has to be made between call and put IV, a rule of thumb can be to choose the lowest of the two.

5 Conclusions

In this paper we have investigated the relation between IV, historical volatility and realised volatility in the DAX index options market. Since IV varies across option type (call versus put), we have run a horse race of different IV estimates: call IV, put IV. Two hypotheses have been tested: unbiasedness and efficiency of the different volatility forecasts. Our results suggest that both IV forecasts contain more information about future realised volatility than historical volatility. In particular, they are unbiased (after a constant adjustment) and efficient forecasts of realised volatility in that they subsume all the information contained in historical volatility. In our sample both IV forecasts obtain almost the same performance, with put IV marginally better than call

IV. This is an interesting result and is a warning against the *a priori* choice of using call IV. The recent turmoil in financial markets caused by the current financial crisis has determined high levels of volatility. High on the research agenda is to test the unbiasedness and efficiency hypotheses using the most recent volatility data.

Acknowledgement. This is a revised version of Muzzioli [16]. The author wishes to thank the two anonymous referees, and Marianna Brunetti, Mario Forni and Giuseppe Marotta for helpful comments and suggestions. The author gratefully acknowledges financial support from MIUR. Usual disclaimer applies.

References

1. Barone-Adesi, G., Whaley, R.: Efficient analytic approximation of American option values. J. Finan. 6, 301–320 (1987)
2. Canina, L., Figlewski, S.: The informational content of implied volatility. Rev. Finan. Stud. 6, 659–681 (1993)
3. Christensen, B.J., Hansen, C.S.: New evidence on the implied-realized volatility relation. Eur. J. Finan. 8, 187–205 (2002)
4. Christensen, B.J., Hansen, C.S., Prabhala, N.R.: The telescoping overlap problem in options data. Working paper, University of Aarhus and University of Maryland (2001)
5. Christensen, B.J., Prabhala, N.R.: The relation between implied and realized volatility. J. Finan. Econ. 50, 125–150 (1998)
6. Day, T.E., Lewis, C.M.: Stock market volatility and the informational content of stock index options. J. Econometrics 52, 267–287 (1992)
7. Dimson, E., Marsh P.: Volatility forecasting without data snooping. J. Bank. Finan. 29, 399–421 (1990)
8. Ederington, L.H., Guan, W.: Is implied volatility an informationally efficient and effective predictor of future volatility? J. Risk 4, 29–46 (2002)
9. Fleming, J.: The quality of market volatility forecasts implied by S&P100 index option prices. J. Emp. Finan. 5, 317–345 (1998)
10. Godbey, J.M., Mahar, J.W.: Forecasting power of implied volatility: evidence from individual equities. B>Quest, University of West Georgia (2007)
11. Hausman, J.: Specification tests in econometrics. Econometrica 46, 1251–1271 (1978)
12. Hentschel, L.: Errors in implied volatility estimation. J. Finan. Quant. Anal. 38, 779–810 (2003)
13. Jiang, G.J., Tian, Y.S.: Model free implied volatility and its information content. Rev. Finan. Stud. 4, 1305–1342 (2005)
14. Lamourex, C.G., Lastrapes, W.D.: Forecasting stock-return variance: toward an understanding of stochastic implied volatilities. Rev. Finan. Stud. 6, 293–326 (1993)
15. Moriggia, V., Muzzioli, S., Torricelli, C.: Call and put implied volatilities and the derivation of option implied trees. Front. Finan. Econ. 4, 35–64 (2007)
16. Muzzioli, S.: The relation between implied and realised volatility: are call options more informative than put options? Evidence from the DAX Index Options Market. CEFIN working paper n. 4 (2007)
17. Poon, S., Granger, C.W.: Forecasting volatility in financial markets: a review. J. Econ. Lit. 41, 478–539 (2003)

18. Romano, J.P., Shaikh, A.M., Wolf, M.: Formalized data snooping based on generalized error rates. Econometric Theory 24, 404–447 (2008)
19. Szakmary, A., Evren, O., Jin, K.K., Wallace N.D.: The predictive power of implied volatility: evidence from 35 futures markets. J. Bank. Finan. 27, 2151–2175 (2003)
20. White, H.: A reality check for data snooping. Econometrica 68, 1097–1126, (2000)

Binomial algorithms for the evaluation of options on stocks with fixed per share dividends

Martina Nardon and Paolo Pianca

Abstract. We consider options written on assets which pay cash dividends. Dividend payments have an effect on the value of options: high dividends imply lower call premia and higher put premia. Recently, Haug et al. [13] derived an integral representation formula that can be considered the exact solution to problems of evaluating both European and American call options and European put options. For American-style put options, early exercise may be optimal at any time prior to expiration, even in the absence of dividends. In this case, numerical techniques, such as lattice approaches, are required. Discrete dividends produce discrete shift in the tree; as a result, the tree is no longer reconnecting beyond any dividend date. While methods based on non-recombining trees give consistent results, they are computationally expensive. In this contribution, we analyse binomial algorithms for the evaluation of options written on stocks which pay discrete dividends and perform some empirical experiments, comparing the results in terms of accuracy and speed.

Key words: options on stocks, discrete dividends, binomial lattices

1 Introduction

We consider options written on assets which pay dividends. Dividends are announced as a pure cash amount D to be paid at a specified ex-dividend date t_D. Empirically, one observes that at the ex-dividend date the stock price drops. Hence dividends imply lower call premia and higher put premia. In order to exclude arbitrage opportunities, the jump in the stock price should be equal to the size of the net dividend. Since we cannot use the proportionality argument, the price dynamics depend on the timing of the dividend payment.

Usually, derivative pricing theory assumes that stocks pay known dividends, both in size and timing. Moreover, new dividends are often supposed to be equal to the former ones. Even if these assumptions might be too strong, in what follows we assume that we know both the amount of dividends and times in which they are paid.

Valuation of options on stocks which pay discrete dividends is a rather hard problem which has received a lot of attention in the financial literature, but there is much confusion concerning the evaluation approaches. Different methods have been proposed for the pricing of both European and American options on dividend paying stocks, which suggest various model adjustments (such as, for example, subtracting the present value of the dividend from the asset spot price). Nevertheless, all such approximations have some drawbacks and are not so efficient (see e.g., Haug [11] for a review).

Haug and Haug [12] and Beneder and Vorst [2] propose a volatility adjustment which takes into account the timing of the dividend. The idea behind the approximation is to leave volatility unchanged before the dividend payment and to apply the adjusted volatility after the dividend payment. This method performs particularly poorly in the presence of multiple dividends. A more sophisticated volatility adjustment to be used in combination with the escrowed dividend model is proposed by Bos et al. [4]. The method is quite accurate for most cases. Nevertheless, for very large dividends, or in the case of multiple dividends, the method can yield significant mispricing. A slightly different implementation (see Bos and Vandermark [5]) adjusts both the stock price and the strike. The dividends are divided into two parts, called "near" and "far", which are used for the adjustments to the spot and the strike price respectively. This approach seems to work better than the approximation mentioned above. Haug et al. [13] derive an integral representation formula that can be considered the exact solution to problems of evaluating both European and American call options and European put options. Recently, de Matos et al. [7] derived arbitrarily accurate lower and upper bounds for the value of European options on a stock paying a discrete dividend.

For American-style put options, it can be optimal to exercise at any time prior to expiration, even in the absence of dividends. Unfortunately, no analytical solutions for both the option price and the exercise strategy are available, hence one is generally forced to numerical solutions, such as binomial approaches. As is well known (see Merton [14]), in the absence of dividends, it is never optimal to exercise an American call before maturity. If a cash dividend payment is expected during the lifetime of the option, it might be optimal to exercise an American call option right before the ex-dividend date, while for an American put it may be optimal to exercise at any point in time until maturity.

Lattice methods are commonly used for the pricing of both European and American options. In the binomial model (see Cox et al. [6]), the pricing problem is solved by backward induction along the tree. In particular, for American options, at each node of the lattice one has to compare the early exercise value with the continuation value.

In this contribution, we analyse binomial algorithms for the evaluation of options written on stocks which pay discrete dividends of both European and American types. In particular, we consider non-recombining binomial trees, hybrid binomial algorithms for both European and American call options, based on the Black-Scholes formula for the evaluation of the option after the ex-dividend date and up to maturity; a binomial method which implements the efficient continuous approximation proposed in [5]; and we propose a binomial method based on an interpolation idea given by Vellekoop and Nieuwenhuis [17], in which the recombining feature is maintained. The model based on the interpolation procedure is also extended to the case of multiple dividends; this feature is very important for the pricing of long-term options and index options. We performed some empirical experiments and compare the results in terms of accuracy and speed.

2 European-style options

Dividends affect option prices through their effect on the underlying stock price. In a continuous time setting, the underlying price dynamics depends on the timing of the dividend payment and is assumed to satisfy the following stochastic differential equation

$$dS_t = rS_t dt + \sigma S_t dW_t \qquad t \neq t_D$$
$$S_{t_D}^+ = S_{t_D}^- - D_{t_D}, \qquad\qquad (1)$$

where $S_{t_D}^-$ and $S_{t_D}^+$ denote the stock price levels right before and after the jump at time t_D, respectively. Due to this discontinuity, the solution to equation (1) is no longer log-normal but in the form[1]

$$S_t = S_0 e^{(r-\sigma^2/2)t+\sigma W_t} - D_{t_D} e^{(r-\sigma^2/2)(t-t_D)+\sigma W_{t-t_D}} I_{\{t \geq t_D\}}. \qquad (2)$$

Recently, Haug et al. [13] (henceforth HHL) derived an integral representation formula for the fair price of a European call option on a dividend paying stock. The basic idea is that after the dividend payment, option pricing reduces to a simple Black-Scholes formula for a non-dividend paying stock. Before t_D one considers the discounted expected value of the BS formula adjusted for the dividend payment. In the geometric Brownian motion setup, the HHL formula is

$$C_{HHL}(S_0, D, t_D) = e^{-rt_D} \int_d^\infty c_E(S_x - D, t_D) \frac{e^{-x^2/2}}{\sqrt{2\pi}} dx, \qquad (3)$$

where $d = \frac{\log(D/S_0)-(r-\sigma^2/2)t_D}{\sigma\sqrt{t_D}}$, $S_x = S_0 e^{(r-\sigma^2/2)t_D+\sigma\sqrt{t_D}x}$ and $c_E(S_x - D, t_D)$ is simply the BS formula with time to maturity $T - t_D$. The integral representation (3) can be considered as the exact solution to the problem of valuing a European call option written on stock with a discrete and known dividend. Let us observe that the well known put-call parity relationship allows the immediate calculation of the theoretical price of a European put option with a discrete dividend.

3 American-style options

Most traded options are of American style. The effect of a discrete dividend payment on American option prices is different than for European options. While for European-style options the pricing problem basically arises from mis-specifying the variance

---

[1] I_A denotes the indicator function of A.

of the underlying process, for American options the impact on the optimal exercise strategy is more important. As is well known, it is never optimal to exercise an American call option on non-dividend paying stocks before maturity. As a result, the American call has the same value as its European counterpart. In the presence of dividends, it may be optimal to exercise the American call and put before maturity. In general, early exercise is optimal when it leads to an alternative income stream, i.e., dividends from the stock for a call and interest rates on cash for a put option. In the case of discrete cash dividends, the call option may be optimally exercised early instantaneously prior to the ex-dividend date,[2] t_D^-; while for a put it may be optimal to exercise at any point in time till maturity. Simple adjustments like subtracting the present value of the dividend from the asset spot price make little sense for American options.

The first approximation to the value of an American call on a dividend paying stock was suggested by Black in 1975 [3]. This is basically the *escrowed* dividend method, where the stock price in the BS formula is replaced by the stock price minus the present value of the dividend. In order to account for early exercise, one also computes an option value just before the dividend payment, without subtracting the dividend. The value of the option is considered to be the maximum of these values.

A model which is often used and implemented in much commercial software was proposed, simplified and adjusted by Roll [15], Geske [8, 10] and Whaley [18] (RGW model). These authors construct a portfolio of three European call options which represents an American call and accounts for the possibility of early exercise right before the ex-dividend date. The portfolio consists of two long positions with exercise prices X and $S^* + D$ and maturities T and t_D^-, respectively. The third option is a short call on the first of the two long calls with exercise price S^*+D-X and maturity t_D^-. The stock price S^* makes the holder of the option indifferent between early exercise at time t_D and continuing with the option. Formally, we have $C(S^*, T-t_D, X) = S^*+D-X$. This equation can be solved if the ex-dividend date is known. The two long positions follow from the BS formula, while for the compound option Geske [9] provides an analytical solution.

The RGW model was considered for more than twenty years as a brilliant solution in closed form to the problem of evaluating American call options on equities that pay a discrete dividend. Although some authoritative authors still consider the RGW formula as the exact solution, the model does not yield good results in many cases of practical interest. Moreover, it is possible to find situations in which the use of the formula RGW allows for arbitrage. Whaley, in a recent monograph [19], presents an example that shows the limits of the RGW model.

Haug et al. [13] derived an integral representation formula for the American call option fair price in the presence of a single dividend D paid at time t_D. Since early exercise is only optimal instantaneously prior to the ex-dividend date, in order to obtain the exact solution for an American call option with a discrete dividend one can

---

[2] Note that after the dividend date t_D, the option is a standard European call which can be priced using the BS formula; this idea can be implemented in a hybrid BS-binomial model.

merely replace relation (3) with

$$C_{HHL}(S_0, D, t_D) = e^{-rt_D} \int_d^\infty \max\{S_x - X, c_E(S_x - D, t_D)\} \frac{e^{-x^2/2}}{\sqrt{2\pi}} dx . \quad (4)$$

4 Binomial models

The evaluation of options using binomial methods is particularly easy to implement and efficient at standard conditions, but it becomes difficult to manage in the case in which the underlying asset pays one or more discrete dividends, due to the fact that the number of nodes grows considerably and entails huge calculations. In the absence of dividends or when dividends are assumed to be proportional to the stock price, the binomial tree reconnects in the sense that the price after a up-down movement coincides with the price after a down-up movement. As a result, the number of nodes at each step grows linearly.

If during the life of the option a dividend of amount D is paid, at each node after the ex-dividend date a new binomial tree has to be considered (see Fig. 1), with the result that the total number of nodes increases to the point that it is practically impossible to consider trees with an adequate number of stages. To avoid such a complication, often it is assumed that the underlying dynamics are characterised by a dividend yield which is discrete and proportional to the stock price. Formally,

$$\begin{cases} S_0 u^j d^{i-j} & j = 0, 1, \ldots i \\ S_0(1 - q)u^j d^{i-j} & j = 0, 1, \ldots i, \end{cases} \quad (5)$$

where the first law applies if the period preceding the ex-dividend date and the second applies after the dividend date, and where S_0 indicates the initial price, q is the dividend yield, and u and d are respectively the upward and downward coefficients, defined by $u = e^{\sigma\sqrt{T/n}}$ and $d = 1/u$. The hypothesis of a proportional dividend yield can be accepted as an approximation of dividends paid in the long term, but it is not acceptable in a short period of time during which the stock pays a dividend in cash and its amount is often known in advance or estimated with appropriate accuracy.

If the underlying asset is assumed to pay a discrete dividend D at time $t_D < T$ (which in a discrete time setting corresponds to the step n_D), the dividend amount is subtracted at all nodes at time point t_D. Due to this discrete shift in the tree, as already noticed, the lattice is no longer recombining beyond the time t_D and the binomial method becomes computationally expensive, since at each node at time t_D a separate binomial tree has to be evaluated until maturity (see Fig. 1). Also, in the presence of multiple dividends this approach remains theoretically sound, but becomes unattractive due to the computational intensity.

Schroder [16] describes how to implement discrete dividends in a recombining tree. The approach is based on the escrowed dividend process idea, but the method leads to significant pricing errors.

The problem of the enormous growth in the number of nodes that occurs in such a case can be simplified if it is assumed that the price has a stochastic component $\tilde{S}$

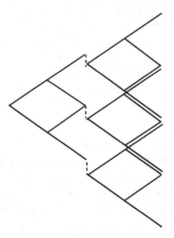

Fig. 1. Non-recombining binomial tree after the dividend payment

given by

$$\tilde{S} = \begin{cases} S - De^{-r(t_D - i)} & i \leq t_D \\ S & i > t_D, \end{cases} \tag{6}$$

and a deterministic component represented by the discounted value of the dividend or of dividends that will be distributed in the future. Note that the stochastic component gives rise to a reconnecting tree. Moreover, you can build a new tree (which is still reconnecting) by adding the present value of future dividends to the price of the stochastic component in correspondence of each node. Hence the tree reconnects and the number of nodes in each period i is equal to $i + 1$.

The recombining technique described above can be improved through a procedure that preserves the structure of the tree until the ex-dividend time and that will force the recombination after the dividend payment. For example, you can force the binomial tree to recombine by taking, immediately after the payment of a dividend, as extreme nodes

$$S_{n_D+1,0} = (S_{n_D,0} - D) d \qquad S_{n_D,n_D} = (S_{n_D,n_D} - D) u, \tag{7}$$

and by calculating the arithmetic average of the values that are not recombining. This technique has the characteristic of being simple from the computational point of view.

Alternatively, you can use a technique, called "stretch", that calculates the extreme nodes as in the previous case; in such a way, one forces the reconnection at the intermediate nodes by choosing the upward coefficients as follows

$$u(i, j) = e^{\lambda \sigma \sqrt{T/n}}, \tag{8}$$

where λ is chosen in order to make equal the prices after an up and down movement. This technique requires a greater amount of computations as at each stage both the coefficients and the corresponding probabilities change.

In this paper, we analyse a method which performs very efficiently and can be applied to both European and American call and put options. It is a binomial method which maintains the recombining feature and is based on an interpolation idea proposed by Vellekoop and Nieuwenhuis [17].

For an American option, the method can be described as follows: a standard binomial tree is constructed without considering the payment of the dividend (with $S_{ij} = S_0 u^j d^{i-j}$, $u = e^{\sigma\sqrt{T/n}}$, and $d = 1/u$), then it is evaluated by backward induction from maturity until the dividend payment; at the node corresponding to an ex-dividend date (at step n_D), we approximate the continuation value V_{n_D} using the following linear interpolation

$$V(S_{n_D,j}) = \frac{V(S_{n_D,k+1}) - V(S_{n_D,k})}{S_{n_D,k+1} - S_{n_D,k}} (S_{n_D,j} - S_{n_D,k}) + V(S_{n_D,k}), \qquad (9)$$

for $j = 0, 1, \ldots, n_D$ and $S_{n_D,k} \le S_{n_D,j} \le S_{n_D,k+1}$; then we continue backward along the tree. The method can be easily implemented also in the case of multiple dividends (which are not necessarily of the same amount).

We have implemented a very efficient method which combines this interpolation procedure and the binomial algorithm for the evaluation of American options proposed by Basso et al. [1].[3]

We performed some empirical experiments and compare the results in terms of accuracy and speed.

5 Numerical experiments

In this section, we briefly report the results of some empirical experiments related to European calls and American calls and puts. In Table 1, we compare the prices provided by the HHL exact formula for the European call, with those obtained with the 2000-step non-combining binomial method and the binomial method based on interpolation (9). We also report the results obtained with the approximation proposed by Bos and Vandermark [5] (BV). For a European call, the non-recombining binomial method requires a couple of seconds, while the calculations with a 2000-step binomial interpolated method are immediate.

Table 2 shows the results for the American call and put options. We have compared the results obtained with non-recombining binomial methods and the 10,000-step binomial method based on the interpolation procedure (9). In the case of the American put, the BV approximation leads to considerable pricing errors.

We also extended the model based on the interpolation procedure to the case of multiple dividends. Table 3 shows the results for the European call with multiple

---

[3] The algorithm exploits two devices: (1) the symmetry of the tree, which implies that all the asset prices defined in the lattice at any stage belong to the set $\{S_0 u^j : j = -n, -n + 1, \ldots, 0, \ldots, n - 1, n\}$, and (2) the fact that in the nodes of the early exercise region, the option value, equal to the intrinsic value, does not need to be recomputed when exploring the tree backwards.

Table 1. European calls with dividend $D = 5$ ($S_0 = 100$, $T = 1$, $r = 0.05$, $\sigma = 0.2$)

t_D	X	HHL	Non-rec. bin. ($n = 2000$)	Interp. bin. ($n = 2000$)	BV
	70	28.7323	28.7323	28.7324	28.7387
0.25	100	7.6444	7.6446	7.6446	7.6456
	130	0.9997	0.9994	1.000	0.9956
	70	28.8120	28.8120	28.8121	28.8192
0.5	100	7.7740	7.7742	7.7742	7.7743
	130	1.0501	1.0497	1.0506	1.0455
	70	28.8927	28.8927	28.8928	28.8992
0.75	100	7.8997	7.8999	7.8999	7.9010
	130	1.0972	1.0969	1.0977	1.0934

Table 2. American call and put options with dividend $D = 5$ ($S_0 = 100$, $T = 1$, $r = 0.05$, $\sigma = 0.2$)

t_D	X	American Call non-rec. hyb. bin. ($n = 5000$)	interp. bin. ($n = 10{,}000$)	American put Non-rec. bin. ($n = 2000$)	interp. bin. ($n = 10{,}000$)	BV
	70	30.8740	30.8744	0.2680	0.2680	0.2630
0.25	100	7.6587	7.6587	8.5162	8.5161	8.5244
	130	0.9997	0.9998	33.4538	33.4540	350112
	70	31.7553	31.7557	0.2875	0.2876	0.2901
0.5	100	8.1438	8.1439	8.4414	8.4412	8.5976
	130	1.0520	1.0522	32.1195	32.1198	35.0112
	70	32.6407	32.6411	0.3070	0.3071	0.2901
0.75	100	9.1027	9.1030	8.2441	8.2439	8.6689
	130	1.1764	1.1767	30.8512	30.8515	35.0012

dividends. We have compared the non-reconnecting binomial method with $n = 2000$ steps (only for the case with one and two dividends) and the interpolated binomial method with $n = 10{,}000$ steps (our results are in line with those obtained by Haug et al. [13]).

Table 4 shows the results for the American call and put options for different maturities in the interpolated binomial method with multiple dividends.

6 Conclusions

The evaluation of the options on stocks that pay discrete dividends was the subject of numerous studies that concerned both closed-form formula and numerical approximate methods. Recently, Haug et al. [13] proposed an integral expression that allows the calculation of European call and put options and American call options in precise

Table 3. European call option with multiple dividends $D = 5$ paid at times $t_D \in$ {0.5, 1.5, 2.5, 3.5, 4.5, 5.5}, for different maturities $T = 1, \ldots, 6$ ($S_0 = 100$, $X = 100$, $r = 0.05$, $\sigma = 0.2$)

T	Non-rec. bin. ($n = 2000$)	Interp. bin. ($n = 10,000$)
1	7.7742	7.7741
2	10.7119	10.7122
3		12.7885
4		14.4005
5		15.7076
6		16.7943

Table 4. American options with multiple dividends in the interpolated 10,000-step binomial method (with parameters $S_0 = 100$, $X = 100$, $r = 0.05$, $\sigma = 0.2$); a cash dividend $D = 5$ is paid at the dates $t_D \in$ {0.5, 1.5, 2.5, 3.5, 4.5, 5.5}, for different maturities $T = 1, \ldots, 6$

T	American call	American put
1	8.1439	8.4412
2	11.2792	11.5904
3	13.3994	13.7399
4	15.0169	15.3834
5	16.3136	16.7035
6	17.3824	17.7938

terms. The formula proposed by Haug et al. requires the calculation of an integral. Such an integral representation is particularly interesting because it can be extended to the case of non-Brownian dynamics and to the case of multiple dividends.

The pricing of American put options written on stocks which pay discrete dividend can be obtained with a standard binomial scheme that produces very accurate results, but it leads to non-recombining trees and therefore the number of nodes does not grow linearly with the number of steps.

In this contribution, we implemented alternative methods to the classical binomial approach for American options: a hybrid binomial-Black-Scholes algorithm, a binomial method which translates the continuous approximation proposed in [5] and a binomial method based on an interpolation procedure, in which the recombining feature is maintained. We performed some empirical experiments and compared the results in terms of accuracy and efficiency. In particular, the efficient implementation of the method based on interpolation yields very accurate and fast results.

References

1. Basso A., Nardon M., Pianca P.: A two-step simulation procedure to analyze the exercise features of American options. Decis. Econ. Finan. 27, 35–56 (2004)

2. Beneder R., Vorst T.: Option on dividends paying stocks. In: Yong, J. (ed.) Recent Developments in Mathematical Finance. World Scientiric Publishing, River Edge, NJ (2002)
3. Black F.: Fact and fantasy in the use of options. Finan. Anal. J. July-August, 36–72 (1975)
4. Bos R., Gairat A., Shepeleva A.: Dealing with discrete dividends. Risk 16, 109–112 (2003)
5. Bos R., Vandermark S.: Finessing fixed dividends. Risk 15, 157–158 (2002)
6. Cox, J., Ross, S., Rubinstein, M.: Option pricing: a simplified approach. J. Finan. Econ. 7, 229–263 (1979)
7. de Matos J.A., Dilao R., Ferreira B.: The exact value for European options on a stock paying a discrete dividend, Munich Personal RePEc Archive (2006)
8. Geske R.: A note on an analytic formula for unprotected American call options on stocks with known dividends. J. Finan. Econ. 7, 375–380 (1979)
9. Geske R.: The valuation of compound options. J. Finan. Econ. 7, 53–56 (1979)
10. Geske R.: Comments on Whaley's note. J. Finan. Econ. 9, 213–215 (1981)
11. Haug, E.S.: The Complete Guide to Option Pricing Formulas. McGraw-Hill, New York (2007)
12. Haug E.S., Haug J.: A new look at pricing options with time varying volatility. Working paper (1998)
13. Haug, E.S., Haug, J., Lewis, A.: Back to basics: a new approach to discrete dividend problem. Wilmott Magazine 9 (2003)
14. Merton R.: The rational theory of options pricing. J. Econ. Manag. Sci. 4, 141–183 (1973)
15. Roll R.: An analytical formula for unprotected American call options on stocks with known dividends. J. Finan. Econ. 5, 251–258 (1977)
16. Schroder M.: Adapting the binomial model to value options on assets with fixed-cash payouts. Finan. Anal. J. 44, 54–62 (1988)
17. Vellekoop, M.H., Nieuwenhuis, J.W.: Efficient pricing of derivatives on assets with discrete dividends. Appl. Math. Finan. 13, 265–284 (2006)
18. Whaley R.E.: On the evaluation of American call options on stocks with known dividends. J. Finan. Econ. 9, 207–211 (1981)
19. Whaley R.E.,: Derivatives, Markets, Valuation, and Risk Management. Wiley-Finance, Danvers 2006

Nonparametric prediction in time series analysis: some empirical results

Marcella Niglio and Cira Perna

Abstract. In this paper a new approach to select the lag p for time series generated from Markov processes is proposed. It is faced in the nonparametric domain and it is based on the minimisation of the estimated risk of prediction of one-step-ahead kernel predictors. The proposed procedure has been evaluated through a Monte Carlo study and in empirical context to forecast the weakly 90-day US T-bill secondary market rates.

Key words: kernel predictor, estimated risk of prediction, subsampling

1 Introduction

One of the aims in time series analysis is forecasting future values taking advantage of current and past knowledge of the data-generating processes. These structures are often summarised with parametric models that, based on specific assumptions, define the relationships among variables. In this parametric context a large number of models have been proposed (among others, [3], [20], [4], and, more recently, [11], which discusses parametric and nonparametric methods) and for most of them the forecast performance has been evaluated.

To overcome the problem of prior knowledge about the functional form of the model, a number of nonparametric methods have been proposed and widely used in statistical applications. In this context, our attention is focused on nonparametric analysis based on kernel methods which have received increasing attention due to their flexibility in modelling complex structures.

In particular, given a Markov process of order p, in this paper a new approach to select the lag p is proposed. It is based on the minimisation of the risk of prediction, proposed in [13], estimated for kernel predictors by using the subsampling.

After presenting some results on the kernel predictors, we discuss, in Section 2, how they can be introduced in the proposed procedure.

In Section 3 we further describe the algorithm whose performance has been discussed in a Monte Carlo study. To evaluate the forecast accuracy of the nonparametric predictor in the context of real data, in Section 4 we present some results on the weekly

90-day US T-bill secondary market rates. Some concluding remarks are given at the end.

2 Nonparametric kernel predictors

Let $\mathbf{X}_T = \{X_1, X_2, \ldots, X_T\}$ be a real-valued time series. We assume that:

A1: $\mathbf{X}_T$ is a realisation of a strictly stationary stochastic process that belongs to the class:

$$X_t = f(X_{t-1}, X_{t-2}, \ldots, X_{t-p}) + \epsilon_t, \qquad 1 \leq t \leq T, \tag{1}$$

where the innovations $\{\epsilon_t\}$ are i.i.d. random variables, independent from the past of X_t, with $E(\epsilon_t) = 0$, $E(\epsilon_t^2) = \sigma^2 < +\infty$, and p is a nonnegative integer.

In class (1), $f(X_{t-1}, X_{t-2}, \ldots, X_{t-p})$ is the conditional expectation of X_t, given $X_{t-1}, \ldots, X_{t-p}$, that can be differently estimated. When the Nadaraya-Watson (N-W)-type estimator (among others see [2]) is used:

$$\hat{f}(x_1, x_2, \ldots, x_p) = \frac{\displaystyle\sum_{t=p+1}^{T} \prod_{i=1}^{p} K\left(\frac{x_i - X_{t-i}}{h_i}\right) X_t}{\displaystyle\sum_{t=p+1}^{T} \prod_{i=1}^{p} K\left(\frac{x_i - X_{t-i}}{h_i}\right)}, \tag{2}$$

where $K(\cdot)$ is a kernel function and h_i is the bandwidth, for $i = 1, 2, \ldots, p$.

Under mixing conditions, the asymptotic properties of the estimator (2) have been widely investigated in [18], and the main properties, when it is used in predictive context, have been discussed in [9] and [2].

When the estimator (2) is used, the selection of the "optimal" bandwidth, the choice of the kernel function and the determination of the autoregressive order p are needed. To solve the latter problem, many authors refer to automatic methods, such as AIC and BIC, or to their nonparametric analogue suggested by [19].

Here we propose a procedure based on one-step-ahead kernel predictors.

The estimator for the conditional mean (2) has a large application in prediction contexts. In fact, when a quadratic loss function is selected to find a predictor for $X_{T+\ell}$, with lead time $\ell > 0$, it is well known that the best predictor is given by $\hat{X}_{T+\ell} = E[X_{T+\ell}|\mathbf{X}_T]$, obtained from the minimisation of

$$\underset{\hat{X}_{T+\ell} \in \mathbb{R}}{\arg\min} \; E[(\hat{X}_{T+\ell} - X_{T+\ell})^2 | \mathbf{X}_T], \qquad \text{with} \quad \ell > 0.$$

It implies that when N-W estimators are used to forecast $X_{T+\ell}$, the least-squares predictor $\hat{X}_{T+\ell}$ becomes:

$$\hat{X}_{T+\ell} = \frac{\displaystyle\sum_{t=p+1}^{T-\ell} \prod_{i=1}^{p} K\left(\frac{x_i - X_{t-i}}{h_i}\right) X_{t+\ell}}{\displaystyle\sum_{t=p+1}^{T-\ell} \prod_{i=1}^{p} K\left(\frac{x_i - X_{t-i}}{h_i}\right)}. \tag{3}$$

The properties of (3) in the presence of strictly stationary and Markovian processes of order p are discussed in [9] and [2].

Under well defined assumptions on the generating process, [9] shows that when $\ell = 1$ the predictor (3) is a strong consistent estimator for $E[X_{T+1}|\mathbf{X}_T]$ and this result has been subsequently generalised in [2] to the case with $\ell \geq 1$.

In presence of real data, [5], [16], [10] and recently [23] evaluate the forecast accuracy of (3) and give empirical criteria to define confidence intervals for $\hat{X}_{T+\ell}$.

In the following, taking advantage of (3), we propose the use of the estimated risk of prediction (ERP), discussed in [13], to select the order p of the the autoregression (1).

In particular we further assume that:

A2: $\mathbf{X}_T$ is a realisation of a strong mixing (or α-mixing) process. Under conditions A1 and A2, the ERP can be estimated through resampling techniques and in particular using the subsampling approach as proposed by [13].

The subsampling has a number of interesting advantages with respect to other resampling techniques: in particular it is robust against misspecified models and gives consistent results under weak assumptions.

This last remark makes the use of subsampling particularly useful in a nonparametric framework and can be properly applied in the context of model selection.

Let $\hat{X}_{T+1}$ be the N-W predictor (3); its mean square error is defined as

$$\Delta_T = E[(X_{T+1} - \hat{X}_{T+1})^2].$$

The algorithm we propose to select p is established on the estimation of Δ_T that, as described in Procedure 1, is based on the overlapping subsampling. Note that in this procedure Step 2 implies the choice of the subsample length b. A large number of criteria have been proposed in the statistical literature to select b (inter alia [17]). Here we refer to the proposal in [14], which describes an empirical rule for estimating the optimal window size in the presence of dependent data of smaller length (m) than the original (T). The details are given in Procedure 2.

Procedure 1: Selection of the autoregressive order p

1. Choose a grid for $p \in (1, \ldots, P)$.
2. Select the subsample length b (*Procedure 2*).
3. For each p, compute the estimated risk of prediction (ERP):

$$\hat{\Delta}_{T,b} = (T - b + 1)^{-1} \sum_{i=0}^{T-b} \left(\hat{X}_{i+b}^{(i)} - X_{i+b} \right)^2,$$

where $\hat{X}_{i+b}^{(i)}$ is the one-step-ahead predictor (3) of X_{i+b}, based on the subsample $(X_{i+1}, X_{i+2}, \ldots, X_{i+b-1})$ of length b.
4. Select $\hat{p}$ which minimises $\hat{\Delta}_{T,b}$.

Procedure 2: Subsample length selection

1. Fix $m < T$ and compute $\hat{X}_{T,m}$, the subsampling estimate from the entire data set $\mathbf{X}_T$.
2. For all $b_m < m$, compute $\hat{X}^{(i)}_{m,b_m}$, the subsampling estimate of the forecast computed from $(X_i, X_{i+1}, \ldots, X_{i+m-1})$.
3. Select the value $\hat{b}_m$ that minimizes the estimated mean square error (EMSE):

$$EMSE(m, b_m) = (T - m + 1)^{-1} \sum_{i=1}^{T-m+1} \left(\hat{X}^{(i)}_{m,b_m} - \hat{X}_{T,m} \right)^2.$$

4. Choose $\hat{b} = (T/m)^\delta * \hat{b}_m$, where $\delta \in (0, 1)$ is a fixed real number.

3 Simulation results

To illustrate the performance of the proposed procedure we have used simulated data sets generated by models with known structure. The aim is to evaluate the ability of our procedure to select a proper value for the autoregressive parameter p in the presence of given data-generating processes.

The simulated time series have been generated by two structures: a linear autoregressive model (AR) and a self-exciting threshold autoregressive model (SETAR) that, as is well known, both belong to the class of Markov processes (1).

More precisely the simulated models are:

Model 1 - AR(1): $X_t = -0.8X_{t-1} + \epsilon_t$, with $\epsilon_t \sim N(0, 1)$;

Model 2 - SETAR(2;1,1):

$$X_t = \begin{cases} 1.5 - 0.9X_{t-1} + \epsilon_t & X_{t-1} \leq 0 \\ -0.4 - 0.6X_{t-1} + \epsilon_t & X_{t-1} > 0, \end{cases} \quad \text{with} \quad \epsilon_t \sim N(0, 1),$$

where *Model 2* has been used in [21] to evaluate the forecast ability of SETAR models.

The simulation study has been implemented defining a grid value for $p = 1, 2, 3, 4$ and using series of length $T = 70$ and $T = 100$.

In order to take into account the two different lengths, we have chosen two grids for m. When $T = 70$, the grid is $m = \{20, 25, 30, 35\}$ whereas for $T = 100$ it is $m = \{25, 35, 40, 50\}$. The two values for T have been chosen to evaluate the proposed procedure in the presence of series of moderate length whereas the grid for m has been defined following [14].

Starting from these values, Procedure 1 has been run in a Monte Carlo study with 100 replications. Following [13] we have fixed the parameter $\delta = 0.4$ whereas the kernel function is Gaussian and the bandwidths h_i $(i = 1, 2, \ldots, p)$ in (3) are selected using a cross-validation criterion.

The results are summarised in Tables 1 and 2 where the distribution of the 100 simulated series is presented for the AR(1) and SETAR(2; 1,1) models respectively, comparing the classes in which $\hat{b}$ lies and the candidate values for p.

Table 1. Distribution of the 100 series of length $T = 70$ and $T = 100$ respectively, simulated from *Model 1*

	$\hat{b}$ ($T = 70$)					$\hat{b}$ ($T = 100$)			
p	[9, 21]	[22, 34]	[35, 43]	Tot.	p	[9,29]	[30,44]	[45,64]	Tot.
1	14	23	51	88	1	19	21	50	90
2	1	4	2	7	2	1	3	1	5
3	4	0	0	4	3	2	2	1	5
4	0	1	0	1	4	0	0	0	0

Table 2. Distribution of the 100 series of length $T = 70$ and $T = 100$ respectively, simulated from *Model 2*

	$\hat{b}$ ($T = 70$)					$\hat{b}$ ($T = 100$)			
p	[17, 26]	[27, 35]	[36, 44]	Tot.	p	[15, 31]	[32, 47]	[48, 64]	Tot.
1	15	11	59	85	1	21	14	54	89
2	1	0	4	5	2	2	2	3	7
3	3	6	1	10	3	1	0	3	4
4	0	0	0	0	4	0	0	0	0

In both cases, the proposed procedure gives satisfactory results on the selection of the autoregressive order in the presence of a Markov process of order 1 that, as expected, improves as T grows.

Note that the good performance is a guarantee for time series of moderate length T, that rises the interest on the procedure.

As expected, most "well selected" models belong to the last class of $\hat{b}$. It should not be surprising because the results used in the proposed procedure are mainly given in asymptotic context.

4 Empirical results on 90-day US T-bill rate

Starting from the theoretical results described in Section 2, the model selection procedure has been applied to generate forecasts from the weekly 90-day US T-bill secondary market rates covering the period 4 January 1957–17 December 1993. The time series, X_t, of length $T = 1929$, has been extracted from the H.15 release of the Federal Reserve System (http://www.federalreserve.gov/releases/h15/data.htm).

The 90-day US T-bill has been widely investigated in nonlinear and nonparametric literature (among others [1] and [15]) and, in particular, the data set under study, plotted in Figure 1, has been analysed in [10] to compare three kernel-based multi-step predictors. The authors, after computing proper unit-root tests, show the nonstationarity of the series X_t, which can be empirically appreciated by observing the correlogram in Figure 2, which presents a very slow decay.

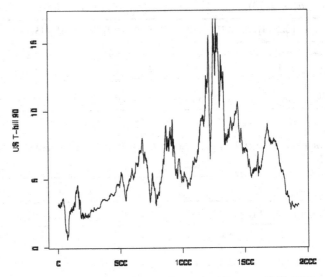

Fig. 1. Weekly 90-day US T-bill secondary market rates: 4 January 1957–17 December 1993

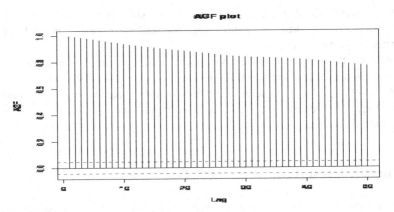

Fig. 2. ACF plot of the weakly 90-day US T-bill

In Table 4 the nonlinearity of r_t is further investigated through the likelihood ratio (LR) test proposed in [6] and [7], where the linearity of the process is tested against threshold nonlinearity. In particular, the test statistic with the corresponding p-value is presented when the null autoregressive model of order p and the threshold delay d (of the alternative hypothesis) allow refusal of the linearity of the data-generating process.

The first differences of X_t (denoted by r_t in the following) are then plotted in Figure 3 where it is evident that the behaviour of the series changes considerably in the time interval taken into account.

Following [10], which further assesses the nonlinearity of the data-generating process of r_t, we examine the conditional mean of r_t, neglecting the conditional heteroschedasticity that gives rise to the volatility clustering that can be clearly seen in Figure 3.

Starting from these results, we firstly evaluate some features of the series using the descriptive indexes presented in Table 4. In particular, the mean, the median, the standard deviation, the skewness and kurtosis (given as third and fourth moment of the standardised data respectively) of r_t, are computed. As expected, the distribution of r_t has null median and shows negative skewness and heavy tails.

It is widely known that when the prediction of the mean level of asymmetric time series needs to be generated, a parametric structure that can be properly applied is the SETAR(2; p_1, p_2) model that treats positive and negative values of r_t differently. This is the reason why the SETAR models have been widely applied to analyse and forecast data related to financial markets (among others: [20] for a wide presentation of the model and [12] for its application to financial data).

Table 3. Descriptive indexes of r_t

	Mean	Median	S.D.	Skewness	Kurtosis
r_t	−6.224e-05	0	0.2342	−0.5801	16.6101

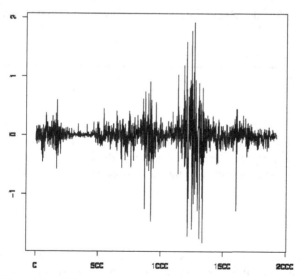

Fig. 3. First differences of the weakly 90-day US T-bill rate (r_t)

Table 4. LR linearity test of r_t

	p	d	Stat	(p-value)
r_t	4	1	30.5996	(1.9343e-05)

The results of the LR test show that a linear structure does not seem to be capabel of catching the structure of the generating process (this explains the poor performance of the autoregressive forecasts in [10]). The threshold autoregressive model fitted to the data is clearly based on a strict parametric structure from which the forecasts are generated.

Here, we alternatively propose the nonparametric predictor (3), which is more flexible than that generated from SETAR models, and whose Markov order is selected following Procedure 1.

For both approaches, we have generated one-step-ahead, out-of-sample forecasts following an expanding window algorithm over the forecast horizon $L = 26$, which corresponds to the last six months of the time interval under analysis.

Further, a null threshold value has been fixed for the SETAR model (with threshold delay given in Table 4) and at each iteration the model has been estimated following [22].

SETAR and nonparametric least-squares forecasts have been evaluated using the mean square error and the mean absolute error, $MSE(L) = L^{-1} \sum_{i=1}^{L} (\hat{X}_{T+i} - X_{T+i})^2$ and $MAE(L) = L^{-1} \sum_{i=1}^{L} |\hat{X}_{T+i} - X_{T+i}|$, whose values are compared in Table 5 where the MSE (and the MAE) of (3) over the MSE (and MAE) of the SETAR predictions are shown.

Table 5. MSE (and MAE) of the nonparametric forecasts over the MSE (and MAE) of the SETAR forecasts

	$MSE(L)_{np}[MSE(L)_{thr}]^{-1}$	$MAE(L)_{np}[MAE(L)_{thr}]^{-1}$
r_t	0.839081	0.944058

The better forecast accuracy, in terms of MSE and MAE, of predictor (3) can be appreciated. It further confirms the good performance of the proposed procedure in the presence of one-step-ahead forecasts. Moreover, the forecast accuracy seems not to be affected when different values, of moderate size, are assigned to m in Procedure 2.

5 Conclusions

We have proposed a procedure to select the order p in the presence of strictly stationary Markov processes (1). It is based on the use of one-step-ahead predictors generated from nonparametric Nadaraya-Watson kernel smoothers.

The selection of p is obtained from the minimisation of a quadratic loss function that makes use of the subsampling estimate of the one-step-ahead forecasts as shown in Procedure 1.

The simulated and empirical results show the good performance of the proposed procedure that can be considered, in the context of model selection, an alternative to more consolidated approaches given in the literature.

Much remains to be done: to investigate the properties of $\hat{p}$; to generalise the procedure to the case with lead time $\ell > 1$; to consider more complex data-generating processes that belong to the Markov class. Further, the procedure could be extended to parametric and/or semiparametric predictors that can be properly considered to minimize $\hat{\Delta}_{T,b}$.

All these tasks need proper evaluation of the computational effort that is requested when computer-intensive methods are selected.

Acknowledgement. The authors would like to thank two anonymous referees for their useful comments.

References

1. Barkoulas, J.T., Baum C.F., Onochie, J.: A nonparametric investigation of the 90-day T-bill rate. Rev. Finan. Econ. 6, 187–198 (1997)
2. Bosq, D.: Nonparametric Statistics for Stochastic Process. Springer, New York (1996)
3. Box, G.E.P., Jenkins, G.M.: Time Series Analysis: Forecasting and Control. Holden-Day, San Francisco (1976)
4. Brockwell, P.J., Davies, R.A.: Time series: theory and methods. Springer-Verlag, New York (1991)
5. Carbon, M., Delecroix M.: Non-parametric vs parametric forecasting in time series: a computational point of view. Appl. Stoch. Models Data Anal. 9, 215–229 (1993)
6. Chan, K.S.: Testing for threshold autoregression. Ann. Stat. 18, 1886–1894 (1990)
7. Chan, K.S., Tong H.: On likelihood ratio test for threshold autoregression, J. R. Stat. Soc. (B) 52, 595–599 (1990)
8. Clements, M.P.: Evaluating Econometric Forecasts of Economic and Financial Variables. Palgave Macmillan, New York (2005)
9. Collomb, G.: Propriétés de convergence presque complète du predicteur à noyau. Zeitschrift für Wahrscheinlichkeitstheorie und verwandte Genbeite 66, 441–460 (1984)
10. De Gooijer, J., Zerom, D.: Kernel-based multistep-ahead predictions of the US short-term interest rate. J. Forecast. 19, 335–353 (2000)
11. Fan, J., Yao, Q.: Nonlinear Time Series. Nonparametric and Parametric Methods. Springer-Verlag, New York (2003)
12. Franses P.H., van Dijk, D.: Non-Linear Time Series Models in Empirical Finance. Cambridge University Press, Cambridge (2000)
13. Fukuchi, J.: Subsampling and model selection in time series analysis. Biometrika 86, 591–604 (1999)
14. Hall, P., Jing B.: On sample reuse methods for dependent data. J. R. Stat. Soc. (B) 58, 727–737 (1996)

Marcella Niglio and Cira Perna

15. Lanne, M., Saikkonen, P.: Modeling the U.S. short-term interest rate by mixture autoregressive processes. J. Finan. Econ. 1, 96–125 (2003)
16. Matzner-Løber, E., De Gooijer, J.: Nonparametric forecasting: a comparison of three kernel-based methods. Commun. Stat.: Theory Methods 27, 1593–1617 (1998)
17. Politis D.N., Romano, J.P., Wolf, M.: Subsampling. Springer-Verlag, New York (1999)
18. Robinson, P.M.: Nonparametric estimators for time series. J. Time Ser. Anal. 4, 185–207 (1983)
19. Tjøstheim, D., Auestad, H.: Nonparametric identification of nonlinear time series: selecting significant lags. J. Am. Stat. Assoc. 89, 1410–1419 (1994)
20. Tong, H.: Nonlinear Time Series: A Dynamical System Approach. Oxford University Press, Oxford (1990)
21. Tong , H., Moeannadin, R.: On multi-step non-linear least squares prediction. Statistician 37, 101–110 (1981)
22. Tsay, R.: Testing and modelling threshold autoregressive processes. J. Am. Stat. Assoc. 84, 231–240 (1989)
23. Vilar-Fernàndez, J.M., Cao, R.: Nonparametric forecasting in time series. A comparative study. Commun. Stat.: Simul. Comput. 36, 311–334 (2007)

On efficient optimisation of the CVaR and related LP computable risk measures for portfolio selection

Włodzimierz Ogryczak and Tomasz Śliwiński

Abstract. The portfolio optimisation problem is modelled as a mean-risk bicriteria optimisation problem where the expected return is maximised and some (scalar) risk measure is minimised. In the original Markowitz model the risk is measured by the variance while several polyhedral risk measures have been introduced leading to Linear Programming (LP) computable portfolio optimisation models in the case of discrete random variables represented by their realisations under specified scenarios. Recently, the second order quantile risk measures have been introduced and become popular in finance and banking. The simplest such measure, now commonly called the Conditional Value at Risk (CVaR) or Tail VaR, represents the mean shortfall at a specified confidence level. The corresponding portfolio optimisation models can be solved with general purpose LP solvers. However, in the case of more advanced simulation models employed for scenario generation one may get several thousands of scenarios. This may lead to the LP model with a huge number of variables and constraints, thus decreasing the computational efficiency of the model. We show that the computational efficiency can be then dramatically improved with an alternative model taking advantages of the LP duality. Moreover, similar reformulation can be applied to more complex quantile risk measures like Gini's mean difference as well as to the mean absolute deviation.

Key words: risk measures, portfolio optimisation, computability, linear programming

1 Introduction

In the original Markowitz model [12] the risk is measured by the variance, but several polyhedral risk measures have been introduced leading to Linear Programming (LP) computable portfolio optimisation models in the case of discrete random variables represented by their realisations under specified scenarios. The simplest LP computable risk measures are dispersion measures similar to the variance. Konno and Yamazaki [6] presented the portfolio selection model with the mean absolute deviation (MAD). Yitzhaki [25] introduced the mean-risk model using Gini's mean (absolute) difference as the risk measure. Gini's mean difference turn out to be a special aggregation technique of the multiple criteria LP model [17] based on the pointwise comparison of the absolute Lorenz curves. The latter leads to the quantile shortfall risk measures that are more commonly used and accepted. Recently, the second-order quantile risk measures have been introduced in different ways by many authors [2,5,15,16,22]. The measure, usually called the Conditional Value at Risk (CVaR) or Tail VaR, represents the mean shortfall at a specified confidence level. Maximisation of the CVaR measures is consistent with the second-degree stochastic dominance [19]. Several empirical analyses confirm its applicability to various financial optimisation problems [1,10]. This paper is focused on computational efficiency of the CVaR and related LP computable portfolio optimisation models.

For returns represented by their realisations under T scenarios, the basic LP model for CVaR portfolio optimisation contains T auxiliary variables as well as T corresponding linear inequalities. Actually, the number of structural constraints in the LP model (matrix rows) is proportional to the number of scenarios T, while the number of variables (matrix columns) is proportional to the total of the number of scenarios and the number of instruments $T + n$. Hence, its dimensionality is proportional to the number of scenarios T. It does not cause any computational difficulties for a few hundred scenarios as in computational analysis based on historical data. However, in the case of more advanced simulation models employed for scenario generation one may get several thousands of scenarios [21]. This may lead to the LP model with a huge number of auxiliary variables and constraints, thus decreasing the computational efficiency of the model. Actually, in the case of fifty thousand scenarios and one hundred instruments the model may require more than half an hour of computation time [8] with the state-of-art LP solver (CPLEX code). We show that the computational efficiency can be then dramatically improved with an alternative model formulation taking advantage of the LP duality. In the introduced model the number of structural constraints is proportional to the number of instruments n, while only the number of variables is proportional to the number of scenarios T, thus not affecting the simplex method efficiency so seriously. Indeed, the computation time is then below 30 seconds. Moreover, similar reformulation can be applied to the classical LP portfolio optimisation model based on the MAD as well as to more complex quantile risk measures including Gini's mean difference [25].

2 Computational LP models

The portfolio optimisation problem considered in this paper follows the original Markowitz' formulation and is based on a single period model of investment. At the beginning of a period, an investor allocates the capital among various securities, thus assigning a nonnegative weight (share of the capital) to each security. Let $J = \{1, 2, \ldots, n\}$ denote a set of securities considered for an investment. For each security $j \in J$, its rate of return is represented by a random variable R_j with a given mean $\mu_j = \mathbb{E}\{R_j\}$. Further, let $x = (x_j)_{j=1,2,\ldots,n}$ denote a vector of decision variables x_j expressing the weights defining a portfolio. The weights must satisfy a set of constraints to represent a portfolio. The simplest way of defining a feasible set $\mathcal{P}$ is by a requirement that the weights must sum to one and they are nonnegative (short sales are not allowed), i.e.,

$$\mathcal{P} = \{x \ : \ \sum_{j=1}^{n} x_j = 1, \quad x_j \geq 0 \quad \text{for } j = 1, \ldots, n\}. \tag{1}$$

Hereafter, we perform detailed analysis for the set $\mathcal{P}$ given with constraints (1). Nevertheless, the presented results can easily be adapted to a general LP feasible set given as a system of linear equations and inequalities, thus allowing one to include short sales, upper bounds on single shares or portfolio structure restrictions which may be faced by a real-life investor.

Each portfolio x defines a corresponding random variable $R_x = \sum_{j=1}^{n} R_j x_j$ that represents the portfolio rate of return while the expected value can be computed as $\mu(x) = \sum_{j=1}^{n} \mu_j x_j$. We consider T scenarios with probabilities p_t (where $t = 1, \ldots, T$). We assume that for each random variable R_j its realisation r_{jt} under the scenario t is known. Typically, the realisations are derived from historical data treating T historical periods as equally probable scenarios ($p_t = 1/T$). The realisations of the portfolio return R_x are given as $y_t = \sum_{j=1}^{n} r_{jt} x_j$.

Let us consider a portfolio optimisation problem based on the CVaR measure optimisation. With security returns given by discrete random variables with realisations r_{jt}, following [1,9,10], the CVaR portfolio optimisation model can be formulated as the following LP problem:

$$\text{maximise } \eta - \frac{1}{\beta} \sum_{t=1}^{T} p_t d_t$$

$$\text{s.t. } \sum_{j=1}^{n} x_j = 1, \quad x_j \geq 0 \quad \text{for } j = 1, \ldots, n \tag{2}$$

$$d_t - \eta + \sum_{j=1}^{n} r_{jt} x_j \geq 0, \ d_t \geq 0 \quad \text{for } t = 1, \ldots, T,$$

where η is an unbounded variable. Except for the core portfolio constraints (1), model (2) contains T nonnegative variables d_t plus a single η variable and T corresponding linear inequalities. Hence, its dimensionality is proportional to the number of scenarios T. Exactly, the LP model contains $T + n + 1$ variables and $T + 1$ constraints. For a few hundred scenarios, as in typical computational analysis based on historical data [11], such LP models are easily solvable. However, the use of more advanced simulation models for scenario generation may result in several thousands of scenarios. The corresponding LP model (2) contains then a huge number of variables and constraints, thus decreasing its computational efficiency dramatically. If the core portfolio constraints contain only linear relations, like (1), then the computational efficiency can easily be achieved by taking advantage of the LP dual model (2). The

248   W. Ogryczak and T. Śliwiński

LP dual model takes the following form:

minimise q

$$\text{s.t. } q - \sum_{t=1}^{T} r_{jt} u_t \geq 0 \quad \text{for } j = 1, \ldots, n$$

$$\sum_{t=1}^{T} u_t = 1, \quad 0 \leq u_t \leq \frac{p_t}{\beta} \quad \text{for } t = 1, \ldots, T. \tag{3}$$

The dual LP model contains T variables u_t, but the T constraints corresponding to variables d_t from (2) take the form of simple upper bounds (SUB) on u_t thus not affecting the problem complexity (c.f., [13]). Actually, the number of constraints in (3) is proportional to the total of portfolio size n, thus it is independent from the number of scenarios. Exactly, there are $T + 1$ variables and $n + 1$ constraints. This guarantees a high computational efficiency of the dual model even for a very large number of scenarios. Note that introducing a lower bound on the required expected return in the primal portfolio optimisation model (2) results only in a single additional variable in the dual model (3). Similarly, other portfolio structure requirements are modelled with a rather small number of constraints, thus generating a small number of additional variables in the dual model.

We have run computational tests on 10 randomly generated test instances developed by Lim et al. [8]. They were originally generated from a multivariate normal distribution for 50 or 100 securities with the number of scenarios of 50,000 just providing an adequate approximation to the underlying unknown continuous price distribution. Scenarios were generated using the Triangular Factorization Method [24] as recommended in [3]. All computations were performed on a PC with a Pentium 4 2.6 GHz processor and 1 GB RAM employing the simplex code of the CPLEX 9.1 package. An attempt to solve the primal model (2) with 50 securities resulted in 2600 seconds of computation (much more than reported in [8]). On the other hand, the dual models (3) were solved in 14.3–27.7 CPU seconds on average, depending on the tolerance level (see Table 1). For 100 securities the optimisation times were longer but still about 1 minute.

Table 1. Computational times (in seconds) for the dual CVaR model (averages of 10 instances with 50,000 scenarios)

Number of securities	$\beta = 0.05$	$\beta = 0.1$	$\beta = 0.2$	$\beta = 0.3$	$\beta = 0.4$	$\beta = 0.5$
$n = 50$	14.3	18.7	23.6	26.4	27.4	27.7
$n = 100$	38.1	52.1	67.9	74.8	76.7	76.0

The SSD consistent [14] and coherent [2] MAD model with complementary risk measure ($\mu_\delta(x) = \mathbb{E}\{\min\{\mu(x), R_x\}\}$) leads to the following LP problem [18]:

$$\text{maximise} \sum_{j=1}^{n} \mu_j x_j - \sum_{t=1}^{T} p_t d_t$$

$$\text{s.t.} \sum_{j=1}^{n} x_j = 1, \quad x_j \geq 0 \quad \text{for } j = 1, \ldots, n \tag{4}$$

$$d_t - \sum_{j=1}^{n} (\mu_j - r_{jt}) x_j \geq 0, \; d_t \geq 0 \quad \text{for } t = 1, \ldots, T.$$

The above LP formulation uses $T + n$ variables and $T + 1$ constraints while the LP dual model then takes the following form:

$$\text{minimise } q$$

$$\text{s.t.} \quad q + \sum_{t=1}^{T} (\mu_j - r_{jt}) u_t \geq \mu_j \quad \text{for } j = 1, \ldots, n \tag{5}$$

$$0 \leq u_t \leq p_t \quad \text{for } t = 1, \ldots, T,$$

with dimensionality $n \times (T + 1)$. Hence, there is guaranteed high computational efficiency even for very large numbers of scenarios. Indeed, in the test problems with 50,000 scenarios we were able to solve the dual model (5) in 25.3 seconds on average for 50 securities and in 77.4 seconds for 100 instruments.

For a discrete random variable represented by its realisations y_t, Gini's mean difference measure $\Gamma(x) = \sum_{t'=1}^{T} \sum_{t'' \neq t'-1} \max\{y_{t'} - y_{t''}, 0\} p_{t'} p_{t''}$ is LP computable (when minimised). This leads us to the following GMD portfolio optimisation model [25]:

$$\max -\sum_{t=1}^{T} \sum_{t' \neq t} p_t p_{t'} d_{tt'}$$

$$\text{s.t.} \sum_{j=1}^{n} x_j = 1, \quad x_j \geq 0 \quad \text{for } j = 1, \ldots, n \tag{6}$$

$$d_{tt'} \geq \sum_{j=1}^{n} r_{jt} x_j - \sum_{j=1}^{n} r_{jt'} x_j, \; d_{tt'} \geq 0 \quad \text{for } t, t' = 1, \ldots, T; \; t \neq t',$$

which contains $T(T - 1)$ nonnegative variables $d_{tt'}$ and $T(T - 1)$ inequalities to define them. This generates a huge LP problem even for the historical data case where the number of scenarios is 100 or 200. Actually, as shown with the earlier experiments [7], the CPU time of 7 seconds on average for $T = 52$ has increased to above 30 s with $T = 104$ and even more than 180 s for $T = 156$. However, similar to the CVaR models, variables $d_{tt'}$ are associated with the singleton coefficient columns. Hence, while solving the dual instead of the original primal, the corresponding dual constraints take the form of simple upper bounds (SUB) which are handled implicitly outside the LP matrix. For the simplest form of the feasible set (1) the dual GMD model takes the following form:

$$
\min v
$$
$$
\text{s.t. } v - \sum_{t=1}^{T} \sum_{t' \neq t} (r_{jt} - r_{jt'}) u_{tt'} \geq 0 \quad \text{for } j = 1, \ldots, n \tag{7}
$$
$$
0 \leq u_{tt'} \leq p_t p_{t'} \quad \text{for } t, t' = 1, \ldots, T; t \neq t',
$$

where original portfolio variables x_j are dual prices to the inequalities. The dual model contains $T(T-1)$ variables $u_{tt'}$ but the number of constraints (excluding the SUB structure) $n + 1$ is proportional to the number of securities. The above dual formulation can be further simplified by introducing variables:

$$
\bar{u}_{tt'} = u_{tt'} - u_{t't} \quad \text{for } t, t' = 1, \ldots, T; t < t', \tag{8}
$$

which allows us to reduce the number of variables to $T(T-1)/2$ by replacing (7) with the following:

$$
\min v
$$
$$
\text{s.t. } v - \sum_{t=1}^{T} \sum_{t' > t} (r_{jt} - r_{jt'}) \bar{u}_{tt'} \geq 0 \quad \text{for } j = 1, \ldots, n \tag{9}
$$
$$
-p_t p_{t'} \leq \bar{u}_{tt'} \leq p_t p_{t'} \quad \text{for } t, t' = 1, \ldots, T; t < t'.
$$

Such a dual approach may dramatically improve the LP model efficiency in the case of a larger number of scenarios. Actually, as shown with the earlier experiments [7], the above dual formulations let us to reduce the optimisation time to below 10 seconds for $T = 104$ and $T = 156$. Nevertheless, the case of really large numbers of scenarios still may cause computational difficulties, due to the huge number of variables ($T(T-1)/2$). This may require some column generation techniques [4] or nondifferentiable optimisation algorithms [8].

3 Conclusions

The classical Markowitz model uses the variance as the risk measure, thus resulting in a quadratic optimisation problem. Several alternative risk measures were introduced, which are computationally attractive as (for discrete random variables) they result in solving linear programming (LP) problems. The LP solvability is very important for applications to real-life financial decisions where the constructed portfolios have to meet numerous side constraints and take into account transaction costs [10]. The corresponding portfolio optimisation models can be solved with general purpose LP solvers, like ILOG CPLEX providing a set of C++ and Java class libraries allowing the programmer to embed CPLEX optimisers in C++ or Java applications.

Unfortunately, in the case of more advanced simulation models employed for scenario generation one may get several thousands of scenarios. This may lead to the LP model with a huge number of variables and constraints, thus decreasing the computational efficiency of the model. We have shown that the computational efficiency can then be dramatically improved with an alternative model taking advantage of the LP duality. In the introduced model the number of structural constraints (matrix rows) is proportional to the number of instruments thus not seriously affecting the simplex method efficiency by the number of scenarios. For the case of 50,000 scenarios, it has resulted in computation times below 30 seconds for 50 securities or below a minute for 100 instruments. Similar computational times have also been achieved for the dual reformulation of the MAD model. Dual reformulation applied to the GMD portfolio optimisation model results in a dramatic problem size reduction with the number of constraints equal to the number of instruments instead of the square of the number of scenarios. Although, the remaining high number of variables (square of the number of scenarios) still generates a need for further research on column-generation techniques or nondifferentiable optimisation algorithms for the GMD model.

Acknowledgement. The authors are indebted to Professor Churlzu Lim from the University of North Carolina at Charlotte for providing the test data.

References

1. Andersson, F., Mausser, H., Rosen, D., Uryasev, S.: Credit risk optimization with conditional value-at-risk criterion. Math. Program. 89, 273–291 (2001)
2. Artzner, P., Delbaen, F., Eber, J.-M., Heath, D.: Coherent measures of risk. Math. Finance 9, 203–228 (1999)
3. Barr, D.R., Slezak, N.L.: A comparison of multivariate normal generators. Commun. ACM 15, 1048–1049 (1972)
4. Desrosiers, J., Luebbecke, M.: A primer in column generation. In: Desaulniers, G., Desrosiers, J., Solomon, M. (eds.) Column Generation, pp. 1–32, Springer, Heidelberg (2005)
5. Embrechts, P., Klüppelberg, C., Mikosch, T.: Modelling Extremal Events for Insurance and Finance. Springer, New York (1997)
6. Konno, H., Yamazaki, H.: Mean-absolute deviation portfolio optimization model and its application to Tokyo stock market. Manag. Sci. 37, 519–531 (1991)
7. Krzemienowski, A., Ogryczak, W.: On extending the LP computable risk measures to account downside risk. Comput. Optim. Appl. 32, 133–160 (2005)
8. Lim, C., Sherali, H.D., Uryasev, S.: Portfolio optimization by minimizing conditional value-at-risk via nondifferentiable optimization. University of North Carolina at Charlotte, Lee College of Engineering, Working Paper 2007
9. Mansini, R., Ogryczak, W., Speranza, M.G.: On LP solvable models for portfolio selection. Informatica 14, 37–62 (2003)
10. Mansini, R., Ogryczak, W., Speranza, M.G.: LP solvable models for portfolio optimization: A classification and computational comparison. IMA J. Manag. Math. 14, 187–220 (2003)
11. Mansini, R., Ogryczak, W., Speranza, M.G.: Conditional value at risk and related linear programming models for portfolio optimization. Ann. Oper. Res. 152, 227–256 (2007)

12. Markowitz, H.M.: Portfolio selection. J. Finan. 7, 77–91 (1952)
13. Maros, I.: Computational Techniques of the Simplex Method. Kluwer, Dordrecht (2003)
14. Müller, A., Stoyan, D.: Comparison Methods for Stochastic Models and Risks. Wiley, Chichester (2002)
15. Ogryczak, W.: Stochastic dominance relation and linear risk measures. In: A.M.J. Skulimowski (ed.) Financial Modelling – Proc. 23rd Meeting EURO WG Financial Modelling, Cracow, 1998, pp. 191–212, Progress & Business Publ., Cracow (1999)
16. Ogryczak, W.: Risk measurement: Mean absolute deviation versus Gini's mean difference. In: W.G. Wanka (ed.) Decision Theory and Optimization in Theory and Practice – Proc. 9th Workshop GOR WG Chemnitz 1999, pp. 33–51, Shaker Verlag, Aachen (2000)
17. Ogryczak, W.: Multiple criteria linear programming model for portfolio selection. Ann. Oper. Res. 97, 143–162 (2000)
18. Ogryczak, W., Ruszczyński, A.: From stochastic dominance to mean-risk models: semideviations as risk measures. Eur. J. Oper. Res. 116, 33–50 (1999)
19. Ogryczak, W., Ruszczyński, A.: Dual stochastic dominance and related mean-risk models. SIAM J. Optim. 13, 60–78 (2002)
20. Pflug, G.Ch.: Some remarks on the value-at-risk and the conditional value-at-risk. In: S. Uryasev (ed.) Probabilistic Constrained Optimization: Methodology and Applications. pp. 272–281, Kluwer, Dordrecht (2000)
21. Pflug, G.Ch.: Scenario tree generation for multiperiod financial optimization by optimal discretization. Math. Program. 89, 251–271 (2001)
22. Rockafellar, R.T., Uryasev, S.: Optimization of conditional value-at-risk. J. Risk 2, 21–41 (2000)
23. Rothschild, M., Stiglitz, J.E.: Increasing risk: I. A definition. J. Econ. Theory 2, 225–243 (1969)
24. Scheuer, E.M., Stoller, D.S.: On the generation of normal random vectors. Technometrics 4, 278–281 (1962)
25. Yitzhaki, S.: Stochastic dominance, mean variance, and Gini's mean difference. Am. Econ. Rev. 72, 178–185 (1982)

A pattern recognition algorithm for optimal profits in currency trading

Danilo Pelusi

Abstract. A key issue in technical analysis is to obtain good and possibly stable profits. Various trading rules for financial markets do exist for this task. This paper describes a pattern recognition algorithm to optimally match training and trading periods for technical analysis rules. Among the filter techniques, we use the Dual Moving Average Crossover (DMAC) rule. This technique is applied to hourly observations of Euro-Dollar exchange rates. The matching method is accomplished using ten chart patterns very popular in technical analysis. Moreover, in order for the results to have a statistical sense, we use the bootstrap technique. The results show that the algorithm proposed is a good starting point to obtain positive and stable profits.

Key words: training sets, trading sets, technical analysis, recognition algorithm

1 Introduction

The choice of the best trading rules for optimal profits is one of the main problems in the use of technical analysis to buy financial instruments. Park and Irwin [31] described various types of filter rules, for instance the Dual Moving Average Crossover family, the Momentum group of rules and the Oscillators. For each of these filter rules we need to find the rule that assures the highest profit. Some good technical protocols, to get optimal profits in the foreign exchange market, have been found by Pelusi et al. [32].

The traders attribute to some chart patterns the property of assessing market conditions (in any financial market) and anticipating turning points. This kind of analysis started with the famous [23], which produced an important stream of literature. However, the popularity of this kind of analysis has been frequently challenged by mainstream financial economists [7, 9, 22, 28–30, 35].

Generally, the success of a rule in actual trading is independent of the type of filter used. It depends on the choice of a so-called "training set", where the maximum profit parameters of a rule are found, and of an independent "trading set" where you apply the optimised filter found in the training phase. In other words, a rule which gives good profits in one period could cause some losses in a different period. This is due to substantial differences in the shapes of the asset price in the two. So, the main issue for the successful application of technical trading rules is to find the best association of a "training sets" (TN-S) and of "trading sets" (TD-S), for the highest and possibly most stable profit streams.

In this paper, we propose a synthesis of the two traditional approaches in technical analysis, outlined above, and use the chart pattern recognition technique for the best association of training and trading phases. Some works [7,9,22,23,29] contain studies on the information content of chart patterns.

Our target is to investigate the existence of non-linear configurations in the hourly observations of the Euro-Dollar (EUR-USD), Dollar-Yen (USD-JPY) and Pound-Dollar (GBP-USD) exchange rates. Our pattern recognition algorithm takes into account ten chart patterns which are traditionally analysed in the literature [7,23,28].

In Section 2 we describe the algorithm. The algorithm results are shown in Section 3, whereas Section 4 contains the conclusions.

2 Pattern recognition algorithm

As outlined above, we consider hourly exchange rates. The first task in the construction of our algorithm is the recognition that some exchange rate movements are significant and others are not. The most significant movements of exchange rates generate a specific pattern. Typically, in order to identify regularities and patterns in the time series of asset prices, it is necessary to extract non-linear patterns from noisy data. This signal extraction can be performed by the human eye, however in our algorithm we use a suitable smoothing estimator. Therefore, to spot the technical patterns in the best way we use the kernel regression. Hardle [16] describes this smoothing method which permits easier analysis of the curve that describes the exchange rate.

Generally, the various chart patterns are quite difficult to quantify analytically (see the technical analysis manuals [2,22,28]). However, to identify a formal way of detecting the appearance of a technical pattern, we have chosen the definitions shown in the paper of Lo et al. [23]. In these definitions, the technical patterns depend on extrema, which must respect certain properties. The use of kernel regression permits easy detection of these extrema because the curve that describes the exchange rate is smoothed. To identify these extrema we use a suitable method described by Omrane and Van Oppens [28].

To detect the presence of technical patterns in the best way, we use a cutoff value as in the work of Osler and Chang [30]. In this manner, the number of maxima and minima identified in the data is inversely related to the value of the cutoff. In other words, an increase or decrease of it generates a different series of maxima and minima, which will result in a different set of chart patterns. For each cutoff value, the algorithm searches the chart patterns HS, IHS, BTOP, BBOT, TTOP, RTOP, RBOT, DTOP, DBOT on the basis of their definitions [23]. Considering a single pattern at a time, the algorithm counts the patterns number of that type, for each cutoff value.

To establish a similarity parameter, we define, for each jth technical pattern ($j = 1, 2, dots, 10$), the coefficient that represents the similarity degree between two different periods. Therefore, our algorithm takes the pattern number for each ith cutoff value and it computes

$$d_{i,j} = \left| n^j_{1,i} - n^j_{2,i} \right|, i = 1, 2, \ldots, 18, \tag{1}$$

where $d_{i,j}$ is the absolute value of the difference between the number of j-type chart patterns of the period 1 and that of period 2, for each cutoff value. So, we are able to define the similarity coefficient S_j as

$$S_j = \begin{cases} \dfrac{\sum_{i=1}^{n_{c_j}} \frac{1}{2^{d_{i,j}}}}{n_{c_j}}, & n_{c_j} \geq 1 \\ 0, & n_{c_j} = 0. \end{cases} \tag{2}$$

The similarity coefficient assumes values that lie between 0 and 1 and n_{c_j} is the number of possible comparisons. At this step, our algorithm gives ten similarity coefficients connected to the ten technical patterns named above. The next step consists of computing a single value that gives informational content on the similarity between the periods. We refer to this value as Global Similarity (GS) and we define it as a weighted average

$$GS = \sum_{j=1}^{10} w_j S_j. \tag{3}$$

The weights w_j are defined as the ratio between the comparisons number of the jth pattern and the sum of comparisons number n_t of all patterns (see formula 4).

$$w_j = \frac{n_{c_j}}{n_t}, n_t = n_{c_1} + n_{c_2} + \ldots + n_{c_{10}} \tag{4}$$

$$\sum_{j=1}^{10} w_j = 1. \tag{5}$$

Moreover, the sum of weights w_j, with j from 0 to 10, is equal to 1 (see formula 5). Computing the global similarity GS through the (3), we assign more weight to the similarity coefficients with greater comparisons number.

The next step is related to the choice of time period amplitude for training and trading phases. For the trading set, we consider the time series of a certain year. Therefore, we consider the exchange rates that start from the end of the time series until the six preceding months. In this manner, we obtain a semester in the year considered. Subsequently, we create the second semester, starting from the end of the year minus a month, until the six preceding months. Thus, we obtain a certain number of semesters. Therefore, we compare these trading semesters with various semesters of the previous years. Subsequently, a selection of training and trading semesters pairs is accomplished, splitting the pairs with positive slopes and those with negative slopes. So, we compute the profits[1] of the trading semesters by considering the optimised

---

[1] To compute the profits, we use the DMAC filter rule [32].

parameters (see [32]) of the corresponding training semesters. For each semesters pair, we compute the GS coefficient through formula 3 and define the following quantity:

$$f_{GS} = \frac{n_{GS}}{n_p}, \tag{6}$$

where n_p is the number of semester pairs with positive profit and n_{GS} is the number of profitable semesters with similarity index GS that lies between two extrema. In particular, we consider the GS ranges: 0.0–0.1, 0.1–0.2, ..., until 0.9–1.0. So, the quantity f_{GS} represents a measure of the frequency of global similarity values based on their memberships at the ranges named above.

In order for the algorithm results to have a statistical sense, we need to apply our technique to many samples that have a trend similar to the exchange rates of the year considered. To solve this problem, we use a technique described by Efron [12], called the bootstrap method [17, 34]. The key idea is to resample from the original data, either directly or via a fitted model,[2] to create various replicate data sets, from which the variability of the quantities of interest can be assessed without long-winded and error-prone analytical calculations. In this way, we create some artificial exchange rate series, each of which is of the same length as the original series.

3 Experimental results

We apply our algorithm to hourly Euro-Dollar exchange rates and consider 2006 as the year for trading. To create samples with trends similar to the Euro-Dollar exchange rate 2006, we use parametric bootstrap methods [3].

In the parametric bootstrap setting, we consider an unknown distribution F to be a member of some prescribed parametric family and obtain a discrete empirical distribution F_n^* by estimating the family parameters from the data. By generating an iid random sequence from the distribution F_n^*, we can arrive at new estimates of various parameters of the original distribution F.

The parametric methods used are based on assuming a specific model for the data. After estimating the model by a consistent method, the residuals are bootstrapped. In this way, we obtain sample sets with the same length of exchange rates as 2006.

Table 1 shows the results with 10, 100, 200, 300, 400 and 500 samples. On the rows, we have the samples number and on the columns we have the global similarity frequency defined in formula (6). We can note that there are no results for the ranges 0.5–0.6, 0.6–0.7, until 0.9–1.0 because they give null contribution, that is there are no global similarity values belonging to the above-named ranges. Moreover, we can observe that for 10 samples, the range with highest frequency is 0.0–0.1, that is, it is more likely that with a similarity coefficient between 0 and 0.1 we have a positive profit than for the other ranges.

The statistics results for 100 samples show that the range with the greatest frequency is 0.1–0.2. For 200, 300, 400 and 500 samples we obtain about the same value.

---

[2] We use a GARCH model (see [4]).

Table 1. Pattern recognition algorithm results

Samples	f_{GS_1}	f_{GS_2}	f_{GS_3}	f_{GS_4}	f_{GS_5}
10	0.4500	0.4440	0.0920	0.0150	0
100	0.3295	0.4859	0.1633	0.0206	0.0006
200	0.3386	0.4722	0.1671	0.0214	0.0007
300	0.3242	0.4770	0.1778	0.0206	0.0005
400	0.3206	0.4833	0.1774	0.0183	0.0003
500	0.3228	0.4813	0.1768	0.0188	0.0003

Therefore, we can infer that from 100 to 500 samples results remain stable.[3] It is most likely that with a number of samples greater than 500, the distribution will be wider.

We also report the results related to the algorithm's application to the real word. To do this, we consider the Euro-Dollar exchange rates of the year 2007. In particular, we choose three semester pairs that could be defined as "similar" by the human eye. We consider the second semester of 2003 the training semester and the second semester of 2007 the trading semester for Euro-Dollar exchange rates. These semesters are shown in Figure 1.

All the figures contain two graphs: the first one shows the half-yearly trend of Euro-Dollar exchange rate, whereas the second one has a double-scale graph. The use of double scale is necessary to study the relationship between exchange rate and profit by the application of technical filter. The trading rule described in this practical example is a long-short strategy. Moreover, the technical filter used is the DMAC rule, which is based on the moving average definition and on Take Profit (TP) and Stop Loss (SL) parameters.

The double-scale graphs have time on the x-axis, the shape of the Euro-Dollar exchange rate on the left y-axis and profit on the right y-axis. We underline that the y-axis values are pure numbers, that is without units of measurement.

The results of Table 2 show that there is a loss of about 13 % with a GS coefficient of 0.61844. From Figure 1 we can note that there are substantial shape differences at the beginning and at the end of semesters and that the profit has a decreasing trend for more than half a semester.

Figure 2 shows the second semester of 2004 (training semester) and the second semester of 2007 (trading semester). As can be seen in Table 2, we obtain a profit of about 26 % with a global similarity index of 0.66299. Observing Figure 2, we can see that the profit is essentially growing, except at the beginning and at the end of the trading semester.

We choose as the third pair the first semester of 2006 and the second semester of 2007 (Fig. 3). In this case, we have a loss of 14 % and a GS of 0.61634. We deduce that the loss is probably due to the substantial shape differences in the various semester sub-periods (see Fig. 3), as happened in the case of Figure 1. From Figure 3 we

---

[3] To perform calculations our algorithm needs about three hours of computer time for each sample. However, in the future we will consider sample numbers greater than 500.

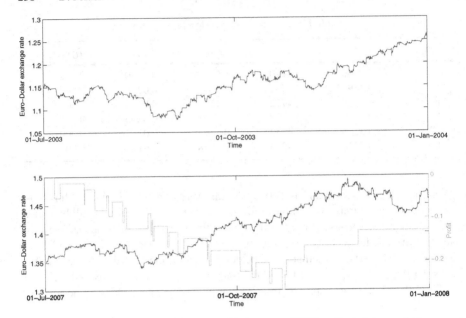

Fig. 1. Trading phrase using the second semester of 2003 as the training period

note that there are considerable losses at the beginning and at the end of the trading semester.

Table 2 summarises the algorithm application results to the semester pairs chosen. From the observation of these results, we infer that there is a global similarity threshold which lies between 0.61844 and 0.66299. For global similarity values greater than this threshold we should obtain profits.

4 Conclusions and future work

In the technical analysis literature, some authors attribute to chart patterns the property of assessing market conditions and anticipating turning points. Some works develop and analyse the information content of chart patterns. Other papers have shown the importance of choosing the best trading rules for maximum and stable profits. There-

Table 2. Profitability and similarity results of the semester pairs

Training semester	Trading semester	Profit	GS
2nd 2003	2nd 2007	−0.1301	0.61844
2nd 2004	2nd 2007	0.2592	0.66299
1st 2006	2nd 2007	−0.1367	0.61634

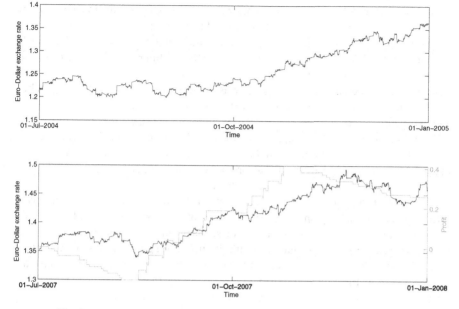

Fig. 2. Trading phase using as training period the second semester of 2004

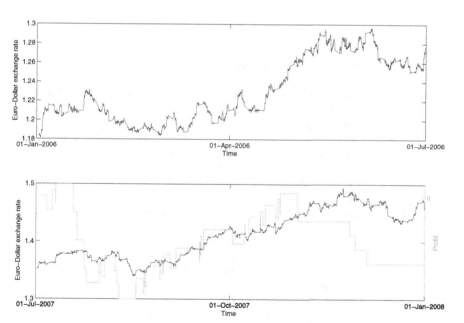

Fig. 3. Trading phase using as training period the first semester of 2006

fore, there are some technical filters that assure the highest profit. In this way, an important issue is the choice of the training and trading sets.

In this paper, we describe a pattern recognition algorithm to optimally match a training period and a trading period in the DMAC filter rule. The trading rule described is a long-short strategy investing in foreign exchange rates. We illustrate a practical example choosing the semester as the testing period and obtaining stable results. This stability is verified also for different periods, such as monthly, yearly and two-yearly periods. Moreover, for these temporal ranges, we realise a statistic on the short and long operations separately. In particular, we compute the mean and the standard error of the operations number, obtaining some interesting information. It might be convenient to also report standard indicators such as performance, volatility and Sharpe ratio, typical of the finance industry.

The aim of this work is to obtain positive profits in accordance with similarity degrees between training and trading periods. Our method gives a similarity index that can be useful to establish how a training set has valuable information for a future trading set. The results show that the similarity index is a good starting point for this kind of study. Therefore, we will need to analyse how differences in shape have an impact on profits for global similarity indexes of comparable magnitude.

References

1. Allen, F., Karjalainen, R.: Using genetic algorithms to find technical trading rules. J. Finan. Econ. 51, 245–271 (1999)
2. Arnold, C., Rahfeldt, D.: Timing the Market: How to Profit in Bull and Bear Markets with Technical Analysis. Probus Publishing, Chicago (1986)
3. Berkowitz, J., Kilian, L.: Recent developments in bootstrapping time series. Discussion Paper, no. 96–45, (1996)
4. Bollersev, T.: Generalized autoregressive conditional heteroskedasticity. J. Econ. 31, 307–327 (1986)
5. Brock, W., Lakonishok, J., LeBaron, B.: Simple technical trading rules and the stochastic properties of stock returns. J. Finan. 47, 1731–1764 (1992)
6. Brooks, C., Clare, A.D., Persand, G.: A word of caution on calculating market-based minimum capital risk requirements. J. Bank. Finan. 14, 1557–1574 (2000)
7. Chang, P., Osler, C.: Methodical madness: technical analysis and the irrationality of exchange-rate forecast. Econ. J. 109, 636–661 (1999)
8. Davison, A.C.: Hinkley, Bootstrap Methods and Their Applications. Cambridge University Press (1997)
9. Dempster, M., Jones, C.: Can technical pattern trading be profitably automated? 2. The head & shoulders. Centre for Financial Research, Judge Institute of Management Studies. University of Cambridge, working paper (1998)
10. Dempster, M.A.H., Jones, C.M.: A real time adaptive trading system using genetic programming. Judge Institute of Management. University of Cambridge, WP n. 36 (2000)
11. Edwards, M.: Technical Analysis of Stock Trends, 5th edn. John Magee, Boston (1966)
12. Efron, B.: Bootstrap methods: another look at the jackknife. Ann. Stat. 7, 1–26 (1979)
13. Efron, B.: The Jackknife, the Bootstrap, and other Resampling Plans. Society for Industrial and Applied Mathematics, Philadelphia (1982)

14. Fan, J.: Local modelling, written for the Encyclopidea of Statistics Science EES update (1995)
15. Gebbie, T.: Kernel regression toolbox. MATLAB Release R14, Version 1.0, URL: http://www.mathworks.com/ (2004)
16. Hardle, W.: Applied Nonparametric Regression. Cambridge University Press, Cambridge (1990)
17. Hardle, W., Horowitz, J., Kreiss, J.: Bootstrap methods for time series. Int. Stat. Rev. 71, 435–459 (2001)
18. Hardy, C.C.: The Investor's Guide to Technical Analysis. McGraw Hill, New York (1978)
19. Hsieh, D.: Implications of nonlinear dynamics for financial risk management. J. Finan. Quant. Anal. 28, 41–64 (1993)
20. Jones, C.M.: Automated technical foreign exchange trading with high frequency data. PhD Thesis, University of Cambridge (1999)
21. Kaufman, P.: Commodity trading systems and methods. Ronald Press, New York (1978)
22. Levy, R.: The predictive significance of five point patterns. J. Bus. 41, 316–323 (1971)
23. Lo, A.W., Mamaysky, H., Wang, J.: Foundations of technical analysis: computational algorithms, statistical inference, and empirical implementation. J. Finan. 55, 1705–1765 (2000)
24. Murphy, J.: Technical Analysis of the Future Market: A Comprehensive Guide to Trading Methods and Applications. Prentice Hall, New York (1986)
25. Murphy, J.: Technical Analysis of the Financial Markets. Institute of Finance, New York (1999)
26. Nadaraya, E.A.: On estimating regression. Theory Prob. Appl. 9, 141–142 (1964)
27. Neely, C.J., Weller, P.A., Ditmar, R.: Is technical analysis of the foreign exchange market profitable? A genetic programming approach. J. Finan. Quant. Anal. 405–426 (1997)
28. Omrane, W.B., Van Oppens, H.: The predictive success and profitability of chart patterns in the Euro/Dollar foreign exchange market. IGA Working Paper bi, 95–03 (2003)
29. Osler, C.: Identifying noise traders: the head and shoulder pattern in U.S. equities. Federal Reserve Bank of New York (1998)
30. Osler, C., Chang, K.: Head and shoulders: not just a flaky pattern. Federal Reserve Bank of New York Staff Reports 4 (1995)
31. Park, C.-H., Irwin, S.H.: The profitability of technical analysis: a review. AgMAS Project Research Report No. 2004-04 (2004)
32. Pelusi, D., Scacciavillani, F., Tivegna, M.: Optimal trading rules at hourly frequency in the foreign exchange markets. Working paper (2007)
33. Pring, M.: Technical Analysis Explained: The Successful Investor's Guide to Spotting Investment Trends and Turning Points, 3rd edn. McGraw-Hill, New York (1985)
34. Ruiz, E., Pascual, L.: Bootstrapping financial time series. J. Econ. Surv. 16, (2002)
35. Savin, G., Weller, P., Zvingelis, J.: The predictive power of head-and-shoulders price patterns in the U.S. stock market. J. Finan. Econ. 5, 243–265 (2007)
36. Sklarew, A.: Techniques of a Professional Commodity Chart Analyst. Commodity Research Bureau, New York (1980)
37. Tsay, R.S.: Analysis of Financial Time Series. Wiley (2002)
38. Watson, G.S.: Smooth regression analysis. Ind. J. Stat. A 26, 359–372 (1964)
39. Zoubir, A.M., Iskander D.R.: Bootstrap matlab toolbox. Curtin University of Technology. http://www.csp.curtin.edu.au/downloads/bootstrap_toolbox.html (May 1998)

Nonlinear cointegration in financial time series

Claudio Pizzi

Abstract. In this paper, the concept of linear cointegration as introduced by Engle and Granger [5] is merged into the local paradigm. Adopting a local approach enables the achievement of a local error correction model characterised by dynamic parameters. Another important result obtained using the local paradigm is that the mechanism that leads the dynamic system back to a steady state is no longer a constant: it is a function not defined a priori but estimated point by point.

Key words: nonlinearity, cointegration, local polynomial model

1 Introduction

One of the aims of the statistical analysis of a time series is to enable the researcher to build a simplified representation of the data-generating process (DGP) and/or the relationship amongst the different phenomena under study. The methods for identifying and estimating these models are based on the assumption of the stationarity of the DGP. Nevertheless, this assumption is often violated when considering financial phenomena, for example stock price, interest rates, exchange rates and so on. The financial time series usually present a non-stationarity of the first order if not higher.

In the case of the construction of a regressive model, the presence of unit roots in the time series means attention should be paid to the possible cointegration amongst the variables.

The cointegration idea, which characterises the long–run relationship between two (or several) time series, can be represented by estimating a vector of parameters and can be used to build a dynamic model that enables both long-run relationships and also some transitional short-run information to be highlighted. This enables the representation of an error correction model that can be considered as a dynamic system characterised by the fact that any departure from the steady state generates a short-run dynamic.

The linear cointegration concept introduced by Engle and Granger [5] has been broadly debated in the literature and much has been published on this topic. The researchers' interest has mainly been tuned to the problem of estimating the cointe-gration relationship (it is worth mentioning, amongst others, Johansen [8], Saikko-nen [18], Stock and Watson [21] Johansen [9] and Strachan and Inder [22]) and building statistical tests to verify the presence of such a relationship.

The first test suggested by Engle and Granger [5] was followed by the tests pro-posed by Stock and Watson [20] to identify common trends in the time series assessed. After them, Phillips and Ouliaris [15] developed a test based on the principal com-ponents method, followed by a test on regression model residuals [14]. Johansen [8] instead proposed a test based on the likelihood ratio.

The idea of linear cointegration then has been extended to consider some kind of nonlinearity. Several research strains can be identified against this background. One suggests that the response mechanism to the departure from the steady state follows a threshold autoregressive process (see for example the work by Balke and Fomby [1]). With regard to the statistical tests to assess the presence of threshold cointegration, see for example Hansen and Byeongseon [7].

The second strain considers the fractional cointegration: amongst the numerous contributions, we would like to recall Cheung and Lai [3], Robinson and Marin-ucci [16], Robinson and Hualde [17] and Caporale and Gil-Alana [2].

Finally, Granger and Yoon [6] introduced the concept of hidden cointegration that envisages an asymmetrical system answer, i.e., the mechanism that guides the system to the steady state is only active in the presence of either positive or negative shocks, but not of both. Schorderet's [19] work follows up this idea and suggests a procedure to verify the presence of hidden cointegration.

From a more general standpoint, Park and Phillips [12] considered non-linear regression with integrated processes, while Lee et al. [11] highlighted the existence of a spurious nonlinear relationship. In the meantime further developments contemplated the equilibrium adjustment mechanism guided by a non-observable weak force, on which further reading is available, by Pellizzari et al. [13].

This work is part of the latter research strain and suggests the recourse to local linear models (LLM) to build a test for nonlinear cointegration. Indeed, the use of local models has the advantage of not requiring the a priori definition of the func-tional form of the cointegration relationship, enabling the construction of a dynamic adjustment mechanism. In other words, a different model can be considered for each instant (in the simplest of cases, it is linear) to guide the system towards a new equi-librium. The residuals of the local model can thus be employed to define a nonlinear cointegration test. The use of local linear models also enables the construction of a Local Error Correction Model (LECM) that considers a correction mechanism that changes in time. The paper is organised as follows. The next section introduces the idea of nonlinear cointegration, presenting the LECM and the unrestricted Local Er-ror Correction Model (uLECM). Section 3 presents an application to real data, to test the nonlinear cointegration assumption. The time series pairs for which the null hypothesis of no cointegration is rejected will be used to estimate both the uLECM and the speed of convergence to equilibrium. The paper will end with some closing remarks.

2 Nonlinear cointegration

Let $X_t, Y_t : t = 1, \ldots, T$ be the realisation of two integrated processes of the order d and consider the following relationship:

$$Y_t = \beta_0 + \beta_1 X_t + u_t. \tag{1}$$

If the vector $\boldsymbol{\beta} = (\beta_0, \beta_1)$ is not null and z_t is an integrated process of the order $b < d$, then the variables X_t and Y_t are linearly cointegrated and $\boldsymbol{\beta}$ is the cointegration vector; as follows, without losing generality, it will be considered that $d = 1$ and consequently $b = 0$. Consider now a general dynamic relationship between Y and X:

$$Y_t = \alpha + \beta X_t + \gamma X_{t-1} + \delta Y_{t-1} + u_t. \tag{2}$$

The parameters restriction $\beta + \gamma = 1 - \delta$ and some algebra lead to the formulation of the following error correction model (ECM):

$$\Delta Y_t = \alpha + \beta \Delta X_t - \phi \hat{z}_{t-1} + v_t, \tag{3}$$

where $\Delta Y_t = Y_t - Y_{t-1}$, $\Delta X_t = X_t - X_{t-1}$ and $\hat{z}_{t-1}$ are residuals of the model estimated by equation (1) and v_t is an error term that satisfies the standard properties.

As an alternative, the unrestricted approach can be considered. The unrestricted error correction model can be specified as:

$$\Delta Y_t = \alpha^* + \beta^* \Delta X_t + \pi_1 Y_{t-1} + \pi_2 X_{t-1} + v_t. \tag{4}$$

In a steady state there are no variations, thus $\Delta Y_t = \Delta X_t = 0$ so that denoting Y^* and X^* the variables of the long-run relationship, the long-run solution can be indicated as:

$$0 = \alpha^* + \pi_1 Y^* + \pi_2 X^* \tag{5}$$

and

$$Y^* = -\frac{\alpha^*}{\pi_1} - \frac{\pi_2}{\pi_1} X^*. \tag{6}$$

The long-run relationship is estimated by π_2/π_1, whereas π_1 is an estimate of the speed of adjustment. The cointegration relationship is interpreted as a mechanism that arises whenever there is a departure from the steady state, engendering a new equilibrium. Both (3) and (6) highlight that the relationship between the variables is linear in nature and that all the parameters are constant with respect to time. If on the one hand this is a convenient simplification to better understand how the system works, on the other its limit consists in it restricting the correction mechanism to a linear and constant answer. To make the mechanism that leads back to a steady-state dynamic, we suggest considering the relationship between the variables of the model described by equation (1) in local terms. In a traditional approach to cointegration, the parameters of (1) are estimated just once by using all available observations and, in the case of homoscedasticity, a constant weight is assigned to each piece of available sample data. On the contrary, the local approach replaces the model represented by (1) with the following equation:

$$\Delta Y_t = \beta_{0,t} + \beta_{1,t} X_t + z_t. \tag{7}$$

The estimation of parameters in the local model is achieved through the following weighted regression:

$$\hat{\beta}_t = \arg \min_{\beta \in \Re^2} \sum_{i=1}^{T} \left[Y_i - \beta_{0,t} - \beta_{1,t} X_i \right]^2 w_{t,i}^2, \tag{8}$$

where $w_{t,i}$ indicates the weight associated to the ith sample point in the estimate of the function at point t. They measure the similarity between the sample points X_t and X_i and are defined as follows:

$$w_{t,i} = \Phi k \left[\left(X_{t-j} - X_{i-j} \right) / h \right], \tag{9}$$

where Φ is an aggregation operator that sums the similarities between ordered pairs of observations. Function k is continuous, positive and achieves its maximum in zero; it is also known as a kernel function. Amongst the different and most broadly used kernel functions aqnd the Epanechnikov kernel, with minimum variance, and the gaussian kernel, which will be used for the application described in the next section. The kernel function in (9) is dependent on parameter h, called bandwidth, which works as a smoother: as it increases, the weight $w_{t,i}$ will be higher and very similar to each other. Parameter h has another interpretation, i.e., to measure the model's "local" nature: the smaller h is, the more the estimates will be based on few sampling data points, very similar to the current one. On the other hand, a higher h value means that many sampling data points are used by the model to achieve an estimate of the parameters. This paper has considered local linear models, but it is also possible to consider other alternative local models models such as, for example, those based on the Nearest Neighbours that resort to constant weights and a subset of fixed size of sample observations. Once the local model has been estimated, a nonlinear cointegration test can be established considering the model's residuals and following the two-stage procedure described by Engle and Granger. Furthermore, with an adaptation from a global to a local paradigm, similar to the one applied to (1), equation (6) becomes:

$$\Delta Y_t = \alpha_t^* + \beta_t^* \Delta X_t + \pi_{1,t} Y_{t-1} + \pi_{2,t} X_{t-1} + v_t. \tag{10}$$

The long-run relationship and the speed of adjustment will also be dependent on time and no longer constant as they depend on the parameters $\pi_{1,t}$ and $\pi_{2,t}$ that are estimated locally.

In the next section both LECM (equation 7) and uLECM (equation 10) will be estimated. The former to test the nonlinear cointegration hypothesis and the latter to estimate the speed of adjustment.

3 An application to a financial time series

To verify whether there is cointegration amongst financial variables the time series of the adjusted closing price of the quotations of 15 US stocks were taken from the S&P 500 basket. For each stock the considered period went from 03.01.2007 to 31.12.2007. Table 1 summarises the 15 stocks considered, the branch of industry they refer to and the results of the Phillips-Perron test performed on the price series to assess the presence of unit roots.

Table 1. *p*-value for the Phillips–Perron test

Code	Name	Industry	Stationarity	Explosive
AIG	American Internat.Group	Insurance	0.565	0.435
CR	Crane Company	Machinery	0.623	0.377
CSCO	Cisco Systems	Communications equipment	0.716	0.284
F	Ford Motor	Automobiles	0.546	0.454
GM	General Motors	Automobiles	0.708	0.292
GS	Goldman Sachs Group	Capital markets	0.536	0.464
JPM	JPMorgan Chase & Co.	Diversified financial services	0.124	0.876
MER	Merrill Lynch	Capital markets	0.585	0.415
MOT	Motorola Inc.	Communications equip.	0.109	0.891
MS	Morgan & Stanley	Investment brokerage	0.543	0.457
NVDA	NVIDIA Corp.	Semiconductor & semiconductor equip.	0.413	0.587
PKI	PerkinElmer	Health care equipment & supplies	0.655	0.345
TER	Teradyne Inc.	Semiconductor & semiconductor equip.	0.877	0.123
TWX	Time Warner Inc.	Media	0.451	0.549
TXN	Texas Instruments	Semiconductor & semiconductor equip.	0.740	0.260

The test was performed both to assess the presence of unit roots vs stationarity (fourth column) and also the presence of unit roots vs explosiveness (last column). The null hypothesis of unit roots was accepted in all the time series considered. A further test, the KPSS test [10], was carried out to assess the null hypothesis that the time series is level or trend stationary. The results have confirmed that all the series are nonstationary. Considering the results from the nonstationarity tests, we proceeded to verify the assumption of cointegration in the time series. The acceptation of the latter assumption is especially interesting: this result can be interpreted in terms of the mechanisms that affect the quotation of the stocks considered. More specifically, the shocks that perturb the quotations of a stock imply departure from the system's steady state, thus inducing variations that depend on the extent of the shock and the estimable speed of convergence towards the new equilibrium. From another standpoint, the presence/absence of cointegration between two stocks may become important when contemplating the implementation of a trading strategy. Recording the variations in a variable (quotation of a stock) enables prediction of the "balancing" response provided by some variables or the purely casual responses of others. Considering the definition of cointegration introduced in the previous section and taking into account the 15 shares contemplated in this application, there are numerous applicable cointegration tests, as each possible *m*-upla of variables can be considered with $m = 2, \ldots, 15$.

In this paper the test of hypothesis of cointegration has been restricted to analysing the relationships between all the different pairs of time series. The Phillips Ouliaris test [15] was performed on each pair, the results of which are summarised in Table 2. The table highlights in bold the p-values lower than 0.05 that show the pairs of stocks for which the hypothesis of no cointegration was rejected. The test was performed resorting to R software. Note that the software outputs 0.15 for any real p-value greater than 0.15. Hence the values 0.15 in the Table 2 mean that the null hypothesis of no cointegration is accepted with a p-value ≥ 0.15.

Table 2. p-value of the Phillips Ouliaris test for the null hypothesis that the time series are not cointegrated

	AIG	CR	CSCO	F	GM	GS	JPM	MER	MOT	MS	NVDA	PKI	TER	TWX
CR	0.15													
CSCO	0.15	0.15												
F	0.15	0.15	0.15											
GM	0.15	0.15	0.15	0.15										
GS	0.15	0.15	0.15	0.15	0.15									
JPM	0.15	0.15	0.15	0.15	0.15	0.15								
MER	0.15	0.15	0.15	0.15	0.15	0.15	0.15							
MOT	0.06	0.15	0.15	0.15	0.15	0.15	0.15	0.07						
MS	0.15	0.15	0.15	0.15	0.15	0.15	0.15	0.15	0.15					
NVDA	**0.01**	0.15	0.15	0.15	0.15	0.15	0.15	0.15	0.15	0.05				
PKI	0.15	0.15	0.15	0.15	0.15	0.15	**0.01**	0.06	0.15	0.15	0.15			
TER	0.15	**0.04**	0.15	0.15	0.15	0.15	0.15	0.15	0.15	0.15	0.15	0.15		
TWX	**0.05**	0.15	**0.04**	0.08	0.15	0.15	0.15	0.15	0.15	0.14	0.07	0.15	0.15	
TXN	0.11	0.15	0.15	0.15	0.15	0.15	0.15	**0.04**	**0.01**	0.11	0.10	0.10	0.15	0.15

Table 3. p-value for the nonlinear cointegration test for the null hypothesis that the time series are not cointegrated

	AIG	CR	CSCO	F	GM	GS	JPM	MER	MOT	MS	NVDA	PKI	TER	TWX
CR	0.19													
CSCO	0.65	**0.00**												
F	0.93	0.97	0.99											
GM	0.95	0.97	0.81	0.59										
GS	**0.03**	0.31	0.06	**0.00**	0.20									
JPM	0.27	0.13	0.98	**0.00**	0.99	**0.00**								
MER	0.23	0.14	1.00	0.99	0.99	0.98	0.28							
MOT	0.08	0.53	0.14	**0.00**	0.99	**0.01**	**0.02**	**0.01**						
MS	**0.00**	**0.00**	0.16	0.94	1.00	0.94	0.59	0.86	0.46					
NVDA	0.13	0.51	1.00	**0.00**	1.00	0.14	0.09	0.77	0.89	**0.00**				
PKI	**0.02**	**0.01**	**0.00**	**0.00**	1.00	**0.00**	**0.00**	**0.00**	0.13	0.89	**0.01**			
TER	0.45	**0.01**	**0.00**	0.99	0.99	0.99	**0.00**	0.05	**0.00**	0.99	**0.00**	**0.00**		
TWX	0.42	0.06	1.00	1.00	1.00	0.99	**0.00**	0.42	**0.01**	1.00	1.00	**0.02**	**0.00**	
TXN	0.41	**0.00**	0.99	1.00	1.00	0.99	**0.00**	0.79	**0.02**	1.00	**0.01**	**0.00**	**0.00**	**0.01**

The figures in bold show that only 7 pairs, out of the total 105 combinations, of time series are linearly cointegrated, highlighting for the majority of cases the lack of a long-run relationship and adjustment mechanism. To assess the presence of nonlinear cointegration, as presented in the previous section, we adapted the two-stage procedure suggested by Engle and Granger to the nonlinear framework. In the first stage, the local linear model was initially estimated amongst the series under investigation; then the stationarity was tested using the residuals of the estimated linear local model. If the null hypothesis of nonstationarity was discarded, the second stage of the procedure was conducted: it consisted in verifying the cointegration hypothesis by performing a second regression as in (3). The results of the two-stage procedure are shown in 3. They highlight that the use of local linear models has enabled the identification of nonlinear cointegration relationships among 40 binary time series combinations. This confirms the initial assumption, i.e., that the time series lacking linear cointegration in fact present a nonlinear relationship.

For the time series that presented nonlinear cointegration, an unrestricted local error correction model was also estimated to obtain both the long-run dynamic relationship and the function of the speed of adjustment. Below, for brevity, only one case is presented. The considered period went from 01/07/2000 to 31/12/2002. has to be interpreted as acceptance of the null hypothesis of the absence of co-integration with a p-value ≥ 0.15.

Fig. 1. Time series of stocks price

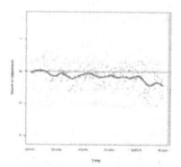

Fig. 2. Speed of adjustment function

Figure 1 shows the time series of the price of the stocks considered, i.e., Ford Motor (F) and Motorola Inc. (MOT). The speed of adjustment is depicted in Figure 2 (points); its behaviour is very rough and to smooth it we estimate the function of speed using the local polynomial regression (LOESS procedure) [4] (line).

It is worth mentioning that the velocity increases when strong market shocks perturb one of the time series disturbing the system from its steady state. As the adjustment mechanism drives the system towards the new equilibrium, the speed of adjustment tends to diminish.

4 Conclusion

The analysis of the time series of the 15 shares has enabled us to highlight that the relationships that bind two stocks in the long run do not always follow a linear error correction structure. To overcome this limit, we have suggested a local error correction model that enables the investigation of the presence of nonlinear cointegration. By applying this local model, it has been shown that, out of all those analysed, several pairs of stocks are bound by a nonlinear cointegration relationship. Furthermore, the LECM, reformulated in terms of an unrestricted local error correction model, has also enabled the determination of the correction speed and the long-run relationship between variables as a function of time, enabling the consideration of a dynamic cointegration relationship.

References

1. Balke, N.S., Fomby, T.B.: Threshold cointegration. International Economic Review 38, 627–645 (1997)
2. Caporale, G.M., Gil-Alana, L.A.: Fractional cointegration and real exchange rates. Review of Financial Economics 13, 327–340 (2004)
3. Cheung, Y.W., Lai, K.S.: A fractional cointegration analysis of purchasing power parity. Journal of Business & Economic Statistics 11, 103–112 (1993)
4. Cleveland, W.S., Grosse, E., Shyu, W.M.: Local regression models. In: Chambers, J.M., Hastie, T.J. (eds) Statistical Models in S. Wadsworth & Brooks (1992)
5. Engle, R.F., Granger, C.W.J.: Cointegration and error correction: representation, estimation and testing. Econometrica 55, 251–276 (1987)
6. Granger, C., Yoon, G.: Hidden cointegration. Working Paper 2002-02, University of California, Department of Economics, San Diego (2002)
7. Hansen, B., Byeongseon, S.: Testing for two-regime threshold cointegration in vector error correction models. Journal of Econometrics 110, 293–318 (2002)
8. Johansen, S.: Statistical analysis of cointegration vectors. Journal of Economic Dynamics and Control 12, 231–254 (1988)
9. Johansen, S.: Likelihood-Based Inference in Cointegrated Autoregressive Models, 2nd edn. Oxford University Press, Oxford (1992)
10. Kwiatkowski, D., Phillips, P.C.B., Schmidt, P., Shin, Y.: Testing the null hypothesis of stationarity against the alternative of a unit root: How sure are we that economic time series have a unit root? Journal of Econometrics 54, 159—178 (1992)

11. Lee, Y.S., Kim, T.H., Newbold, P.: Spurious nonlinear regressions in econometrics. Economics Letters 87, 301–306 (2005)
12. Park, J.Y., Phillips, P.C.B.: Nonlinear regressions with integrated time series. Econometrica 69, 117–161 (2001)
13. Pellizzari, P., Pizzi, C., Salmasi, L.: Dynamical cointegration in economic time series. In: Proceedings of Complex Models and Computational Methods for Estimation and Prediction, pp. 389–394, Cleup, Padua (2005)
14. Phillips, P.C.B., Ouliaris, S.: Asymptotic properties of residual based tests for cointegration. Econometrica 58, 165–193 (1990)
15. Phillips, P.C.B., Ouliaris, S.: Testing for cointegration using principal components methods. Journal of Economic Dynamics and Control 12, 205–230 (1988)
16. Robinson, P., Marinucci, D.: Semiparametric frequency domain analysis of fractional cointegration. In: P. Robinson (ed.) Time Series with Long Memory. Oxford University Press, Oxford (2003)
17. Robinson, P.M., Hualde, J.: Cointegration in fractional systems with unknown integration orders. Econometrica 71, 1727–1766 (2003)
18. Saikkonen, P.J.: Asymptotically efficient estimation of cointegration regressions. Econometric Theory 7, 1–12 (1991)
19. Schorderet, Y.: Asymmetric cointegration. Working Paper 2003-01, University of Geneve, Department of Econometrics (2003)
20. Stock, J.H., Watson, M.W.: Testing for common trends. Journal of the American Statistical Association 83, 1097–1107 (1988)
21. Stock, J.H., Watson, M.W.: A simple estimator of cointegrating vectors in higher order integrated systems. Econometrica 61, 783–820 (1993)
22. Strachan, R.W., Inder, B.: Bayesian analysis of the error correction model. Journal of Econometrics 123, 307–325 (2004)

Optimal dynamic asset allocation in a non–Gaussian world

Gianni Pola

Abstract. Asset Allocation deals with how to combine securities in order to maximize the investor's *gain*. We consider the Optimal Asset Allocation problem in a multi-period investment setting: the optimal portfolio allocation is synthesised to maximise the joint probability of the portfolio fulfilling some target returns requirements. The model does not assume any particular distribution on asset returns, thus providing an appropriate framework for a non–Gaussian environment. A numerical study clearly illustrates that an optimal total-return fund manager is *contrarian* to the market.

Key words: asset allocation, portfolio management, multi-period investment, optimal control, dynamic programming

1 Introduction

In the finance industry, portfolio allocations are usually achieved by an optimization process. Standard approaches for Optimal Asset Allocation are based on the Markowitz model [15]. According to this approach, return stochastic dynamics are mainly driven by the first two moments, and asymmetry and fat-tails effects are assumed to be negligible. The model does not behave very well when dealing with non–Gaussian-shaped asset classes, like Hedge Funds, Emerging markets and Commodities. Indeed it has been shown that sometimes minimizing the second order moment leads to an increase in kurtosis and a decrease in skewness, thus increasing the probability of extreme negative events [3, 10, 22]. Many works have appeared recently in the literature that attempt to overcome these problems: these approaches were based on an optimization process with respect to a cost function that is sensitive to higher-order moments [2, 11, 12], or on a generalisation of the Sharpe [21] and Lintner [14] CAPM model [13, 16].

The second aspect of the Markowitz model is that it is static in nature. It permits the investor to make a one-shot allocation to a given time horizon: portfolio re–balancing during the investment lifetime is not faced. Dynamic Asset Allocation models address the portfolio optimisation problem in multi-period settings [4, 7, 17, 20, 23].

In this paper we consider the Optimal Dynamic Asset Allocation (ODAA) problem from a Control System Theory perspective. We will show that the ODAA problem can be reformulated as a suitable optimal control problem. Given a sequence of target sets, which represent the portfolio specifications, an optimal portfolio allocation strategy is synthesized by maximizing the probability of fulfilling the target sets requirements. The proposed optimal control problem has been solved by using a Dynamic Programming [6] approach; in particular, by using recent results on the Stochastic Invariance Problem, established in [1, 18]. The proposed approach does not assume any particular distribution on the stochastic random variables involved and therefore provides an appropriate framework for non–Gaussian settings. Moreover the model does not assume stationarity in the stochastic returns dynamics. The optimal solution is given in a closed algorithmic form.

We applied the formalism to a case study: a 2-year trade investing in the US market. The objective of the strategy is to beat a fixed target return at the end of the investment horizon. This study shows markedly that an (optimal) total return fund manager should adopt a *contrarian* strategy: the optimal solution requires an increase in risky exposure in the presence of market drawdowns and a reduction in the bull market. Indeed the strategy is a *concave* dynamic strategy, thus working pretty well in oscillating markets. We contrast the ODAA model to a *convex* strategy: the Constant-Proportional-Portfolio-Insurance (CPPI) model.

Preliminary results on the ODAA problem can be found in [19].

The paper is organised as follows. In Section 2 we give the formal statement of the model and show the optimal solution. Section 3 reports a case study. Section 4 contains some final remarks.

2 The model: formal statement and optimal solution

Consider an investment universe made of m asset-classes. Given $k \in \mathcal{N}$, define the vector:

$$w_k = \left[\, w_k(1) \ w_k(2) \ \cdots \ w_k(m) \, \right]^T \in \mathcal{R}^m,$$

where the entries are the returns at time k. Let

$$u_k = [\, u_k(1) \ u_k(2) \ \ldots \ u_k(m) \,]^T \in \mathcal{R}^m$$

be the portfolio allocation at time $k \in \mathcal{N}$. Usually some constraints are imposed on u_k in the investment process: we assume that the portfolio u_k is constrained to be in a given set $U_k \subset \mathcal{R}^m$. The portfolio time evolution is governed by the following stochastic dynamical control system:

$$x_{k+1} = x_k(1 + u_k^T w_{k+1}), \ k \in \mathcal{N}, \tag{1}$$

where:

- $x_k \in X = \mathcal{R}$ is the state, representing the portfolio value at time k;

- $u_k \in U_k \subseteq \mathcal{R}^m$ is the control input, representing the portfolio allocation at time k; and

- $w_k \in \mathcal{R}^m$ is a random vector describing the asset classes' returns at time $k \in \mathcal{N}$.

Equation (1) describes the time evolution of the portfolio value: $u_k^T w_{k+1}$ quantifies the percentage return of the portfolio allocation u_k at time k in the time interval $[k, k+1)$ due to market performances w_{k+1}.

Let $(\Omega, \mathcal{F}, P)$ be the probability space associated with the stochastic system in (1). Portfolio value x_k at time $k = 0$ is assumed to be known and set to $x_0 = 1$. The mathematical model in (1) is characterised by no specific distribution on the asset classes' returns. We model asset classes' returns by means of Mixtures of Multivariate Gaussian Models (MMGMs), which provide accurate modelling of non–Gaussian distributions while being computationally simple to be implemented for practical issues[1]. We recall that a random vector Y is said to be distributed according to a MMGM if its probability density function p_Y can be expressed as the convex combination of probability density functions p_{Y_i} of some multivariate Gaussian random variables Y_i, i.e.,

$$p_Y(y) = \sum_{i=1}^{N} \lambda_i p_{Y_i}(y), \quad \lambda_i \in [0, 1], \quad \sum_{i=1}^{N} \lambda_i = 1.$$

Some further constraints are usually imposed on coefficients λ_i so that the resulting random variable Y is well behaved, by requiring, for example, semi-definiteness of the covariance matrix and unimodality in the marginal distribution. The interested reader can refer to [8] for a comprehensive exposition of the main properties of MMGMs.

The class of control inputs that we consider in this work is the one of Markov policies [6]. Given a finite time horizon $N \in \mathcal{N}$, a Markov policy is defined by the sequence

$$\pi = \{u_0, u_1, \ldots, u_{N-1}\}$$

of measurable maps $u_k : X \to U_k$. Denote by $\mathcal{U}_k$ the set of measurable maps $u_k : X \to U_k$ and by Π_N the collection of Markov policies. For further purposes let $\pi^k = \{u_k, u_{k+1}, \ldots, u_{N-1}\}$.

Let us consider a finite time horizon N which represents the lifetime of the considered investment. Our approach in the portfolio construction deals with how to select a Markov policy π in order to fulfill some specifications on the portfolio value x_k at times $k = 1, \ldots, N$. The specifications are defined by means of a sequence of target sets $\{\Sigma_1, \Sigma_2, \ldots, \Sigma_N\}$ with $\Sigma_i \subseteq X$. The investor wishes to have a portfolio value x_k at time k that is in Σ_k. Typical target sets Σ_k are of the form $\Sigma_k = [\underline{x}_k, +\infty[$ and aim to achieve a performance that is downside bounded by $\underline{x}_k \in \mathcal{R}$. This formulation of specifications allows the investor to have a portfolio evolution control during its lifetime, since target sets Σ_k depend on time k.

The portfolio construction problem is then formalized as follows:

---

[1] We stress that MMGM modelling is only one of the possible choices: formal results below hold without any assumptions on the return stochastic dynamics.

Problem 1. *(Optimal Dynamic Asset Allocation (ODAA))* Given a finite time horizon $N \in \mathcal{N}$ and a sequence of target sets

$$\{\Sigma_1, \Sigma_2, ..., \Sigma_N\}, \tag{2}$$

where Σ_k are Borel subsets of X, find the optimal Markov policy π that maximizes the joint probability quantity

$$P(\{\omega \in \Omega : x_0 \in \Sigma_0, x_1 \in \Sigma_1, \ldots, x_N \in \Sigma_N\}). \tag{3}$$

The ODAA problem can be solved by using a dynamic programming approach [6] and in particular by resorting to recent results on stochastic reachability (see e.g., [18]).

Since the solution of Problem 1 can be obtained by a direct application of the results in the work of [1, 18], in the following we only report the basic facts which lead to the synthesis of the optimal portfolio allocation. Given $x \in X$ and $u \in \mathcal{R}^m$, denote by $p_{f(x,u,w_k)}$ the probability density function of random variable:

$$f(x, u, w_{k+1}) = x(1 + u^T w_{k+1}), \tag{4}$$

associated with the dynamics of the system in (1). Given the sequence of target sets in (2) and a Markov policy π, we introduce the following cost function V, which associates a real number $V(k, x, \pi^k) \in [0, 1]$ to a triple (k, x, π^k) by:

$$V(k, x, \pi^k) = \begin{cases} I_{\Sigma_k}(x), & \text{if } k = N, \\ \displaystyle\int_{\Sigma_{k+1}} V(k+1, z, \pi^{k+1}) p_f(z) dz, & \text{if } k = N-1, N-2, \ldots, 0, \end{cases} \tag{5}$$

where $I_{\Sigma_N}(x)$ is the indicator function of the Borel set Σ_N (i.e. $I_{\Sigma_N}(x) = 1$ if $x \in \Sigma_N$ and $I_{\Sigma_N}(x) = 0$, otherwise) and p_f stands for $p_{f(x,u_k,w_{k+1})}$. Results in [18] show that cost function V is related to the probability quantity in (3) as follows:

$$P(\{\omega \in \Omega : x_0 \in \Sigma_0, x_1 \in \Sigma_1, \ldots, x_N \in \Sigma_N\}) = V(0, x_0, \pi).$$

Hence the ODAA problem can be reformulated, as follows:

Problem 2. (Optimal Dynamic Asset Allocation) Given a finite time horizon $N \in \mathcal{N}$ and the sequence of target sets in (2), compute:

$$\pi^* = \arg \sup_{\pi \in \Pi_N} V(0, x_0, \pi).$$

The above formulation of the ODAA problem is an intermediate step towards the solution of the optimal control problem under study which can now be reported hereafter.

Theorem 1. *The optimal value of the ODAA Problem is equal to [18]*

$$p^* = J_0(x_0),$$

where $J_0(x)$ is given by the last step of the following algorithm,

$$J_N(x) = I_{\Sigma_N}(x),$$

$$J_k(x) = \sup_{u_k \in \mathcal{U}_k} \int_{\Sigma_{k+1}} J_{k+1}(z) p_{f(x,u_k,w_{k+1})}(z) dz, \ k = N-1, N-2, \ldots, 1, 0. \quad (6)$$

The algorithm proceeds as follows. Suppose that the time horizon of our investment is N. First, the optimisation algorithm (6) is solved for $k = N - 1$. This optimisation problem can be automatically solved by using a wealth of optimisation packages in many computer programs, for example, MATLAB, Mathematica and Maple. The solution to (6) provides the optimal strategy $\hat{u}_k(x)$ to be applied to the investment when the value of the portfolio is x at time k. Once the optimization problem (6) is solved for $k = N - 1$, function $J_{N-1}(x)$ is also known. Hence, on the basis of the knowledge of function $J_{N-1}(x)$, one can proceed one step backwards and solve the optimisation problem (6) at step $j = N - 2$. This algorithmic optimisation terminates when $j = 0$. The outcome of this algorithm is precisely the optimal control strategy that solves (3), as formally stated in Theorem 1.

3 Case study: a total return portfolio in the US market

In this section we apply the proposed methodology to the synthesis of a total return product. The investment's universe consists of 3 asset classes: the money market, the US bond market and the US equity market. Details on the indices used in the analysis are reported below:

Label	Asset	Index
C	Money market	US Generic T-bills 3 months
B	US bond	JP Morgan US Government Bond All Maturity
E	US equity	S&P500

Time series are in local currency and weekly based from January 1st 1988 to December 28th 2007. The total return product consists of a 2-year trade. The investor objective is to beat a target return of 7% (annualised value) at maturity; his budget risk corresponds to 7% (ex ante) monthly Value at Risk at 99% (VaR99m) confidence level.[2]

The portfolio allocation will be synthesized applying the results presented in the previous section. We first consider an ODAA problem with a quarter rebalancing ($N = 8$).

---

[2] This budget risk corresponds to an ex ante (annual) volatility of 10.42%.

Table 1. Probabilistic model assumptions

	C	B	E
return (ann)	3.24%	5.46%	10.62%
vol (ann)	0%	4.45%	14.77%
skewness	0	−0.46	−0.34
kurtosis	3	4.25	5.51
corr to C	1	0	0
corr to B	0	1	0.0342
corr to E	0	0.0342	1

The first step consists in building up a probabilistic model that describes the asset classes' return dynamics. Risk figures and expected returns[3] are reported in Table 1. Asset classes present significant deviations to the Gaussian nature (Jarque–Bera test; 99% confidence level): bond and equity markets are leptokurtic and negative skewed. We assume stationarity in the dynamics of the returns distribution. This market scenario has been modelled with a 2-states Mixture of Multivariate Gaussian Models (MMGM), as detailed in the Appendix. The proposed MMGM modelling *exactly* fits up to the fourth-order the asset-classes' performance and risk figures, and up to the second order the correlation pattern.

The investment requirements are translated into the model as follows. The optimisation criterion consists in maximising the probability $P(x_8 > 1.07^2)$, x_8 being the portfolio value at the end of the second year. The target sets Σs formalisation is given below:

$$\Sigma_0 = \{1\}, \quad \Sigma_k = [0, +\infty), \quad \forall k = 1, 2, ..., 7, \quad \Sigma_8 = [1.07^2, +\infty). \qquad (7)$$

More precisely, the optimisation problem consists in determining the (optimal) dynamic allocation grids u_k ($k = 0, 1, ..., 7$) in order to maximise the joint probability $P(x_1 \in \Sigma_1, ..., x_8 \in \Sigma_8)$ subjected to the Value-at-Risk budget constraint. By applying Theorem 1 we obtain the optimal control strategy that is illustrated in Figure 1. (Budget and long-only constraints have been included in the optimisation process.)

The allocation at the beginning of the investment (see Figure 1, upper-left panel) is 46% Bond and 54% Equity market. After the first quarter, the fund manager revises the portfolio allocation (see Figure 1, upper-right panel). Abscissas report the portfolio value x_1 at time $k = 1$. For each portfolio realisation x_1, the map gives the corresponding portfolio allocation. As the portfolio strategy delivers higher and higher performance in the first quarter, the optimal rebalancing requires a reduction in the risky exposure. If x_1 reaches a value around 1.0832, a 100% cash allocation guarantees the target objective will be reached at maturity. Conversely, a portfolio

---

[3] In the present work we do not face the problem of returns and risk-figures forecasting. Volatility, skewness, kurtosis and the correlation pattern have been estimated by taking the historical average. Expected returns have been derived by assuming a constant Sharpe ratio (0.50), and a cash level given by the US Generic T-bills 3 months in December 31st 2007.

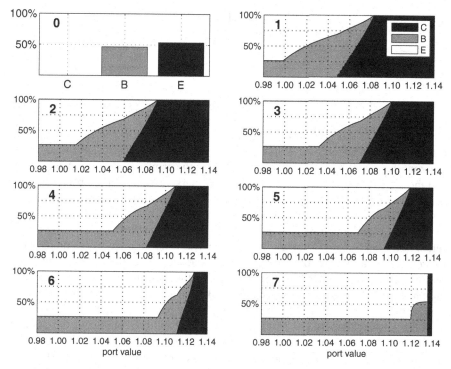

Fig. 1. ODAA optimal solution

value of $x_1 = 0.9982$ moves the optimal strategy to the maximum allowed risky exposure (i.e. 7% VaR99m).

Maps for $k = 2, 3, 4, 5, 6, 7$ exhibit similar characteristics as for $k = 1$. The main difference is that as k increases the portfolio rebalancing gets sharper and sharper.

The ODAA maximal probability p^* is 68.40%. In order to make a comparison with more standard approaches, we run the same exercise for a Markowitz constant-mix investor: in this case the optimal solution requires the full budget-risk to be invested, with a maximal probability of 61.90%. The ODAA model gets more and more efficient as the rebalancing frequency increases. Table 2 reports the maximal probabilities for rebalancing frequency of three months ($N = 8$), one month ($N = 24$), two weeks ($N = 52$) and one week ($N = 104$). In fact, this result is a direct consequence of the Dynamic Programming approach pursued in this paper. It is worth emphasising that the Markowitz constant-mix approach does not produce similar results: in fact the probability is rather insensitive to the rebalancing frequency.

The allocation grids reported in Figure 1 clearly show that an (optimal) total return fund manager should adopt a *contrarian* rebalancing policy [9]: the investor should increase the risky exposure in the presence of market drawdowns and reduce it in case of positive performance. The contrarian attitude of the model is a peculiarity of

Table 2. Maximal probabilities p^*

	3 m	1 m	2 w	1 w
Probability	68.40%	72.50%	76.28%	77.76%

concave strategies.[4] This feature makes the model particularly appealing in oscillating markets.

We conclude this section by making a comparison with a *convex* strategy. We run a Constant–Proportional–Portfolio–Insurance (CPPI) model [5] with a 2-years horizon: the model has been designed to protect the capital at maturity. We assume that the risky basket is composed by the equity market (multiplier has been set to 6) and trading is weekly based. In order to make a comparison, we assume the same Value-at-Risk budget risk as in the previous case.

Table 3 offers a comparison between the two approaches. (The ODAA results refers to a weekly based strategy; $N = 104$.) The CPPI strategy works pretty well to protect the capital (99.60%), but it presents a lower probability of achieving large returns. (The probability of beating a 7% target return at maturity is 33.64%.) Distribution of performance at maturity is positive skewed and platykurtic, thus revealing a very stable strategy. Conversely, the ODAA strategy presents a higher probability of delivering its target return (77.76%), but a lower probability of protecting the capital. ODAA performance distribution is negative skewed and leptokurtic. Higher-order risk is paid off by the the large probability of achieving more ambitious returns.

The applications presented in this section should be considered for illustrating the methodology. The views expressed in this work are those of the author and do not necessarily correspond to those of Crédit Agricole Asset Management.

Table 3. Comparison between the ODAA and CPPI strategy

	ODAA	CPPI
Mean perf N (ann)	5.68%	6.15 %
Median perf N (ann)	7.03%	3.64%
Skewness perf N	−2.80	1.35
Kurtosis perf N	11.08	4.44
Vol (ann)	2.70%	3.05%
Sharpe (ann)	1.51	0.44
Prob 0%	91.40%	99.60%
Prob cash	85.58%	53.21%
Prob 7%	77.76%	33.64%

---

[4] The exposure diagram reports on the X–axis the portfolio value and on the Y–axis the risky exposure. Concave (resp. convex) strategies are characterised by a concave (resp. convex) exposure diagram.

4 Conclusions

In this paper we considered the Optimal Dynamic Asset Allocation problem. Given a sequence of target sets that the investor would like his portfolio to stay within, the optimal strategy is synthesised in order to maximise the joint probability of fulfilling the investment requirements. The approach does not assume any specific distributions for the asset classes' stochastic dynamics, thus being particularly appealing to treat non-Gaussian asset classes. The proposed optimal control problem has been solved by leveraging results on stochastic invariance. The optimal solution exhibits a contrarian attitude, thus performing very well in oscillating markets.

Acknowledgement. The author would like to thank Giordano Pola (University of L'Aquila, Center of Excellence DEWS, Italy), Roberto Dopudi and Sylvie de Laguiche (Crédit Agricole Asset Management) for stimulating discussions on the topic of this paper.

Appendix: Markets MMGM modeling

Asset classes used in the case study present significant deviation to gaussianity. This market scenario has been modelled by a 2-state MMGM. States 1 and 2 are characterised by the following univariate statistics:[5]

$$\{\mu_1(i)\}_i = [0.000611; 0.001373; 0.002340],$$
$$\{\sigma_1(i)\}_i = [0.000069; 0.005666; 0.019121],$$
$$\{\mu_2(i)\}_i = [0.000683; -0.016109; -0.017507],$$
$$\{\sigma_2(i)\}_i = [0.000062; 0.006168; 0.052513],$$

and correlation matrix:[6]

	C	B	E
corr to C	1	0.0633	0.0207
corr to B	0.0633	1	−0.0236
corr to E	0.0207	−0.0236	1

Transition probabilities are uniform and the unconditional probability of State 1 is 98%. The above MMGM model correctly represents the univariate statistics of the asset classes up to the fourth order (as detailed in Table 1) and up to the second order concerning the correlation patterns.

---

[5] $\mu_s(i)$ and $\sigma_s(i)$ indicate (resp.) the performance and volatility of asset i in the state s. (Values are weekly based.)

[6] We assume the same correlation matrix for the above gaussian models.

References

1. Abate, A.: Probabilistic Reachability for Stochastic Hybrid Systems: Theory, Computations, and Applications. PhD Thesis. EECS Department, University of California, Berkeley (2007)
2. Agarwal, V., Naik, N.: Risks and portfolio decisions involving hedge funds. Ref. Finan. Stud. 17, 63–98 (1990)
3. Amin, G., Kat, H.: Stocks, bonds and hedge funds: not a free lunch! J. Portfolio Man. 29 4, 113–120 (2003)
4. Barberis, N.: Investing for the Long Run when Returns are Predictable. J. Finan. 55, 225–264 (2000)
5. Bertrand, P., Prigent, J. L.: Portfolio Insurance Strategies: OBPI Versus CPPI. University of CERGY Working Paper No. 2001-30. GREQAM Working Paper
6. Bertsekas, D.P.: Dynamic Programming and Optimal Control, 2nd Edn, Vols. 1 and 2. Athena Scientific, Belmont, MA (2001)
7. Brennan, M. J., Xia, Y.: Dynamic Asset Allocation under Inflation. The Journal of Finance 57(3), 1201–1238 (2002)
8. Buckley, I., Saunders, D., Seco, L.: Portfolio optimization when asset returns have the Gaussian mixture distribution. Eur. J. Oper. Res. 185, 1434–1461 (2008)
9. Chan, K. C.: On the Contrarian Investment Strategy. Journal of Business 61, 147–163 (1988)
10. Cremers, J. H., Kritzman, M., Page, S.: Optimal Hedge Fund Allocations: Do Higher Moments the Matter? The Journal of Portfolio Management 31(3), 70–81 (2005)
11. DeSouza, C., Gokcan, S.: Allocation Methodologies and Customizing Hedge Fund Multi-Manager Multi-Strategy Products. Journal of Alternative Investments 6(4), 7–21 (2004)
12. Favre-Bulle, A., Pache, S.: The Omega Measure: Hedge Fund Portfolio Optimization. MBF master's thesis. University of Lausanne (2003)
13. Hwang, S., Satchell, S.: Modeling Emerging Risk Premia Using Higher Moments. Int. J. Finan. Econ. 4(4), 271–296 (1999)
14. Lintner, J.: The Valuation of Risk Assets and the Selection of Risky Investments in Stock Portfolio and Capital Budgets. Rev. Econ. Stat. 47(1), 13–37 (1965)
15. Markowitz, H.: Portfolio Selection. The Journal of Finance 7(1), 77–91 (1952)
16. Martellini, L., Ziemann, V.: Extending Black-Litterman Analysis Beyond the Mean-Variance Framework. The Journal of Portfolio Management 33(4), 33–44 (2007)
17. Merton, R. C.: Lifetime Portfolio Selection Under Uncertainty: the Continuous–Time Case. Review of Economics and Statistics 51, 247–257 (1969)
18. Pola, G., Lygeros, J., Di Benedetto, M. D.: Invariance in Stochastic Dynamical Systems. 17th International Symposium on Mathematical Theory of Network and Systems (MTNS 2006). Kyoto, Japan, July 24th – 28th (2006)
19. Pola, G., Pola G.: Optimal Dynamic Asset Allocation: A Stochastic Invariance Approach. 45th IEEE Conference on Decision and Control (CDC 2006). San Diego, USA, December 12nd – 25th, pp. 2589–2594 (2006)
20. Samuelson, P. A.: Lifetime Portfolio Selection by Dynamic Stochastic Programming. Review of Economics and Statistics 51, 239–246 (1969)
21. Sharpe, W.: Capital Asset Prices: A Theory of Market Equilibrium under Conditions of Risk. Journal of Finance 19(3), 425–442 (1964)
22. Sornette, D., Andersen, J. V., Simonetti, P.: Portfolio Theory for Fat Tails. International Journal of Theoretical and Applied Finance 3(3), 523–535 (2000)
23. Xia, Y.: Learning about Predictability: the effect of parameter uncertainty on dynamic asset allocation. The Journal of Finance 56(1), 205–246 (2001)

Fair costs of guaranteed minimum death benefit contracts

François Quittard-Pinon and Rivo Randrianarivony

Abstract. The authors offer a new perspective on the domain of guaranteed minimum death benefit contracts. These products have the particular feature of offering investors a guaranteed capital upon death. A complete methodology based on the generalised Fourier transform is proposed to investigate the impacts of jumps and stochastic interest rates. This paper thus extends Milevsky and Posner (2001).

Key words: life insurance contracts, variable annuities, guaranteed minimum death benefit, stochastic interest rates, jump diffusion models, mortality models

1 Introduction

The contract analysed in this article is a Guaranteed Minimum Death Benefit contract (GMDB), which is a life insurance contract pertaining to the class of variable annuities (VAs). For an introduction to this subject, see Hardy [4] and Bauer, Kling and Russ [2]. The provided guaranty, only in effect upon death, is paid by continuously deducting small amounts from the policyholder's subaccount. It is shown in this chapter how these fees can be endogenously determined. Milevsky and Posner [8] found these fees overpriced by insurance companies with respect to their model fair price. To answer this overpricing puzzle, the effects of jumps in financial prices, stochastic interest rates and mortality are considered. For this purpose, a new model is proposed which generalises Milevsky and Posner [8].

2 General framework and main notations

2.1 Financial risk and mortality

Financial risk is related to market risk firstly because the policyholder's account is linked to a financial asset or an index, and secondly via interest rates. We denote by r

the stochastic process modelling the instantaneous risk-free rate. The discount factor is thus given by:

$$\delta_t = e^{-\int_0^t r_s\,ds}. \tag{1}$$

The policyholder's account value is modelled by the stochastic process S. In that model, ℓ stands for the fees associated with the Mortality and Expense (M&E) risk charge.

The future lifetime of a policyholder aged x is the r.v. T_x. For an individual aged x, the probability of death before time $t \geq 0$ is

$$P(T_x \leq t) = 1 - \exp\left(-\int_0^t \lambda(x+s)ds\right), \tag{2}$$

where λ denotes the force of mortality. As usual, F_x and f_x denote respectively the c.d.f. and the p.d.f. of the r.v. T_x. To ease notation, we generally omit the x from the future lifetime and write T when no confusion is possible. We assume stochastic independence between mortality and financial risks.

2.2 Contract payoff

The insurer promises to pay upon the policyholder's death the contractual amount $\max\{S_0 e^{gT}, S_T\}$, where g is a guaranteed rate, S_0 is the insured initial investment and S_T is the subaccount value at time of death $x + T$. We can generalise this payoff further by considering a contractual expiry date $x + \Theta$. The contract only provides a guarantee on death. If the insured is otherwise still alive after time Θ passes, she will receive the account value by that time. For the sake of simplicity, we keep the first formulation, and we note that:

$$\max\{S_0 e^{gT}, S_T\} = S_T + \left[S_0 e^{gT} - S_T\right]^+. \tag{3}$$

Written in this way, the contract appears as a long position on the policyholder account plus a long position on a put option written on the insured account. Two remarks are in order: firstly, the policyholder has the same amount as if she invested in the financial market (kept aside the fees), but has the insurance to get more, due to the put option. Secondly, because T is a r.v., her option is not a vanilla one but an option whose exercise date is itself random (the policyholder's death).

The other difference with the option analogy lies in the fact that in this case there is no upfront payment. In this contract, the investor pays the guarantee by installments. The paid fees constitute the so-called M&E risk charges. We assume they are continuously deducted from the policyholder account at the contractual proportional rate ℓ. More precisely, we consider that in the time interval $(t, t + dt)$, the life insurance company receives $\ell S_t\,dt$ as instantaneous earnings. We denote by F the cumulative discounted fees. F_τ is the discounted accumulated fees up to time τ, which can be a stopping time for the subaccount price process S. The contract can also be designed in order to cap the guaranteed rate g; in the VA literature, this is known as capping the rising floor.

2.3 Main Equations

Under a chosen risk-neutral measure Q, the GMDB option fair price is thus

$$G(\ell) = E_Q\big[\delta_T(S_0 e^{gT} - S_T)^+\big],$$

and upon conditioning on the insured future lifetime,

$$G(\ell) = E_Q\Big[E_Q\big[\delta_T(S_0 e^{gT} - S_T)^+ | T = t\big]\Big], \qquad (4)$$

which – taking into account a contractual expiry date – gives:

$$G(\ell) = \int_0^\Theta f_x(t) E_Q\big[\delta_T(S_0 e^{gT} - S_T)^+ | T = t\big] dt. \qquad (5)$$

If F_T denotes the discounted value of all fees collected up to time T, the fair value of the M&E charges can be written

$$ME(\ell) = E_Q[F_T],$$

which after conditioning also gives:

$$ME(\ell) = E_Q\big[E_Q[F_T | T = t]\big]. \qquad (6)$$

Because the protection is only triggered by the policyholder's death, the endogenous equilibrium price of the fees is the solution in ℓ, if any, of the following equation

$$G(\ell) = ME(\ell). \qquad (7)$$

This is the key equation of this article. To solve it we have to define the investor account dynamics, make assumptions on the process S, and, of course, on mortality.

3 Pricing model

The zero-coupon bond is assumed to obey the following stochastic differential equation (SDE) in the risk-neutral universe:

$$\frac{dP(t, T)}{P(t, T)} = r_t\, dt + \sigma_P(t, T)\, dW_t, \qquad (8)$$

where $P(t, T)$ is the price at time t of a zero-coupon bond maturing at time T, r_t is the instantaneous risk-free rate, $\sigma_P(t, T)$ describes the volatility structure and W is a standard Brownian motion.

In order to take into account a dependency between the subaccount and the interest rates, we suggest the introduction of a correlation between the diffusive part of the subaccount process and the zero-coupon bond dynamics. The underlying account price process S is supposed to behave according to the following SDE under the chosen equivalent pricing measure Q:

$$\frac{dS_t}{S_{t^-}} = (r_t - \ell) \, dt + \rho\sigma \, dW_t + \sigma\sqrt{1 - \rho^2} \, dZ_t + (Y - 1) \, d\tilde{N}_t. \quad (9)$$

Again, r_t is the instantaneous interest rate, ℓ represents the fixed proportional insurance risk charge, σ is the asset's volatility, ρ is the correlation between the asset and the interest rate, W and Z are two independent standard Brownian motions, and the last part takes into account the jumps. $\tilde{N}$ is a compensated Poisson process with intensity λ, while Y, a r.v. independent from the former processes, represents the price change after a jump. The jump size is defined by $J = \ln(Y)$.

Let us emphasise here that the non-drift part M, defined by $dM_t = \rho\sigma \, dW_t + \sigma\sqrt{1 - \rho^2} \, dZ_t + (Y - 1) \, d\tilde{N}_t$, is a martingale in the considered risk-neutral universe.

3.1 Modelling stochastic interest rates and subaccount jumps

Denoting by N_t the Poisson process with intensity λ and applying Itō's lemma, the dynamics of S writes as:

$$S_t = S_0 \, e^{\int_0^t r_s \, ds - (\ell + \frac{1}{2}\sigma^2 + \lambda\kappa)t + \rho\sigma \, W_t + \sigma\sqrt{1-\rho^2} \, Z_t + \sum_{i=1}^{N_t} \ln\left((Y)_i\right)}, \quad (10)$$

where $\kappa = E(Y - 1)$. The zero-coupon bond price obeys the following equation:

$$P(t, T) = P(0, T) \, e^{\int_0^t \sigma_P(s,T) \, dW_s - \frac{1}{2}\int_0^t \sigma_P^2(s,T) \, ds + \int_0^t r_s \, ds}.$$

The subaccount dynamics can be written as:

$$S_t = \frac{S_0}{P(0, t)} \, e^{-(\ell + \frac{1}{2}\sigma^2 + \lambda\kappa)t + \frac{1}{2}\int_0^t \sigma_P^2(s,t) \, ds + \int_0^t [\rho\sigma - \sigma_P(s,t)] dW_s + \sigma\sqrt{1-\rho^2} \, Z_t + \sum_{i=1}^{N_t} \ln\left((Y)_i\right)}.$$

Let us introduce the T-forward measure Q_T defined by

$$\frac{dQ_T}{dQ}\bigg|_{\mathcal{F}_t} = \frac{\delta_t \, P(t, T)}{P(0, T)} = e^{\int_0^t \sigma_P(s,T) \, dW_s - \frac{1}{2}\int_0^t \sigma_P^2(s,T) \, ds}, \quad (11)$$

where δ_t is the discount factor defined in (1). Girsanov's theorem states that the stochastic process W^T, defined by $W_t^T = W_t - \int_0^t \sigma_P(s, T) \, ds$, is a standard Brownian motion under Q_T. Hence, the subaccount price process can be derived under the T-forward measure:

$$S_t = \frac{S_0}{P(0, t)} e^{X_t}, \quad (12)$$

where X is the process defined by

$$X_t = -(\ell + \tfrac{1}{2}\sigma^2 + \lambda\kappa)t + \int_0^t \left(\sigma_P(s, T)(\rho\sigma - \sigma_P(s, t)) + \tfrac{1}{2}\sigma_P^2(s, t)\right) ds$$

$$+ \int_0^t (\rho\sigma - \sigma_P(s, t)) dW_s^T + \sigma\sqrt{1 - \rho^2} \, Z_t + \sum_{i=1}^{N_t} \ln((Y)_i). \quad (13)$$

A lengthy calculation shows that the characteristic exponent $\phi_T(u)$ of X_T under the T-forward measure, defined by $E_{Q_T}\left[e^{iuX_T}\right] = e^{\phi_T(u)}$, writes:

$$\phi_T(u) = -iu\ell T - \frac{iu}{2}\Sigma_T^2 - \frac{u^2}{2}\Sigma_T^2 + \lambda T\left(\phi_J(u) - \phi_J(-i)\right), \qquad (14)$$

where $\phi_J(u)$ denotes the characteristic function of the i.i.d. r.v.'s $J_i = \ln((Y)_i)$ and

$$\Sigma_T^2 = \int_0^T \left(\sigma^2 - 2\rho\sigma\sigma_P(s, T) + \sigma_P^2(s, T)\right)ds. \qquad (15)$$

3.2 Present value of fees

Using the definition of F_t and (6), it can be shown that:

$$ME(\ell) = 1 - \int_0^\infty e^{-\ell t} f_x(t)dt, \qquad (16)$$

where f_x is the p.d.f. of the r.v. T. A very interesting fact is that only the mortality model plays a role in the computation of the present value of fees as seen in (16).

Taking into account the time to contract expiry date Θ, we have:

$$ME(\ell) = 1 - \int_0^\Theta e^{-\ell t} f_x(t)dt - \left(1 - F_x(\Theta)\right)e^{-\ell\Theta}. \qquad (17)$$

3.3 Mortality models

Two mortality models are taken into account, namely the Gompertz model and the Makeham model. Another approach could be to use the Lee-Carter model, or introduce a mortality hazard rate as in Ballotta and Haberman (2006). In the case of the Gompertz mortality model, the force of mortality at age x follows

$$\lambda(x) = B.C^x, \qquad (18)$$

where $B > 0$ and $C > 1$. It can also be written as $\lambda(x) = \frac{1}{b}\exp\left(\frac{x-m}{b}\right)$, where $m > 0$ is the modal value of the Gompertz distribution and $b > 0$ is a dispersion parameter.

Starting from (2), it can be shown that the present value of fees[1] in the case of a Gompertz-type mortality model amounts to:

$$\begin{aligned} ME(\ell) =&\, 1 - e^{b\lambda(x)}e^{(x-m)\ell}\left[\Gamma\left(1 - \ell b, b\lambda(x)\right) - \Gamma\left(1 - \ell b, b\lambda(x)e^{\frac{\Theta}{b}}\right)\right]\\ &- e^{b\lambda(x)\left(1-e^{\frac{\Theta}{b}}\right)}e^{-\ell\Theta}, \end{aligned} \qquad (19)$$

---

[1] It is to be noted that formula (19) corrects typos in Milevsky and Posner's (2001) original article.

where $\Gamma(a, x) = \int_x^\infty e^{-t} t^{a-1} dt$ is the upper incomplete gamma function where a must be positive. This condition entails an upper limit on the possible value of the insurance risk charge ℓ:

$$\ell < \frac{1}{b}. \tag{20}$$

The Makeham mortality model adds an age-independent component to the Gompertz force of mortality (18) as follows:

$$\lambda(x) = A + B.C^x, \tag{21}$$

where $B > 0$, $C > 1$ and $A \geq -B$.

In this case, a numerical quadrature was used to compute the M&E fees.

3.4 Valuation of the embedded GMDB option

The valuation of this embedded GMDB option is done in two steps:

First, taking the conditional expectation given the policyholder's remaining lifetime, the option is valued in the context of a jump diffusion process with stochastic interest rates, with the assumption that the financial asset in the investor subaccount is correlated to the interest rates.

More precisely, let us recall the embedded GMDB option fair price, as can be seen in (4):

$$G(\ell) = E_Q \Big[E_Q \big[\delta_T (S_0 e^{gT} - S_T)^+ | T = t \big] \Big].$$

Using the zero-coupon bond of maturity T as a new numéraire, the inner expectation I_T can be rewritten as:

$$I_T = E_Q \big[\delta_T (S_0 e^{gT} - S_T)^+ \big] = P(0, T) E_{Q_T} \big[(K - S_T)^+ \big].$$

Then this expectation is computed using an adaptation[2] of the generalised Fourier transform methodology proposed by Boyarchenko and Levendorskiĭ [3].

4 Empirical study

This section gives a numerical analysis of jumps, stochastic interest rates and mortality effects. To study the impacts of jumps and interest rates, a numerical analysis is performed in a first section while a second subsection examines all these risk factors together.

---

[2] A detailed account is available from the authors upon request.

4.1 Impact of jumps and interest rates

The GMDB contract expiry is set at age 75. A guaranty cap of 200 % of the initial investment is also among the terms of the contract.

The Gompertz mortality model is used in this subsection. The Gompertz parameters used in this subsection and the next one are those calibrated to the 1994 Group Annuity Mortality Basic table in Milevsky and Posner [8]. They are recalled in Table 1.

Table 1. Gompertz distribution parameters

	Female		Male	
Age (years)	m	b	m	b
30	88.8379	9.213	84.4409	9.888
40	88.8599	9.160	84.4729	9.831
50	88.8725	9.136	84.4535	9.922
60	88.8261	9.211	84.2693	10.179
65	88.8403	9.183	84.1811	10.282

A purely diffusive model with a volatility of 20 % serves as a benchmark throughout the study. It corresponds to the model used by Milevsky and Posner [8].

The particular jump diffusion model used in the following study is the one proposed by Kou [5]. Another application in life insurance can be seen in Le Courtois and Quittard-Pinon [6]. In this model, jump sizes $J = \ln(Y)$ are i.i.d. and follow a double exponential law:

$$f_J(y) = p\lambda_1 e^{-\lambda_1 y} 1_{y>0} + q\lambda_2 e^{\lambda_2 y} 1_{y\leq 0}, \qquad (22)$$

with $p \geq 0, q \geq 0, p + q = 1, \lambda_1 > 0$ and $\lambda_2 > 0$.

The following Kou model parameters are set as follows: $p = 0.4$, $\lambda_1 = 10$ and $\lambda_2 = 5$. The jump arrival rate is set to $\lambda = 0.5$. The diffusive part is set so that the overall quadratic variation is 1.5 times the variation of the no-jump case.

Table 2 shows the percentage of premium versus the annual insurance risk charge in the no-jump case and the Kou jump diffusion model case for a female policyholder. A flat interest rate term structure was taken into account in this table and set at $r = 6$ %.

The initial yield curve $y(0, t)$ is supposed to obey the following parametric equation: $y(0, t) = \alpha - \beta e^{-\gamma t}$ where α, β and γ are positive numbers. The yield is also supposed to converge towards r for longer maturities. The initial yield curve equation is set as follows:

$$y(0, t) = 0.0595 - 0.0195 \exp(-0.2933\, t). \qquad (23)$$

As stated earlier, the interest rate volatility structure is supposed to be of exponential form. Technically, it writes as follows:

$$\sigma_P(s, T) = \frac{\sigma_P}{a}\left(1 - e^{-a(T-s)}\right), \qquad (24)$$

Table 2. Jumps impact, female policyholder; $r = 6\%$, $g = 5\%$, 200 % cap

Purchase age	No-jump case		Kou model	
(years)	(%)	(bp)	(%)	(bp)
30	0.76	1.77	1.16	2.70
40	1.47	4.45	2.04	6.19
50	2.52	10.85	3.21	13.86
60	2.99	21.58	3.55	25.74
65	2.10	22.56	2.47	26.59

Gompertz mortality model. In each case, the left column displays the relative importance of the M&E charges given by the ratio $ME(\ell)/S_0$. The right column displays the annual insurance risk charge ℓ in basis points (bp).

where $a > 0$. In the sequel, we will take $\sigma_P = 0.033333$, $a = 1$ and the correlation between the zero-coupon bond and the underlying account will be set at $\rho = 0.35$. Plugging (24) into (15) allows the computation of Σ_T^2:

$$\Sigma_T^2 = \left(\frac{2\rho\sigma\sigma_P}{a^2} - \frac{3}{2}\frac{\sigma_P^2}{a^3}\right) + \left(\sigma^2 + \frac{\sigma_P^2}{a^2} - \frac{2\rho\sigma\sigma_P}{a}\right)T + \left(\frac{2\sigma_P^2}{a^3} - \frac{2\rho\sigma\sigma_P}{a^2}\right)e^{-aT} - \frac{\sigma_P^2}{2a^3}e^{-2aT}. \quad (25)$$

The results displayed in Table 3 show that stochastic interest rates have a tremendous impact on the fair value of the annual insurance risk charge across purchase age. Table 3 shows that a 60-year-old male purchaser could be required to pay a risk charge as high as 88.65 bp for the death benefit in a stochastic interest rate environment.

Thus, the stochastic interest rate effect is significantly more pronounced than the jump effect. Indeed, the longer the time to maturity, the more jumps tend to smooth out, hence the lesser impact. On the other hand, the stochastic nature of interest rates are felt deeply for the typical time horizon involved in this kind of insurance contract.

It is to be noted that the annual insurance risk charge decreases after age 60. This decrease after a certain purchase age will be verified again with the figures provided in the next section. Indeed, the approaching contract termination date, set at age 75 as previously, explains this behaviour.

4.2 Impact of combined risk factors

The impact of mortality models on the fair cost of the GMDB is added in this subsection. Melnikov and Romaniuk's [17] Gompertz and Makeham parameters, estimated from the Human mortality database 1959–1999 mortality data, are used in the sequel. As given in Table 4, no more distinction was made between female and male policyholders. Instead, the parameters were estimated in the USA.

In the following figure, the circled curve corresponds to the no-jump model with a constant interest rate. The crossed curve corresponds to the introduction of Kou jumps but still with a flat term structure of interest rates. The squared curve adds jumps and stochastic interest rates to the no-jump case. These three curves are built with a Gompertz mortality model. The starred curve takes into account jumps and

Table 3. Stochastic interest rates impact, male policyholder; $g = 5\%$, 200% cap

Purchase age	Kou model (flat rate)		Kou model (stochastic rates)	
(years)	(%)	(bp)	(%)	(bp)
30	2.01	4.86	8.87	22.27
40	3.46	10.99	11.38	37.81
50	5.35	24.46	13.38	64.07
60	5.81	44.82	11.14	88.65
65	4.08	46.31	6.82	78.55

Gompertz mortality model. In each case, the left column displays the relative importance of the M&E charges given by the ratio $ME(\ell)/S_0$. The right column displays the annual insurance risk charge ℓ.

Table 4. Gompertz (G) and Makeham (M) mortality model parameters for the USA [7]

	A	B	C
G_{US}		6.148×10^{-5}	1.09159
M_{US}	9.566×10^{-4}	5.162×10^{-5}	1.09369

Table 5. Mortality impact on the annual insurance risk charge (bp), USA; $g = 5\%$, 200% cap

Age	No jumps	Gompertz Kou (flat)	Kou (stoch.)	Makeham Kou (stoch.)
30	4.79	6.99	30.23	32.20
40	11.16	15.15	50.86	52.34
50	24.88	31.50	82.50	83.03
60	44.45	52.97	105.27	104.77
65	45.20	53.18	90.41	89.78

stochastic interest rates but changes the mortality model to a Makeham one. Figure 1 displays the annual risk insurance charge with respect to the purchase age in the USA. From 30 years old to around 60 years old, the risk charge is steadily rising across all models. It decreases sharply afterwards as the contract expiry approaches.

The two lower curves correspond strikingly to the flat term structure of the interest rate setting. The jump effect is less pronounced than the stochastic interest rate effect as represented by the two upper curves. The thin band in which these upper curves lie shows that the change of mortality model has also much less impact than the stochastic nature of interest rates.

As is reported in Table 5, and displayed in Figure 1, the behaviour of the insurance risk charge with respect to age is of the same type whatever the considered model. However, within this type, differences can be seen. First, the jump effect alone does not change the fees very much but there are more differences when stochastic interest rates

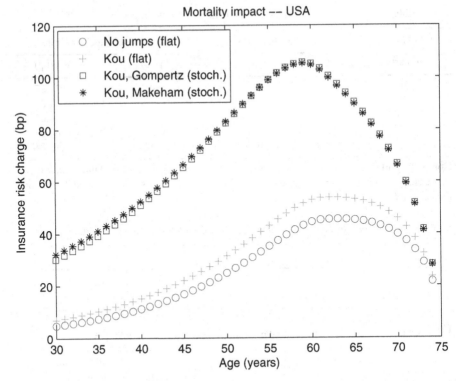

Fig. 1. Annual risk insurance charge in basis points, USA

are introduced. In this case, fees are notably higher. Second, the choice of mortality model does not have a significant impact.

5 Conclusions

To analyse the effects of jump risk, stochastic interest rate and mortality on GMDBs, this paper assumes a particular jump diffusion process, namely a Kou process, for the return of the policyholder subaccount and a Vasicek term structure of interest rate, while the mortality is of a Gompertz or a Makeham type. The contract fair value is obtained using a methodology based on generalised Fourier analysis. It is shown that the largest impact among the three risk factors on the GMDB fees is due to stochastic interest rate. Jumps and mortality have smaller influence. The fair insurance risk charges are found to be significantly higher than Milevsky and Posner [8] reported, but still below the fees required by insurance companies.

References

1. Ballotta, L., Haberman, S.: The fair valuation problem of guaranteed annuity options: the stochastic mortality environment case. Insur. Math. Econ. 38, 195–214 (2006)
2. Bauer, D., Kling, A., Russ, J.: A universal pricing framework for guaranteed minimum benefits in variable annuities. ASTIN Bull. 38, 621–651 (2008)
3. Boyarchenko, S., Levendorkiĭ, S.: Non-Gaussian Merton-Black-Scholes Theory, vol. 9 of Advanced Series on Statistical Science and Applied Probability. World Scientific, London (2002)
4. Hardy, M.: Investment Guarantees: Modeling and Risk Management for Equity-Linked Life Insurance. John Wiley, New Jersey (2003)
5. Kou, S.G.: A jump-diffusion model for option pricing. Manag. Sci. 48, 1086–1101 (2002)
6. Le Courtois, O., Quittard-Pinon, F.: Fair valuation of participating life insurance contracts with jump risk. Geneva Risk Insur. Rev. 33, 106–136 (2008)
7. Melnikov, A., Romaniuk, Y.: Evaluating the performance of Gompertz, Makeham and Lee-Carter mortality models for risk-management with unit-linked contracts. Insur. Math. Econ. 39, 310–329 (2006)
8. Milevsky, M.A., Posner, S.E.: The titanic option: valuation of the guaranteed minimum death benefit in variable annuities and mutual funds. J. Risk Insur. 68, 91–126 (2001)

Solvency evaluation of the guaranty fund at a large financial cooperative

Jean Roy

Abstract. This paper reports on a consulting project whose objective was to evaluate the solvency of the guaranty fund of the Mouvement Desjardins, a large federation of financial cooperatives based in Quebec, Canada. The guaranty fund acts as an internal mutual insurance company; it collects premiums from the 570 local credit unions of the federation and would provide funds to any of these local credit unions facing financial difficulties. At the time of the study, the assets of the insured credit unions totalled 79 billion CA$ and the fund had a capital of 523 million CA$. The purpose of the study was to estimate the probability of insolvency of the fund over various horizons ranging from one to 15 years. Two very different approaches were used to obtain some form of cross-validation. Firstly, under the highly aggregated approach, three theoretical statistical distributions were fitted on the 25 historical yearly rates of subsidy. Secondly, a highly disaggregated Monte-Carlo simulation model was built to represent the financial dynamics of each credit union and the guaranty fund itself, taking into account some 150 parameters for each credit union. Both approaches converged to similar probabilities of insolvency for the fund, which indicated that the fund was well within an implicit AAA rating. The study had several significant financial impacts both internally and externally.

Key words: solvency analysis, financial cooperatives, guaranty fund, Monte Carlo method, credit risk

1 Introduction

The regulatory context brought by the Basel II accord has given a new impetus to the evaluation of the solvency of financial institutions. Although internationally active commercial banks have been at the forefront, other financial institutions, such as large financial cooperatives, are also strongly involved. The decentralised nature of financial cooperatives brings new challenges to the process, as the case study presented here will show. Specifically, this paper will report on a consulting project whose objective was to evaluate the solvency of the guaranty fund of the Mouvement Desjardins, a large federation of credit unions based in Quebec, Canada. The paper will proceed as follows. Section 2 will describe the institutional and technical context of the study. Section 3 will present the preliminary analysis that was conducted to identify and eventually select the methods to be applied. The two polar approaches in terms of aggregation of data were implemented to obtain some form of cross-validation. Section 4 will document the aggregated approach, whereas Section 5 will describe the highly disaggregated approach implemented through a Monte-Carlo simulation model. Section 6 will compare the results of the two approaches, whereas Section 7 will report on the several significant impacts of the study for the organisation. Section 8 will provide the conclusion.

2 Context of the study

The context of this study had two main dimensions, one internal, namely the institutional context, and the other external, namely the technical context. Both will now be addressed.

The institution involved in this study is the "Fonds de sécurité Desjardins" or Desjardins Guaranty Fund (DGF), which is a wholly owned incorporated affiliate of the "Mouvement Desjardins". The Mouvement Desjardins is a federation of some 570 local "caisses populaires" or credit unions. DGF acts as an internal mutual insurance company. It collects annual premiums from the local credit unions and would provide funds to any ot these credit unions in a situation of financial distress. In 2004, DGF had a capital of 523 million CA\$, whereas the insured credit unions had total assets of 79 billion CA\$, leading to a capitalisation ratio of 66.2 basis points. DGF had, at that point, a capitalisation target of 100 basis points. However, management wanted a formal evaluation of the solvency of the fund in order to confirm or review this target. To perform the analysis, data for the 25 years of formal operation of the fund were available. These showed that the fund had played a major role in keeping the local credit unions solvent. Indeed, 964 subsidies were granted to 372 different credit unions for a total of some 220 million un-indexed CA\$ from 1980 to 2004. It also needs to be mentioned that the federation played a key role in managing a consolidation of the network, bringing the total number of credit unions down from its peak of 1465 credit unions in 1982 to 570 in 2004.

A first look at the data showed that the annual average subsidy rate, total subsidies to credit unions divided by the total assets of these, was 3.245 basis points, such that the current capital was equivalent to a 20.4-year reserve at the average rate. However, a cursory analysis of the time series showed a high volatility (4.75 basis points) and also high asymmetry and kurtosis. More precisely, two peaks could be identified in the series: one at 20.8 basis points in 1982 and another one at 5.8 basis points in 1995. Overall, five high values above 5 basis points could be observed during the period of 25 years. Aside from the historical data of the fund itself, extensive data were also available on the insured credit unions over the latest period of seven years, allowing contemplation of highly disaggregated models. At that point, a review of the literature was conducted to survey the status of current practices for the solvency evaluation of similar organisations.

DGF is an organisation that shares many similarities with public deposit insurers. Thus the literature on the solvency evaluation of the US and Canadian deposit insurers was reviewed: the three studies of the Federal Deposit Corporation by Sheehan [5], Bennett [1] and Kuritzkes et al. [2] and the study of the Canada Deposit Insurance Corporation by McIntyre [3]. Overall, the survey of this literature showed that the dominant approach to estimate the solvency of a deposit insurer was the use of a credit risk simulation model and it appeared natural to follow this practice. However, before proceeding, it seemed appropriate to identify the methodological options and make a deliberate choice.

3 Analysis and selection of methodologies

The process consisted in identifying the various possible approaches, evaluating them and eventually selecting one or several for implementation. After some analysis, four dimensions emerged to characterise the possible approaches, namely, the level of aggregation of the data, the estimation technique, the depth of historical data and the horizon considered for the future. The following sections will look at each in turn.

The demand for subsidies by the credit unions depends on the losses that these incur, which depends in turn on the risks they bear. This observation leads to consider the credit unions as a single aggregated entity or to proceed to a disaggregated analysis of each credit union one by one. Similarly, the total risk can be analysed as an aggregate of all risks or risks can be analysed individually (e.g., credit risk, market risk and operational risk). Finally, credit risk itself can be analysed at the portfolio level or can be analysed by segments according to the type, the size and the risk of loans.

To estimate the distribution of the demand for subsidies by credit unions, two techniques appeared possible. If aggregate data were used, it would be possible to fit theoretical statistical distributions to the historical distribution. However if disaggregated data were used, a Monte Carlo simulation model would be more appropriate to estimate the distribution of the demand for subsidies.

If aggregated data were used, 25 years of data would be available. On the other hand, if disaggregated data were used, only seven years of data would be available.

If theoretical statistical distribution were used, the model would be static and the horizon of one year would logically follow from the yearly period of observation of historical data. If a simulation model was used, the financial dynamics of credit unions and of the guaranty fund itself could be modelled and trajectories over time could be built. In this case, a horizon of 15 years was considered relevant.

As the analysis above has shown, even though the four dimensions were not independent, several combinations of choices could be implemented and these could be viewed as forming a spectrum mainly according to the level of aggregation of data, which seemed to be the dimension that had the strongest impact on conditioning the other choices. In this light, it was decided to move forward with the implementation of two polar choices, namely a highly aggregated approach and a highly disaggregated approach. Table 1 summarises the characteristics of each of these two approaches.

It was deemed interesting to implement two very different approaches and to observe whether they would converge or not to similar results. If similarity was obtained, then a cross-validation effect would increase confidence in the results. If dissimilarity

Table 1. Characterisation of the two approaches selected for implementation

	Aggregated	Disaggregated
Aggregation of data	High	Low
Estimation technique	Theoretical distributions	Monte Carlo Simulation
Depth of historical data	25 years	7 years
Horizon for projection	1 year – static	15 years – dynamic

was obtained, then an analysis would have to be conducted to understand the sources of differences and eventually decide on which approach seems more reliable. The next two sections will describe the implementation and the results obtained under the two approaches.

4 The aggregated approach

Under this approach, the aggregate demand for subsidies by the credit unions will be estimated using 25 historical observations of the aggregate rate of subsidy defined as the sum of the subsidies for the year divided by the total assets of the credit unions at the beginning of the year. Three theoretical distributions were selected, namely the Weibull, the Gamma and the Lognormal. Each of these distributions was fitted to the historical cumulative distribution of the rate of subsidy. These three distributions have some common features: they are characterised by two parameters and they accommodate asymmetry.

Before providing the results of the estimation process, it is appropriate to mention that this approach implies two strong hypotheses. First, one must assume that the demand for subsidy has had and will continue to have a distribution that is stable in time. This is indeed a strong assumption as both internal and external structural conditions have evolved significantly. Internally, a strong consolidation of credit unions has taken place which resulted in more than halving their total number giving rise to bigger and hopefully stronger units. Externally, the monetary policy of the Bank of Canada has changed over the years and the strong emphasis now put on the control of inflation will avoid the high nominal interest rates that took place in the early 1980s and which generated massive credit problems. Second, this approach also assumes implicitly that there is no serial correlation, which is most likely contrary to reality as there were clearly periods of good times and bad times that extended over several years. Overall, the first assumption points to overestimating the current demand, whereas the second may lead to underestimating demand in extended periods of difficulties. One may hope that the net effect of the two biases is small. Finally, the depth of the historical data allows the inclusion of two periods of difficulties, which may represent other unknown difficulties that may arise in the future.

With these considerations in mind, we estimated the parameters of the three distributions using non-biased OLS; the results are shown in Table 2 together with various statistics. Overall, the statistics seem to show a reasonably good fit of the distributions to the historical data. With these distributions on hand, one now needs to evaluate the probability that the demand for subsidies is inferior to the current level of capital. It must be remembered that historical values have occurred in the interval ranging from 0 to 20.8 basis points. On the other hand the current level of capital was 66.2 bp. Thus one must extrapolate the distributions to a point that is more than three times bigger than the biggest value ever observed. With the awareness of this fact, we proceeded to evaluate the three estimated distributions at the point corresponding to the current level of capital. Table 3 presents the results that were obtained together with the implied rating according to the scales of S&P and Moody's.

Table 2. Three estimated distributions of the aggregate demand for subsidies

Distribution	Weibull	Gamma	Log-normal
Parameter 1	$\alpha = 0.742$	$\alpha = 0.527$	$\mu = -8.95$
Parameter 2	$\beta = 0.000225$	$\beta = 0.000616$	$\sigma = 1.36$
Mean value	0.03245	0.03245	0.03245
R^2	99.14%	98.37%	99.06%
Kolmogorov-Smirnov Test	0.943	0.964	0.964
Chi squared test	0.139	0.505	0.405
Chi squared test w/o the greatest contribution	0.964	0.948	0.954

Table 3. Solvency estimates of the guaranty fund using statistical distributions

Distribution	Probability of solvency	Probability of default (in bp)	S&P rating	Moody's rating
Weibull	99.9995%	0.05	AAA	Aaa
Gamma	99.9996%	0.04	AAA	Aaa
Log-normal	99.8142%	18.58	BBB	Baa2

As can be observed, the Weibull and the Gamma distributions give very similar results, whereas the log-normal distribution points to a higher probability of default and accordingly a lower credit risk rating. Under the first two distributions, the guaranty fund achieves the implied triple A rating very easily because a probability of default of less than 1 basis point is enough to obtain this rating.

Thus, the aggregated approach has allowed the estimation of the solvency of the fund. Accepting their admittedly strong assumptions, two out of three theoretical distributions lead to a very strong evaluation of the solvency of the fund, the third distribution showing a somewhat weaker position of the fund.

5 The disaggregated approach

Under the disaggregated approach, the income statement of each credit union is simulated using six stochastic variables: namely, the net interest income, loan losses, other income, other expenses, operating expenses and operational losses. Overall, these six variables are meant to capture interest rate risk, market risk, credit risk and operational risk. Once the income statements are simulated, the balance sheets of the credit unions are computed. Should the capital be under the regulatory requirement, then a demand for subsidy at the guaranty fund will be generated. After the total demand for subsidies is computed, the financial statements of the guaranty fund are simulated. Trajectories simulating 15 consecutive years are run using 50 000 trials in order to obtain the distribution of the level of capital of the fund and thus estimate its solvency over the 15-year horizon.

Let us now examine in more details how the six stochastic variables were modelled. Special attention was devoted to credit risk, as it is believed to be the most important source of risk. Thus, the loan portfolio of each credit union was in turn divided into ten types of loans, namely: consumer loans, mortgages, investment loans, commercial loans, agricultural loans, institutional loans, lines of credit to individuals, commercial lines of credit, agricultural line of credit and institutional lines of credit. In turn, each of these ten types of loans was divided into three size categories, namely: below $100 000, between $100 000 and $1 000 000, and above $1 000 000. Now, for each of these 30 credit segments, five historical parameters were available: the probability of default (PD), the exposition at default (EAD), the loss given default (LGD), the number of loans in the segment (N) and the correlation factor with a global latent economic factor (ρ). A Merton-type model is then used. First, the value of the latent economic factor is drawn and then the PD of each segment is conditioned on its value. Secondly, the number of defaults (ND) in a segment is generated with a binomial distribution using the number of loans in the segment N and the conditional PD. Finally, the loan losses are obtained as the product of the number of default ND, the exposure at default EAD and the loss given default LGD. The five other stochastic variables are simulated using normal distributions using expected values and standard deviations. Two sets of assumptions are used: a base case using historical values and a stressed case where the standard deviations were increased by 50% to represent higher risks. Finally, a correlation structure was modelled. Serial correlation factors were assumed for the latent economic factor (0.5) to represent business/credit cycles, for the operating expenses (0.5) and for net interest revenue (0.5) to represent the inertia of these. A negative cross correlation factor (−0.55) was also introduced between net interest revenues and operational losses.

Following the simulation of the financial statements of the credit unions, those of the guaranty fund are generated. The fund has two types of revenues: revenues obtained from investing its assets and premiums collected from the credit unions. It has three main types of expenses: fixed administrative expenses, payment of subsidies to the needing credit unions and taxes on its profit. Stochastic values were generated for investment income, which were correlated to the latent economic factor, and for the subsidies paid through the simulation described in the above subsection. Finally, two policies for premiums were simulated: the current policy which is expressed as 1/14 of 1% of the risky assets of the insured credit union and a hypothetical relaxed policy of 1/17 of 1% of the risky assets. The latter policy was simulated because it was anticipated that given the excellent capital of the fund it could maintain a good solvency level while easing its burden on its insured members.

In total four combinations of scenarios were simulated according to whether the parameters had base-case or stressed values and to whether the policy for premiums was modelled as base-case or relaxed. Table 4 below shows the results for each of the four cases. The probabilities estimated over the 15-year horizon were converted to a one-year horizon as it is this latter horizon that is used for reference by rating agencies such as S&P and Moody's.

It is striking that under the two base case scenarios, the level of insolvency is much lower than one basis point, thus allowing an implied credit rating of AAA to be granted to the fund. Under the two stressed cases, the level of solvency is close to the threshold needed to get a triple A rating. Overall, the simulation model leads to the belief that the solvency of the fund is indeed excellent.

Table 4. Solvency estimates of the guaranty fund by Monte Carlo simulation

Parameters	Base case		Stressed case	
Policy for premiums	1/14%	1/17%	1/14%	1/17%
Nb of cases of insolvency	6	10	74	101
Nb of cases of solvency	49994	49990	49926	49899
Total number of cases	50000	50000	50000	50000
Solvency over 15 years	99.9880%	99.9800%	99.8520%	99.7980%
Solvency over 1 year	99.9992%	99.9987%	99.9901%	99.9865%
Insolvency over 15 years	0.0120%	0.0200%	0.1480%	0.2020%
Insolvency over 1 year	0.0008%	0.0013%	0.0099%	0.0135%
Implied rating	AAA	AAA	AAA	AAA

6 Comparison of the two approaches

It is now interesting to compare the results obtained under the aggregated and disaggregated approaches. Table 5 summarises these results. Overall, Table 5 provides a sensitivity analysis of the solvency estimates while varying methods and hypotheses about distributions, parameters and premium policies. Obviously, there is a wide margin between the best and the worst estimates. However, apart from the log-normal distribution, all other results are basically in the same range. One clear advantage of the simulation model is that it allows the analysis of hypothetical cases, as was done in the last three scenarios considered. So, to make a fair comparison between the statistical distribution approach and the simulation approach one must use the first

Table 5. Summary of solvency estimates under the two approaches

Approach	Probability of solvency	Probability of default (bp)	Implied S&P rating
Statistical distribution			
Weibull	99.9995 %	0.05	AAA
Gamma	99.9996 %	0.04	AAA
Log-normal	99.8142 %	18.58	BBB
Monte Carlo simulation			
Base case and premiums at 1/14 %	99.9992 %	0.08	AAA
Base case and premiums at 1/17 %	99.9987 %	0.13	AAA
Stressed case and premiums at 1/14 %	99.9901 %	0.99	AAA
Stressed case and premiums at 1/17 %	99.9865 %	1.35	AAA

base case scenario. Then one observes that the probability of insolvency obtained (0.08 bp) is quite consistent with the values obtained using the Weibull and Gamma distributions (0.05 bp and 0.04 bp). This observation is quite comforting as we interpret it as each result being reinforced by the other. It is striking indeed that the two very different approaches did in fact converge to basically similar results. Overall, it can be concluded that the solvency of the guaranty fund was excellent and that this conclusion could be taken with a high level of confidence considering the duplication obtained.

7 Financial impact of the study

The study led the management of the fund to take several significant actions. First, the target capital ratio, which was previously set to 1% of the aggregate assets of the credit unions, was brought down to an interval between 0.55% and 0.65%, basically corresponding to the current level of capital of 0.62%. Secondly, management decided to lower the premiums charged to the credit unions. Finally, the deposits made at any of the credit unions of the Mouvement Desjardins are also guaranteed by a public deposit insurer managed by the Financial Market Authority of Quebec (FMA) to which the Mouvement Desjardins has to pay an annual premium. The solvency study that we have described above was presented to the FMA to request a reduction of the premium and after careful examination the FMA granted a very significant reduction. Thus, one could argue that the study achieved its goals. First, it provided management of the guaranty fund with the information it requested, that is a well grounded estimation of the solvency of the fund. Secondly, the very favourable assessment of the solvency allowed management to take several actions to reap the benefits of the excellent state of solvency of the fund.

8 Conclusions

This study faced the challenge of estimating the solvency of the guaranty fund of 523 million CA\$ insuring a network of 570 credit unions totalling some 79 billion CA\$ in assets invested in a broad spectrum of personal, commercial, agricultural and institutional loans. As a preliminary step, the array of possible approaches according to the aggregation of data, the estimation technique, the depth of historical data and the projection horizon was examined. After analysis, two polar approaches in terms of the aggregation of data were selected for implementation. Under the aggregated approach three statistical distributions were fitted to twenty five yearly observations of the total rate of subsidy. Under the disaggregated approach, an elaborate Monte Carlo simulation model was set up whereby the financial statements of each credit union and those of the guaranty fund itself were generated, integrating four types of risks and using more than 7500 risk parameters, mainly representing credit risk at a very segmented level. The Monte Carlo simulation was also used to evaluate the impact of stressed values of the risk parameters and of a relaxation of the policy for the premiums charged by the fund. Overall, both approaches converged to similar estimates of the solvency of the fund, thus reinforcing the level of confidence in the results. Accordingly, the solvency of the fund could be considered as excellent, being well within an implied triple A rating under the base case scenario, and still qualifying, although marginally, for this rating under stressed hypotheses. The detailed analysis of the solvency of the fund and the good evaluation it brought had three significant financial impacts: the target capital ratio of the fund was revised downward, the premiums charged to credit unions were reduced and the Mouvement Desjardins itself obtained a sizable reduction of the premium it pays to the public deposit insurer. Needless to say, management of the guaranty fund was quite satisfied with these outcomes. Finally, it was decided to update the study every five years. From this perspective, several improvements and extensions, namely regarding a more refined modelling of the risk factors other than credit risk and a more adaptive premium policy, are already envisaged.

Acknowledgement. I would like to thank Miguel Mediavilla, M.Sc., of Globevest Capital for programming the simulation model.

References

1. Bennett, R.L.: Evaluating the Adequacy of the Deposit Insurance Fund: A Credit-Risk Modeling Approach. Federal Deposit Insurance Corporation (2002)
2. Kuritzkes A., Schuerman T., Weiner S.: Deposit Insurance and Risk Management of the U.S. Banking System: How Much? How Safe Who Pays? Wharton Financial Institutions Center (2002)
3. McIntyre M.L.: Modelling Deposit Insurance Loss Distributions – A Monte Carlo Simulation Approach. Confidential report to Canada Deposit Insurance Corporation (2003)
4. Roy J.: Funding Issues for Deposit Insurers. International Association of Deposit Insurers (2001)

5. Sheehan K.P.: Capitalization and the Bank Insurance Fund. Federal Deposit Insurance Corporation (1998)

A Monte Carlo approach to value exchange options using a single stochastic factor

Giovanni Villani

Abstract. This article describes an important sampling regarding modification of the Monte Carlo method in order to minimise the variance of simulations. In a particular way, we propose a generalisation of the antithetic method and a new a-sampling of stratified procedure with $a \neq \frac{1}{2}$ to value exchange options using a single stochastic factor. As is well known, exchange options give the holder the right to exchange one risky asset V for another risky asset D and therefore, when an exchange option is valued, we generally are exposed to two sources of uncertainity. The reduction of the bi-dimensionality of valuation problem to a single stochastic factor implies a new stratification procedure to improve the Monte Carlo method. We also provide a set of numerical experiments to verify the accuracy derived by a-sampling.

Key words: exchange options, Monte Carlo simulations, variance reduction

1 Introduction

Simulations are widely used to solve option pricing. With the arrival of ever faster computers coupled with the development of new numerical methods, we are able to numerically solve an increasing number of important security pricing models. Even where we appear to have analytical solutions it is often desirable to have an alternative implementation that is supposed to give the same answer. Simulation methods for asset pricing were introduced in finance by Boyle [3]. Since that time simulation has been successfully applied to a wide range of pricing problems, particularly to value American options, as witnessed by the contributions of Tilley [10], Barraquand and Martineau [2], Broadie and Glasserman [4], Raymar and Zwecher [9].

The aim of this paper is to improve the Monte Carlo procedure in order to evaluate exchange options generalizing the antithetic variate methodology and proposing a new stratification procedure. To realise this objective, we price the most important exchange options using a single stochastic factor P that is the ratio between the underlying asset V and the delivery one D. For this reason, we need a particular sampling to concentrate the simulations in the range in which the function P is more sensitive.

The most relevant models that value exchange options are given in Margrabe [7], McDonald and Siegel [8], Carr [5,6] and Armada et al. [1]. Margrabe [7] values an European exchange option that gives the right to realise such an exchange only at expiration. McDonald and Siegel [8] value an European exchange option considering that the assets distribute dividends and Carr [5] values a compound European exchange option in which the underlying asset is another exchange option. However, when the assets pay sufficient large dividends, there is a positive probability that an American exchange option will be exercised strictly prior to expiration. This positive probability induced additional value for an American exchange option as given in Carr [5,6] and Armada et al. [1].

The paper is organised as follows. Section 2 presents the estimation of a Simple European Exchange option, Section 3 introduces the Monte Carlo valuation of a Compound European Exchange option and Section 4 gives us the estimation of a Pseudo American Exchange option. In Section 5, we apply new techniques that allow reduction of the variance concerning the above option pricing and we also present some numerical studies. Finally, Section 6 concludes.

2 The price of a Simple European Exchange Option (SEEO)

We begin our discussion by focusing on a SEEO to exchange asset D for asset V at time T. Denoting by $s(V, D, T - t)$ the value of SEEO at time t, the final payoff at the option's maturity date T is $s(V, D, 0) = \max(0, V_T - D_T)$, where V_T and D_T are the underlying assets' terminal prices. So, assuming that the dynamics of assets V and D are given by:

$$\frac{dV}{V} = (\mu_v - \delta_v)dt + \sigma_v dZ_v, \tag{1}$$

$$\frac{dD}{D} = (\mu_d - \delta_d)dt + \sigma_d dZ_d, \tag{2}$$

$$Cov\left(\frac{dV}{V}, \frac{dD}{D}\right) = \rho_{vd}\sigma_v\sigma_d \, dt, \tag{3}$$

where μ_v and μ_d are the expected rates of return on the two assets, δ_v and δ_d are the corresponding dividend yields, σ_v^2 and σ_d^2 are the respective variance rates and Z_v and Z_d are two Brownian standard motions with correlation coefficient ρ_{vd}, Margrabe [7] and McDonald and Siegel [8] show that the value of a SEEO on dividend-paying assets, when the valuation date is $t = 0$, is given by:

$$s(V, D, T) = Ve^{-\delta_v T} N(d_1(P, T)) - De^{-\delta_d T} N(d_2(P, T)), \tag{4}$$

where:

- $P = \frac{V}{D}$; $\quad \sigma = \sqrt{\sigma_v^2 - 2\rho_{vd}\sigma_v\sigma_d + \sigma_d^2}$; $\quad \delta = \delta_v - \delta_d$;
- $d_1(P, T) = \frac{\log P + \left(\frac{\sigma^2}{2} - \delta\right)T}{\sigma\sqrt{T}}$; $\quad d_2(P, T) = d_1(P, T) - \sigma\sqrt{T}$;

- $N(d)$ is the cumulative standard normal distribution.

The typical simulation approach is to price the SEEO as the expectation value of discounted cash-flows under the risk-neutral probability $\mathbb{Q}$. So, for the risk-neutral version of Equations (1) and (2), it is enough to replace the expected rates of return μ_i by the risk-free interest rate r plus the premium-risk, namely $\mu_i = r + \lambda_i \sigma_i$, where λ_i is the asset's market price of risk, for $i = V, D$. So, we obtain the risk-neutral stochastic equations:

$$\frac{dV}{V} = (r - \delta_v)dt + \sigma_v(dZ_v + \lambda_v dt) = (r - \delta_v)dt + \sigma_v dZ_v^*, \tag{5}$$

$$\frac{dD}{D} = (r - \delta_d)dt + \sigma_d(dZ_d + \lambda_d dt) = (r - \delta_d)dt + \sigma_d dZ_d^*. \tag{6}$$

The Brownian processes $dZ_v^* \equiv dZ_v + \lambda_v dt$ and $dZ_d^* \equiv dZ_d + \lambda_d dt$ are the new Brownian motions under the risk-neutral probability $\mathbb{Q}$ and $Cov(dZ_v^*, dZ_d^*) = \rho_{vd} dt$. Applying Ito's lemma, we can reach the equation for the ratio-price simulation $P = \frac{V}{D}$ under the risk-neutral measure $\mathbb{Q}$:

$$\frac{dP}{P} = (-\delta + \sigma_d^2 - \sigma_v \sigma_d \rho_{vd}) \, dt + \sigma_v dZ_v^* - \sigma_d dZ_d^*. \tag{7}$$

Applying the log-transformation for D_T, under the probability $\mathbb{Q}$, it results in:

$$D_T = D_0 \exp\{(r - \delta_d)T\} \cdot \exp\left(-\frac{\sigma_d^2}{2} T + \sigma_d Z_d^*(T)\right). \tag{8}$$

We have that $U \equiv \left(-\frac{\sigma_d^2}{2} T + \sigma_d Z_d^*(T)\right) \sim \mathcal{N}\left(-\frac{\sigma_d^2}{2}T, \sigma_d\sqrt{T}\right)$ and therefore $\exp(U)$ is a log-normal whose expectation value is $E_{\mathbb{Q}}\left[\exp(U)\right] = \exp\left(-\frac{\sigma_d^2}{2}T + \frac{\sigma_d^2}{2}T\right) = 1$. So, by Girsanov's theorem, we can define the new probability measure $\widetilde{\mathbb{Q}}$ equivalent to $\mathbb{Q}$ and the Radon-Nikodym derivative is:

$$\frac{d\widetilde{\mathbb{Q}}}{d\mathbb{Q}} = \exp\left(-\frac{\sigma_d^2}{2} T + \sigma_d Z_d^*(T)\right). \tag{9}$$

Hence, using Equation (8), we can write:

$$D_T = D_0 \, e^{(r-\delta_d)T} \cdot \frac{d\widetilde{\mathbb{Q}}}{d\mathbb{Q}}. \tag{10}$$

By the Girsanov theorem, the processes:

$$d\hat{Z}_d = dZ_d^* - \sigma_d dt, \tag{11}$$

$$d\hat{Z}_v = \rho_{vd} d\hat{Z}_d + \sqrt{1 - \rho_{vd}^2} \, dZ', \tag{12}$$

are two Brownian motions under the risk-neutral probability measure $\widetilde{Q}$ and Z' is a Brownian motion under $\widetilde{Q}$ independent of $\hat{Z}_d$. By the Brownian motions defined in Equations (11) and (12), we can rewrite Equation (7) for the asset P under the risk-neutral probability $\widetilde{Q}$. So it results that:

$$\frac{dP}{P} = -\delta\, dt + \sigma_v\, d\hat{Z}_v - \sigma_d\, d\hat{Z}_d. \tag{13}$$

Using Equation (12), it results that:

$$\sigma_v d\hat{Z}_v - \sigma_d d\hat{Z}_d = (\sigma_v \rho_{vd} - \sigma_d)\, d\hat{Z}_d + \sigma_v \left(\sqrt{1 - \rho_{vd}^2}\right) dZ', \tag{14}$$

where $\hat{Z}_v$ and Z' are independent under $\widetilde{Q}$. Therefore, as $(\sigma_v d\hat{Z}_v - \sigma_d d\hat{Z}_d) \sim \mathcal{N}(0, \sigma\sqrt{dt})$, we can rewrite Equation (13):

$$\frac{dP}{P} = -\delta\, dt + \sigma dZ_p, \tag{15}$$

where $\sigma = \sqrt{\sigma_v^2 + \sigma_d^2 - 2\sigma_v\sigma_d\rho_{vd}}$ and Z_p is a Brownian motion under $\widetilde{Q}$. Using the log-transformation, we obtain the equation for the risk-neutral price simulation P:

$$P_t = P_0 \exp\left\{\left(-\delta - \frac{\sigma^2}{2}\right)t + \sigma Z_p(t)\right\}. \tag{16}$$

So, using the asset D_T as numeraire given by Equation (10), we price a SEEO as the expectation value of discounted cash-flows under the risk-neutral probability measure:

$$\begin{aligned} s(V, D, T) &= e^{-rT} E_Q[\max(0, V_T - D_T)] \\ &= D_0 e^{-\delta_d T} E_{\widetilde{Q}}[g_s(P_T)], \end{aligned} \tag{17}$$

where $g_s(P_T) = \max(P_T - 1, 0)$. Finally, it is possible to implement the Monte Carlo simulation to approximate:

$$E_{\widetilde{Q}}[g_s(P_T)] \approx \frac{1}{n}\sum_{i=1}^{n} g_s^i(\hat{P}_T^i), \tag{18}$$

where n is the number of simulated paths effected, $\hat{P}_T^i$ for $i = 1, 2 \ldots n$ are the simulated values and $g_s^i(\hat{P}_T^i) = \max(0, \hat{P}_T^i - 1)$ are the n simulated payoffs of SEEO using a single stochastic factor.

3 The price of a Compound European Exchange Option (CEEO)

The CEEO is a derivative in which the underlying asset is another exchange option. Carr [5] develops a model to value the CEEO assuming that the underlying asset is a SEEO $s(V, D, \tau)$ whose time to maturity is $\tau = T - t_1$ with $t_1 < T$, the exercise price is a ratio q of asset D at time t_1 and the expiration date is t_1. So, considering that the valuation date is $t = 0$, the final payoff of CEEO at maturity date t_1 is:

$$c(s(V, D, \tau), qD, 0) = \max[0, s(V, D, \tau) - qD].$$

Assuming that the evolutions of assets V and D are given by Equations (1) and (2), under certain assumptions, Carr [5] shows that the CEEO price at evaluation date $t = 0$ is:

$$
\begin{aligned}
c(s(V, D, \tau), qD, t_1) = & \, V e^{-\delta_v T} N_2 \left(d_1 \left(\frac{P}{P_1^*}, t_1 \right), d_1 (P, T); \rho \right) \\
& - D e^{-\delta_d T} N_2 \left(d_2 \left(\frac{P}{P_1^*}, t_1 \right), d_2 (P, T); \rho \right) \\
& - q D e^{-\delta_d t_1} N \left(d_2 \left(\frac{P}{P_1^*}, t_1 \right) \right),
\end{aligned}
\tag{19}
$$

where $N_2(x_1, x_2; \rho)$ is the standard bivariate normal distribution evaluated at x_1 and x_2 with correlation $\rho = \sqrt{\frac{t_1}{T}}$ and P_1^* is the critical price ratio that makes the underlying asset and the exercise price equal and solves the following equation:

$$P_1^* e^{-\delta_v \tau} N(d_1(P_1^*, \tau)) - e^{-\delta_d \tau} N(d_2(P_1^*, \tau)) = q. \tag{20}$$

It is obvious that the CEEO will be exercised at time t_1 if $P_{t_1} \geq P_1^*$. We price the CEEO as the expectation value of discounted cash-flows under the risk-neutral probability $\mathbb{Q}$ and, after some manipulations and using D_{t_1} as numeraire, we obtain:

$$
\begin{aligned}
c(s, qD, t_1) &= e^{-rt_1} E_{\mathbb{Q}}[\max(s(V_{t_1}, D_{t_1}, \tau) - qD_{t_1}, 0)] \\
&= D_0 e^{-\delta_d t_1} E_{\tilde{\mathbb{Q}}}[g_c(P_{t_1})],
\end{aligned}
\tag{21}
$$

where

$$g_c(P_{t_1}) = \max[P_{t_1} e^{-\delta_v \tau} N(d_1(P_{t_1}, \tau)) - e^{-\delta_d \tau} N(d_2(P_{t_1}, \tau)) - q, 0]. \tag{22}$$

Using a Monte Carlo simulation, it is possible to approximate the value of CEEO as:

$$c(s, qD, t_1) \approx D_0 e^{-\delta_d t_1} \left(\frac{\sum_{i=1}^n g_c^i(\hat{P}_{t_1}^i)}{n} \right), \tag{23}$$

where n is the number of simulated paths and $g_c^i(\hat{P}_{t_1}^i)$ are the n simulated payoffs of CEEO using a single stochastic factor.

4 The price of a Pseudo American Exchange Option (PAEO)

Let $t = 0$ be the evaluation date and T be the maturity date of the exchange option. Let $S_2(V, D, T)$ be the value of a PAEO that can be exercised at time $\frac{T}{2}$ or T. Following

Carr [5,6], the payoff of PAEO can be replicated by a portfolio containing two SEEOs and one CEEO. Hence, the value of PAEO is:

$$
\begin{aligned}
S_2(V, D, T) = & \ V e^{-\delta_v T} N_2 \left(-d_1 \left(\frac{P}{P_2^*}, \frac{T}{2} \right), d_1(P, T); -\rho \right) \\
& + V e^{-\delta_v \frac{T}{2}} N \left(d_1 \left(\frac{P}{P_2^*}, \frac{T}{2} \right) \right) \\
& - D e^{-\delta_d T} N_2 \left(-d_2 \left(\frac{P}{P_2^*}, \frac{T}{2} \right), d_2(P, T); -\rho \right) \\
& - D e^{-\delta_d \frac{T}{2}} N \left(d_2 \left(\frac{P}{P_2^*}, \frac{T}{2} \right) \right),
\end{aligned}
\tag{24}
$$

where $\rho = \sqrt{\frac{T/2}{T}} = \sqrt{0.5}$ and P_2^* is the unique value that makes the PAEO exercise indifferent or note at time $\frac{T}{2}$ and solves the following equation:

$$
P_2^* e^{-\delta_v \frac{T}{2}} N \left(d_1 \left(P_2^*, \frac{T}{2} \right) \right) - e^{-\delta_d \frac{T}{2}} N \left(d_2 \left(P_2^*, \frac{T}{2} \right) \right) = P_2^* - 1.
$$

The PAEO will be exercised at mid-life time $\frac{T}{2}$ if the cash flows $(V_{T/2} - D_{T/2})$ exceed the opportunity cost of exercise, i.e., the value of the option $s(V, D, T/2)$:

$$
V_{T/2} - D_{T/2} \geq s(V, D, T/2).
\tag{25}
$$

It is clear that if the PAEO is not exercised at time $\frac{T}{2}$, then it's just the value of a SEEO with maturity $\frac{T}{2}$, as given by Equation (4). However, the exercise condition can be re-expressed in terms of just one random variable by taking the delivery asset as numeraire. Dividing by the delivery asset price $D_{T/2}$, it results in:

$$
P_{T/2} - 1 \geq P_{T/2} e^{-\delta_v \frac{T}{2}} N(d_1(P_{T/2}, T/2)) - e^{-\delta_d \frac{T}{2}} N(d_2(P_{T/2}, T/2)).
\tag{26}
$$

So, if the condition (26) takes place, namely, if the value of P is higher than P_2^* at moment $\frac{T}{2}$, the PAEO will be exercised at time $\frac{T}{2}$ and the payoff will be $(V_{T/2} - D_{T/2})$; otherwise the PAEO will be exercised at time T and the payoff will be $\max[V_T - D_T, 0]$. So, using the Monte Carlo approach, we can value the PAEO as the expectation value of discounted cash flows under the risk-neutral probability measure:

$$
\begin{aligned}
S_2(V, D, T) = & \ e^{-r \frac{T}{2}} E_{\mathbb{Q}}[(V_{T/2} - D_{T/2}) \mathbf{1}_{(P_{T/2} \geq P_2^*)}] \\
& + e^{-rT} E_{\mathbb{Q}}[\max(0, V_T - D_T) \mathbf{1}_{(P_{T/2} < P_2^*)}].
\end{aligned}
\tag{27}
$$

Using assets $D_{T/2}$ and D_T as numeraires, after some manipulations, we can write that:

$$
S_2(V, D, T) = D_0 \left(e^{-\delta_d \frac{T}{2}} E_{\widetilde{\mathbb{Q}}}[g_s(P_{T/2})] + e^{-\delta_d T} E_{\widetilde{\mathbb{Q}}}[g_s(P_T)] \right),
\tag{28}
$$

where $g_s(P_{T/2}) = (P_{T/2} - 1)$ if $P_{T/2} \geq P_2^*$ and $g_s(P_T) = \max[P_T - 1, 0]$ if $P_{T/2} < P_2^*$.

So, by the Monte Carlo simulation, we can approximate the PAEO as:

$$S_2(V, D, T) \simeq D_0 \left(\frac{\sum_{i \in A} g_s^i(\hat{P}_{T/2}^i)e^{-\delta_d T/2} + \sum_{i \in B} g_s^i(\hat{P}_T^i)e^{-\delta_d T}}{n} \right), \qquad (29)$$

where $A = \{i = 1..n \text{ s.t. } \hat{P}_{T/2}^i \geq P_2^*\}$ and $B = \{i = 1..n \text{ s.t. } \hat{P}_{T/2}^i < P_2^*\}$.

5 Numerical examples and variance reduction techniques

In this section we report the results of numerical simulations of SEEO, CEEO and PAEO and we propose a generalisation of the antithetic method and a new a-stratified sampling in order to improve on the speed and the efficiency of simulations. To compute the simulations we have assumed that the number of simulated paths n is equal to 500 000. The parameter values are $\sigma_v = 0.40$, $\sigma_d = 0.30$, $\rho_{vd} = 0.20$, $\delta_v = 0.15$, $\delta_d = 0$ and $T = 2$ years. Furthermore, to compute the CEEO we assume that $t_1 = 1$ year and the exchange ratio $q = 0.10$. Table 1 summarises the results of SEEO simulations, while Table 2 shows the CEEO's simulated values. Finally, Table 3 contains the numerical results of PAEO.

Using Equation (16), we can observe that $Y = \ln(\frac{P_t}{P_0})$ follows a normal distribution with mean $(-\delta - \frac{\sigma^2}{2})t$ and variance $\sigma^2 t$. So, the random variable Y can be generated by the inverse of the normal cumulative distribution function $Y = F^{-1}(u; (-\delta - \frac{\sigma^2}{2})t, \sigma^2 t)$ where u is a function of a uniform random variable $U[0, 1]$. Using the Matlab algorithm, we can generate the n simulated prices $\hat{P}_t^i$, for $i = 1...n$, as:

```
Pt=P0*exp(norminv(u,-d*t-0.5*sig^2*t,sig*sqrt(t))),
```

where $u = rand(1, n)$ are the n random uniform values between 0 and 1. As the simulated price $\hat{P}_t^i$ depends on random value u_i, we write henceforth that the SEEO, CEEO and PAEO payoffs g_k^i, for $k = s, c$ using a single stochastic factor depend

Table 1. Simulation prices of Simple European Exchange Option (SEEO)

V_0	D_0	SEEO (true)	SEEO (sim)	$\hat{\sigma}_n^2$	ε_n	$\hat{\sigma}_{av}^2$	Eff$_{av}$	$\hat{\sigma}_{st}^2$	Eff$_{st}$	$\hat{\sigma}_{gst}^2$	Eff$_{gst}$
180	180	19.8354	19.8221	0.1175	0.0011	0.0516	1.13	0.0136	4.32	1.02e-8	22.82
180	200	16.0095	16.0332	0.0808	8.98e-4	0.0366	1.10	0.0068	5.97	8.08e-9	19.98
180	220	12.9829	12.9685	0.0535	7.31e-4	0.0258	1.03	0.0035	7.56	5.89e-9	18.15
200	180	26.8315	26.8506	0.1635	0.0013	0.0704	1.16	0.0253	3.23	1.27e-8	25.54
200	200	22.0393	22.0726	0.1137	0.0011	0.0525	1.08	0.0135	4.19	1.03e-8	21.97
200	220	18.1697	18.1746	0.0820	9.05e-4	0.0379	1.08	0.0072	5.65	8.37e-9	19.58
220	180	34.7572	34.7201	0.2243	0.0015	0.0939	1.19	0.0417	2.68	1.54e-8	28.94
220	200	28.9873	28.9479	0.1573	0.0013	0.0695	1.13	0.0238	3.30	1.23e-8	25.45
220	220	24.2433	24.2096	0.1180	0.0011	0.0517	1.14	0.0135	4.35	1.03e-8	22.88

Table 2. Simulation prices of Compound European Exchange Option (CEEO)

V_0	D_0	CEEO (true)	CEEO (sim)	$\hat{\sigma}_n^2$	ε_n	$\hat{\sigma}_{av}^2$	Eff$_{av}$	$\hat{\sigma}_{st}^2$	Eff$_{st}$	$\hat{\sigma}_{gst}^2$	Eff$_{gst}$
180	180	11.1542	11.1590	0.0284	2.38e−4	0.0123	1.15	0.0043	3.30	2.23e−9	25.44
180	200	8.0580	8.0830	0.0172	1.85e−4	0.0078	1.10	0.0019	4.64	1.61e−9	21.28
180	220	5.8277	5.8126	0.0103	1.43e−4	0.0048	1.06	0.0008	6.30	1.15e−9	17.89
200	180	16.6015	16.6696	0.0464	3.04e−4	0.0184	1.25	0.0089	2.60	2.99e−9	30.94
200	200	12.3935	12.4010	0.0283	2.37e−4	0.0124	1.14	0.0043	3.28	2.22e−9	25.40
200	220	9.2490	9.2226	0.0179	1.89e−4	0.0080	1.11	0.0020	4.42	1.67e−9	21.37
220	180	23.1658	23.1676	0.0684	3.69e−4	0.0263	1.30	0.0158	2.15	3.83e−9	35.71
220	200	17.7837	17.7350	0.0439	2.96e−4	0.0180	1.21	0.0083	2.65	2.91e−9	30.07
220	220	13.6329	13.6478	0.0285	2.38e−4	0.0122	1.17	0.0043	3.33	2.22e−9	25.66

Table 3. Simulation prices of Pseudo American Exchange Option (PAEO)

V_0	D_0	PAEO (true)	PAEO (sim)	$\hat{\sigma}_n^2$	ε_n	$\hat{\sigma}_{av}^2$	Eff$_{av}$	$\hat{\sigma}_{st}^2$	Eff$_{st}$	$\hat{\sigma}_{gst}^2$	Eff$_{gst}$
180	180	23.5056	23.5152	0.0833	9.12e−4	0.0333	1.26	0.0142	2.93	3.29e−8	25.31
180	200	18.6054	18.6699	0.0581	7.62e−4	0.0250	1.16	0.0083	3.51	3.96e−8	14.65
180	220	14.8145	14.8205	0.0411	6.41e−4	0.0183	1.12	0.0051	4.00	3.72e−8	11.03
200	180	32.3724	32.3501	0.1172	0.0011	0.0436	1.34	0.0247	2.36	3.44e−8	24.86
200	200	26.1173	26.1588	0.0839	9.16e−4	0.0328	1.27	0.0142	2.95	3.27e−8	25.64
200	220	21.1563	21.1814	0.0600	7.74e−4	0.0253	1.18	0.0053	3.43	3.83e−8	15.63
220	180	42.5410	42.5176	0.1571	0.0013	0.0536	1.46	0.0319	2.46	3.97e−8	32.82
220	200	34.9165	34.9770	0.1134	0.0011	0.0422	1.34	0.0233	2.43	2.36e−8	27.90
220	220	28.7290	28.7840	0.0819	9.04e−4	0.0338	1.21	0.0142	2.87	3.35e−8	24.41

on u_i. We can observe that the simulated values are very close to true ones. In a particular way, the Standard Error $\varepsilon_n = \frac{\hat{\sigma}_n}{\sqrt{n}}$ is a measure of simulation accurancy and it is usually estimated as the realised standard deviation of the simulations $\hat{\sigma}_n = \sqrt{\frac{\sum_{i=1}^{n}\left(g_k^i(u_i)\right)^2}{n} - \left(\frac{\sum_{i=1}^{n} g_k^i(u_i)}{n}\right)^2}$ divided by the square root of simulations. Moreover, to reduce the variance of results, we propose the Antithetic Variates (AV), the Stratified Sample with two intervals (ST) and a general stratified sample (GST). The Antithetic Variates consist in generating n independent pairwise averages $\frac{1}{2}(g_k^i(u_i)+g_k^i(1-u_i))$ with $u_i \sim U[0, 1]$. The function $g_k^i(1 - u_i)$ decreases whenever $g_k^i(u_i)$ increases, and this produces a negative covariance $cov[g_k^i(u_i), g_k^i(1 - u_i)] < 0$ and so a variance reduction. For instance, we can rewrite the Monte Carlo pricing given by Equation (18) as:

$$E_{\underset{\mathbb{Q}}{\sim}}^{AV}[g_s(P_T)] \approx \frac{1}{n}\left(\sum_{i=1}^{n}\frac{1}{2}g_s^i(u_i) + \frac{1}{2}g_s^i(1 - u_i))\right). \tag{30}$$

We can observe that the variance $\hat{\sigma}_{av}^2$ is halved, but if we generate $n = 500\,000$ uniform variates u and we also use the values of $1-u$, it results in a total of $1\,000\,000$ function evaluations. Therefore, in order to determine the efficiency Eff$_{av}$, the variance $\hat{\sigma}_n^2$ should be compared with the same number ($1\,000\,000$) of function evaluations.

We can conclude that efficiency $\text{Eff}_{av} = \frac{\hat{\sigma}_n^2/2n}{\hat{\sigma}_{av}^2/n}$ and the introduction of antithetic variates has the same effect on precision as doubling the sample size of Monte Carlo path-simulations.

Using the Stratified Sample, we concentrate the sample in the region where the function g is positive and, where the function is more variable, we use larger sample sizes. First of all, we consider the piecewise $ag_k^i(u_1) + (1-a)g_k^i(u_2)$ where $u_1 \sim U[0, a]$ and $u_2 \sim U[a, 1]$, as an individual sample. This is a weighted average of two function values with weights a and $1-a$ proportional to the length of the corresponding intervals. If u_1 and u_2 are independent, we obtain a dramatic improvement in variance reduction since it becomes $a^2 var[g_k^i(u_1)] + (1-a)^2 var[g_k^i(u_2)]$. For instance, the payoff of SEEO $g_s^i(P_T^i) = \max[0, P_T^i - 1]$ with $V_0 = 180$ and $D_0 = 180$ has a positive value starting from $a_s = 0.60$, as shown in Figure 1(a), while the CEEO will be exercised when $P_{t_1} \geq 0.9878$ and the payoff will be positive from $a_c = 0.50$, as illustrated in Figure 1(b). Assuming $a = 0.90$, Tables 1, 2 and 3 show the variance using the Stratified Sample (ST) and the efficiency index. For the same reason as before, we should to compare this result with the Monte Carlo variance with the same number ($1\,000\,000$) of path simulations. The efficiency index $\text{Eff}_{st} = \frac{\hat{\sigma}_n^2/2n}{\hat{\sigma}_{st}^2/n}$ shows that the improvement is about 4. We can assert that it is possibile to use one fourth the sample size by stratifying the sample into two regions: $[0, a]$ and $[a, 1]$.

Finally, we consider the general stratified sample subdividing the interval $[0, 1]$ into convenient subintervals. Then, if we use the stratified method with two strata $[0.80, 0.90], [0.90, 1]$, Tables 1, 2 and 3 show the variance and also the efficiency gain $\text{Eff}_{gst} = \frac{\hat{\sigma}_n^2}{\sum_{i=1}^k n_i \hat{\sigma}_{gst}^2}$. Moreover, for the first simulation of SEEO we have that the optimal choice sample size is $n = 66\,477, 433\,522$, for the first simulation of CEEO we obtain that $n = 59\,915, 440084$, while for the PAEO it results that $n = 59\,492, 440\,507$. It's plain that the functions g_k^i are more variables in the the interval

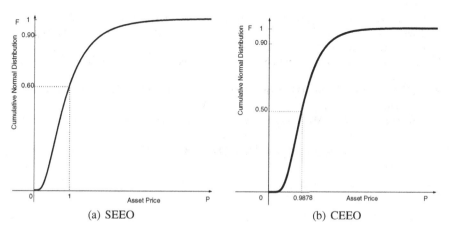

(a) SEEO   (b) CEEO

Fig. 1. Cumulative normal distribution of asset P

[0.90, 1] and so the sample size is about 440 000. We can observe that this stratified sample can account for an improvement in efficiency of about 23.

6 Conclusions

In this paper we have shown a generalisation of the antithetic method and an a-sampling procedure to value exchange options improving on the Monte Carlo simulation. Using the delivery asset D as numeraire, we have reduced the bi-dimensionality of evaluation to one stochastic variable P that is the ratio between assets V and D. But the particular evolution of asset P requires a new sampling procedure to concentrate the simulations in the range in which P is more sensitive in order to reduce the variance. The paper can be improved choosing a^* in order to minimise the variance of simulation through an endogenic process. To realise this objective, a short simulation, to estimate some optimal a^*, and then the a^*-stratification, may be used.

Acknowledgement. Many thanks to the anonymous reviewers for their constructive comments.

References

1. Armada, M.R., Kryzanowsky, L., Pereira, P.J.: A modified finite-lived American exchange option methodology applied to real options valuation. Global Finan. J. 17, 419–438 (2007)
2. Barraquand, J., Martineau. D.: Numerical valuation of high dimensional multivariate american securities. J. Finan. Quant. Anal. 30, 383–405 (1995)
3. Boyle, P.: Options: a Monte Carlo approach. J. Finan. Econ. 4, 323–338 (1977)
4. Broadie, M., Glasserman, P.: Pricing American-style securities using simulation. J. Econ. Dyn. Control 21, 1323–1352 (1997)
5. Carr, P.: The valuation of sequential exchange opportunities. J. Finan. 43, 1235–1256 (1988)
6. Carr, P.: The valuation of American exchange options with application to real options. In: Trigeorgis, L. (ed.) Real Options in Capital Investment: Models, Strategies and Applications. Praeger, Westport, CT, London (1995)
7. Margrabe, W.: The value of an exchange option to exchange one asset for another. J. Finan. 33, 177–186 (1978)
8. McDonald, R.L., Siegel, D.R.: Investment and the valuation of firms when there is an option to shut down. Int. Econ. Rev. 28, 331–349 (1985)
9. Raymar, S., Zwecher, M.: A Monte Carlo valuation of american call options on the maximum of several stocks. J. Derivatives 5, 7–23 (1997)
10. Tilley, J.: Valuing american options in a path simulation model. Trans. Soc. Actuaries 45, 83–104 (1993)

Printed in the United States
By Bookmasters